VARIATIONAL ANALYSIS:

Critical Extremals and Sturmian Extensions

VARIATIONAL ANALYSIS:

Critical Extremals and Sturmian Extensions

MARSTON MORSE
Institute for Advanced Study

A WILEY-INTERSCIENCE PUBLICATION

JOHN WILEY & SONS New York . London . Sydney . Toronto

To

LOUISE

Library of Congress Cataloging in Publication Data

Morse, Marston, 1892–
Variational analysis.

(Pure and applied mathematics)
Bibliography: p.
1. Calculus of variations. 2. Maxima and minima. I. Title.

QA316.M58 515′.64 72-8368
ISBN 0-471-61700-8

Printed in the United States of America

10 9 8 7 6 5 4 3 2 1

PREFACE

The first concern of Euler, Legendre, Jacobi, Weierstrass, and other mathematicians who pioneered in the study of the calculus of variations was to characterize differentiable mappings $x \to g(x)$ of a finite interval $[a,b]$ into R^m ($m = 1$, or 2) which afforded a minimum to an integral of the form

$$\int_a^b f(x, y_1, \ldots, y_m; y_1', \ldots, y_m')\, dx \tag{1}$$

relative to values of this integral along mappings $x \to y(x)$ neighboring g which satisfied simple boundary conditions. Under conditions on f, g and the mappings $x \to y(x)$ neighboring g, which will be made explicit, the mapping $x \to g(x)$ satisfies the Euler equations and has a graph that is called a *minimizing extremal.*

However, it has become increasingly evident that the restriction of the calculus of variations to a study of *minimizing extremals* is not only unnatural but fails to respond to present needs of mathematics in analysis, differential geometry, mathematical physics, and engineering. This is transparent in global and differential topology. It is hoped that the theorems and methods of this book will make this equally clear in analysis, in particular in that branch of analysis which is concerned with extensions of Sturm separations, comparison, and oscillation theorems.

Apart from the development of Sturm-like theorems in my Colloquium Lectures, most applications of variational methods to the Sturm theory have been restricted to the case $m = 1$, where m is the number of variables $y_1, \ldots, y_m$ in (1). This is the case of one linear second-order differential equation. In the case $m = 1$, A. Kneser, Leighton, Nehari, Reid, Howard, and others have effectively used variational methods. In the case $m = 1$, much attention has also been given to linear, homogeneous DE (differential equations) of the third or fourth order. A book by Swanson summarizes these results and contains an extensive bibliography. It is my belief that when the systematic carrying-over of the methods of this book to the general Bolza problem is completed (as it can be), much light will be thrown on DE of orders higher than the second.

Among the exceptions* to the special emphasis on the case $m = 1$ was the comprehensive paper by Birkhoff and Hestenes in 1935. These authors confirmed some of the principal new theorems that were presented in my Colloquium Lectures of 1932. An exception is a well-known theorem that identifies the least characteristic root λ_1 of quadratic functionals $\eta \to I_r^\lambda(\eta)$, associated with the second variation, with the minimum value of $I_r^0(\eta)$, under the condition that the norm $\|\eta\| = 1$. See Hilbert-Courant [1]. We shall replace† this characterization of λ_1 by our Index Theorem 15.2. This theorem effectively characterizes not only λ_1 but each characteristic root λ_n. In particular the classical characterization of λ_1 is confirmed in Corollary 16.2 as a consequence of the Index Theorem and *extended* to the case in which the coefficients of the underlying quadratic form

$$2\omega(x,\eta,\zeta) = R_{ij}(x)\eta_i'\eta_j' + 2Q_{ij}(x)\eta_i'\eta_j + P_{ij}(x)\eta_i\eta_j \tag{2}$$

are required to be no more than continuous.

From my point of view, the principal object of study in variational analysis should be the study of the origin and nature of *critical extremals*, including the study of minimizing extremals as a special case. *A critical extremal is defined as a finite extremal arc that satisfies the transversality conditions associated with the prescribed boundary conditions.* See Definition 9.1.

In addition to the study of minimizing extremals we shall emphasize two principal applications of variational analysis:

1. *Variational topology*
2. *Quadratic functionals*

Variational topology includes and extends differential topology. In variational topology extremals replace the geodesics of differential topology and the topology of the underlying space enters strongly.

The quadratic functionals studied extend the functionals given by the classical second variation. In the "free" form that we give them in Section 15 they are more general than the classical or "derived" second variation. It is a main thesis of this book that Sturm-like separation, comparison, and oscillation theorems are best understood and extended as by-products of a variational study of the relevant quadratic functionals. The study of quadratic

* Also exceptional are paper [2] and book [3] by W. T. Reid.

† In Part V a general structure underlying the theory of characteristic roots is introduced. We follow Sturm and Liouville (see Bôcher, *Méthodes de Sturm*) in allowing the characteristic parameter λ to enter very generally, both in the integrand of the quadratic functional and in the boundary conditions. An Index Theorem 33.5 still holds, even when λ does not enter linearly, and serves to characterize roots λ_n, including the least characteristic root λ_1. A characterization of λ_1 with the aid of a simple isoperimetric condition does not seem possible in the general case.

functionals, as we shall restrict them, will turn out to be the study of a general system of m nonsingular, selfadjoint, linear, homogeneous, second-order DE under the most general selfadjoint boundary conditions. See Sections 30 and 31.

It would not be possible to present an extended variational theory had not the classical theory been so well built by men such as Bolza, Bliss, Hadamard, Kneser, Mayer, von Escherich, Carathéodory, Tonelli, and their pupils. To these names we add the names of present-day expositors such as Hestenes and Gelfand. The contributions to the theory "in the large" by mathematicians such as Deheuvels, Milnor, Serre, Pitcher, Herman, Palais, Bott, Ljusternik, Snirelman, and Smale should be recalled. The lectures by L. C. Young on the "Calculus of Variations and Optimal Control Theory" cover important novel aspects of the theory which, for the sake of brevity and clarity, are not included. In my second volume on *Variational Topology* other names will be cited.

After summarizing and refining the relevant classical theorems on the minimizing extremal this book extends and organizes in a new theory the theorems on "critical" primary and secondary extremals. Theorems on critical secondary extremals take the form of a variational study of quadratic functionals that includes extensions of the Sturm theory. An indication of the methods and technical innovations is given in the Introduction.

I am indebted to Professor George Kozlowski, Professor Ping-Fun Lam, and Professor Everett Pitcher for their critical reading of the text and helpful suggestions, and to Miss Caroline Underwood for her very skillful preparation of the manuscript.

MARSTON MORSE

Princeton, New Jersey
May 1972

CONTENTS

APPENDIX I FREE LINEAR CONDITIONS

APPENDIX II SUBORDINATE QUADRATIC FORMS AND THEIR COMPLEMENTARY FORMS

APPENDIX III MODEL FAMILIES OF QUADRATIC FORMS

Introduction

We shall single out from each chapter special theorems or concepts which are particularly significant and indicate in what respect they are significant. These comments should be read after reading the chapter in question but before reading the following chapter.

CHAPTER 1. Under the conditions of Theorem 7.4

$$\int_{a^1}^{a^2} \omega(x,\eta(x),\eta'(x))\,dx > 0. \tag{1}$$

Classical proofs of this theorem require that the mappings,

$$x \to R_{ij}(x), \qquad x \to Q_{ij}(x), \qquad x \to P_{ij}(x) \tag{2}$$

of $[a^1,a^2]$ into R be of class C^1. It suffices that the mappings (2) be continuous. Similar remarks should be made concerning the remaining theorems of Section 7. These Theorems affect the whole body of Sturm-like theorems.

CHAPTER 2. Let C_r be general end point conditions of Section 8 with $r > 0$. The functional (J,C_r) whose values $J(\chi,\alpha)$ are given by (8.19), is the sum of an integral and an "external" function Θ whose values $\Theta(\alpha)$ are determined by the end points $(X^1(\alpha), Y^1(\alpha))$ and $(X^2(\alpha), Y^2(\alpha))$ of the graph of χ.

Theorem 10.1 shows that the second variation of (J,C_r) whose values are given by the right member of (10.13) has a similar structure. The second variation is the sum of an integral and of an *external* quadratic form $b_{hk}u_hu_k$ where the r-tuple $u = (u_1, \ldots, u_r)$ is uniquely determined, as (10.14) shows, by the end points of the graph of the "variation" η.

This similarity of structure of (J,C_r) and its second variation motivated the choice of the representation (8.19) of the functional (J,C_r) and our definition (10.23) of the basic quadratic functional $\eta \to \mathscr{I}_r(\eta)$.

CHAPTER 3. With each critical extremal g of (J^λ, C_r), Definition 14.2 associates a quadratic *index form* Q^λ whose index (Theorem 14.1) equals the "count" of negative characteristic roots of Conditions (11.8), when $r > 0$, and of conditions (12.2) when $r = 0$. This index form is termed *derived* because its definition depends on an antecedent critical extremal g. It is replaced in Section 15 by a similarly structured *free index form* whose definition is independent of any antecedent critical extremal g.

In the problem of relating the geodesics g joining two fixed points A and B, on a differentiable manifold M_n, to the homological characteristics of M_n, a derived index form is associated with each geodesic g joining A to B, and enables one to assign local homological characteristics to each g. This will be elaborated in our second volume on "variational topology." See the analogous treatment of critical points of a nondegenerate function f on a differentiable manifold. Morse and Cairns [1].

CHAPTER 4. Conditions $W_r(\lambda)$ combine the Jacobi differential equations with boundary conditions $0 < r \leq 2m$. For each $\lambda \in R$, conditions $W_r(\lambda)$: (15.0) are uniquely determined by giving $\omega^\lambda(x,\eta,\zeta)$ as in Appendix I, by prescribing a $2m \times r$ matrix $\|c^s_{ih}\|$ of rank r and an r-square symmetric "*comparison*" *matrix** $\|b_{hk}\|$. If λ ranges over R, a system of conditions $W_r(\lambda)$ in which the matrices $\|c^s_{ih}\|$ and $\|b_{hk}\|$ remain invariable, is called a *canonical system* W_r of dimension r.

With each system W_r and value $\sigma \in R$ there is associated, by Definitions 15.1 and 15.2, respectively, a quadratic functional I^σ_r : (15.2) and a *free* index form Q^σ, whose index and nullity are equal (Theorem 15.3) and are given by Index Theorem 15.2. A nonnull solution of conditions $W_r(\sigma)$ has a graph which is a "critical extremal" of I^σ_r. Sturm-like theorems result from a study of critical extremals of I^σ_r. The index form Q^σ is a technical aid in this study. The exceptional case $r = 0$ is also considered.

CHAPTER 5. The focal points of free focal conditions V_r : (17.6) and (17.7) are introduced in Definition 17.4. The Focal Point Theorem† 17.3 gives the "count" of focal points of V_r, $r \geq 0$, in terms of the characteristic roots of a special canonical system $\mathscr{W}_r$ determined by V_r in Definition 17.6. When $r = 0$, Theorem 17.3 (restated as Theorem 17.4) gives the count of conjugate points of $x = a^1$, at $x = a^2$ or in (a^1,a^2) in terms of characteristic roots of the canonical system W_0 of Conditions (15.1).

In Section 18 the theory of focal points is identified with the theory of von

* The matrix $\|b_{hk}\|$ is called a *comparison matrix* because of the basic role it plays in the extended Sturmian Comparison Theorems.

† The Focal Point Theorem was first given on page 61 in Morse [1].

Escherich families of "mutually conjugate" solutions of the system of DE

$$(3) \qquad \frac{d}{dx}\,\omega_{\zeta_i}(x,\eta,\eta') = \omega_{\eta_i}(x,\eta,\eta') \qquad (i = 1, \ldots, m)$$

and leads to our Separation Theorem.

CHAPTER 6. The "Extended Separation* Theorem" 20.1 implies the classical Sturm Separation Theorem when $m = 1$ and replaces it when $m > 1$. It affirms the following. If two von Escherich families F and $\hat{F}$ of the DE (3) have exactly ρ linearly independent solutions in common, then the "count" of focal points of F in any relatively compact subinterval τ of $\aleph$ differs from the corresponding count for $\hat{F}$ by at most $m - \rho$.

In Section 21 we compare the focal points in an interval $(a^1,d]$ of two sets of focal conditions V_r and $\hat{V}_\rho$, as defined in (17.6) and (17.7). When $r = \rho = 0$ the comparison reduces to a comparison of conjugate points. In our first comparison theorem, Theorem 21.1, the hypotheses, when $m = 1$, are similar to classical hypotheses. When $m = 1$ certain classical theorems such as Reid's Theorem (Theorem 1.29 of Swanson) are extended in Section 22. Our second comparison theorem is the Nuclear Comparison Theorem 21.4. It has hypotheses which are definitely weaker than those of Comparison Theorem 21.1 but imply the same conclusions. When $m = 1$ Leighton's Theorem (Theorem 1.4, page 4 of Swanson) is of the same nature as our Nuclear Comparison Theorem. Leighton's Example 1, on page 6 of Swanson, is used by us for the same end. See Section 21.

CHAPTER 7. The Oscillation Theorem 24.1 differs in character from the classical oscillation theorems to which we shall turn in Section 37. Theorem 24.1 gives the exact value of the difference

$$(4) \qquad \text{index } W_r - \text{index } W_0 \qquad (r > 0)$$

where index W_r denotes the count of negative characteristic roots of a canonical system W_r for which $\lambda = 0$ is not a characteristic root, and index W_0 denotes the count of conjugate points of $x = a^1$ in (a^1,a^2) of the underlying DE (3). The proof of Theorem 24.1 depends upon an auxiliary theorem, Theorem 25.1, on quadratic forms, proved in Appendix II. Many other applications of Theorem 25.1 exist. We apply Theorem 24.1 to the "periodic case," as defined in Section 27.

* The Extended Separation Theorem was first given as Theorem 7 in Morse [2] and later as Theorem 8.3 on page 104 of Morse [1].

CHAPTER 8. In Section 29, general selfadjoint BC (boundary conditions) associated with prescribed selfadjoint DE are defined, as well as the *equivalence* (Definition 29.1) of any two such sets of BC. Theorem 31.1 then implies that an arbitrary set of selfadjoint BC with an accessory r-plane χ_r, (Definition 31.1), is equivalent, when $r > 0$, to some set of BC of form (29.1). When $r = 0$ it is trivial that selfadjoint BC are equivalent to the conditions, $\eta(a^1) = \eta(a^2) = \mathbf{0}$. The Conditions (15.0) and (15.1) are selfadjoint, and (up to an equivalence of BC) are general in form, as we show in Section 31.

CHAPTER 9. The general systems $\dot{W}_r$ of canonical conditions $\dot{W}_r(\lambda)$ introduced in Section 32, include the systems W_r of canonical conditions $W_r(\lambda)$ of Section 15 as special cases. Theorem 34.2 gives necessary and sufficient conditions that a system $\dot{W}_r$ have infinitely many characteristic roots. It is a corollary that the canonical systems W_r of Section 15 have infinitely many such roots.

In Section 37 Oscillation Criteria are given in a set of theorems whose proofs will be presented in a separate paper. A set of DE of the form (3) is given for $x \in (0, \infty)$ and conditioned as in Appendix I. The DE are termed *oscillatory* if the point $x = 1$ has infinitely many conjugate points following $x = 1$. A D^1-mapping $x \to \gamma(x) : [1,\infty) \to R^m$ is called a *thread* and the condition that

$$\liminf_{x\uparrow\infty} \int_1^x \omega(x,\gamma(x),\gamma'(x))\, dx = -\infty \tag{5}$$

a *thread condition*.

One of the corollaries of the general theorems presented in Section 37 for $m \geq 1$, concerns the DE

$$\eta'' + a(x)\eta = 0 \qquad (0 < x < \infty;\ a(x) \geq 0) \tag{6}$$

where the mapping $x \to a(x)$ is continuous. See Swanson, Chapter 2. The corresponding Thread Condition has the form

$$\liminf_{x\uparrow\infty} \int_1^x (\gamma'^2(x) - a(x)\gamma^2(x))\, dx = -\infty. \tag{7}$$

The corollary takes the form:

COROLLARY. *A necessary and sufficient condition that the DE* (6) *be oscillatory is that the Thread Condition* (7) *be satisfied by some thread* γ.

The Leighton-Wintner Theorem that the DE (6) is oscillatory if

$$\int_1^\infty a(x)\, dx = \infty,$$

is implied by the corollary, on taking the thread as the mapping $x \rightarrow \gamma(x) \equiv 1$.

Note. The major class of admissible curves is that of x-parameterized curves of class D^1, including broken extremals. This class of curves is adequate for a first simple presentation of the new theorems and structure. It is preferred to a choice of Hilbert Space from the different Banach Spaces which could be profitably used in variational analysis, because it leaves that choice fully open, while preserving a close connection with the classical theory. "Index forms" are definable in terms of broken extremals and permit an easy transition from the Sturmian properties of a critical extremal to properties which are purely topological. In our presentation of *Variational Topology* which is to follow, admissible curves will range from the merely continuous to the real analytic and include curves definable in the terminology of Lebesgue.

Part I

CRITICAL EXTREMALS

CHAPTER 1

Minimizing Extremals g: Fixed Endpoints

This chapter gives a review of classical results found in varying forms in the works of Bliss, Carathéodory, Bolza, Hadamard, Tonelli, and other writers on variational theory. Present day notation is introduced and classical proofs are freely modified.

1. The Euler Equations

To properly condition admissible curves several definitions are needed.

Mappings of Class D^0. Let $[a^1,a^2]$ be a closed interval of the axis R of real numbers. Two intervals of the R-axis are termed *nonoverlapping* if their intersection is empty or a point. A mapping $x \to h(x) : [a^1,a^2] \to R$ will be said to be of *class* D^0 if $[a^1,a^2]$ is the union of a finite set of closed, nonoverlapping subintervals I, on the interior $\mathring{I}$ of each of which h is continuous and at each endpoint x_0 of which h has a finite limit as x_0 is approached from $\mathring{I}$.

Mappings of Class D^1. A mapping $x \to h(x) : [a^1,a^2] \to R$ will be said to be of *class* D^1 if h is continuous and if $[a^1,a^2]$ is the union of a finite set of nonoverlapping closed subintervals I on the interior $\mathring{I}$ of each of which h' exists and is continuous, and at each endpoint x_0 of which h' has a finite limit when x_0 is approached from $\mathring{I}$.

The Preintegrand f. Let $(x, y_1, \ldots, y_m)$ be written as (x,y) and be the set of rectangular coordinates of a point (x,y) in a euclidean space E_{m+1} of dimension $m + 1$. Let X be an open connected subset of E_{m+1}. Let $p = (p_1, \ldots, p_m)$ be an arbitrary point in the m-fold Cartesian product R^m of R. For brevity we write

$$(x, y_1, \ldots, y_m, p_1, \ldots, p_m) = (x,y,p) \tag{1.1}$$

and consider a real-valued function*

$$(x,y,p) \to f(x,y,p) : X \times R^m \to R$$

such that f and its partial derivatives f_{y_i}, f_{p_i} are of class C^2. These conditions are satisfied if f is of class C^3. It is not sufficient that f be merely of class C^2 if the solution $Y(x : c,a,b)$: (2.14) of the Euler equations is to have continuous second order partial derivatives with respect to the initial parameters c, a, b. Such differentiability is essential in several places.

The "Mapping" g *and the "Curve"* g. Let there be given mappings

$$x \to g_i(x) : [a^1,a^2] \to R \qquad (i = 1, \ldots, m) \tag{1.2}$$

each of class D^1. A mapping

$$x \to g(x) : [a^1,a^2] \to R^m$$

is thereby defined, and said to be of class D^1. The graph of g is given in X and will be called the "curve" g. On the curve g

$$y_i = g_i(x) \qquad (x \in [a^1,a^2],\ i = 1, \ldots, m). \tag{1.3}$$

Let there be given a second mapping

$$x \to \gamma(x) : [a^1,a^2] \to R^m \tag{1.3$'$}$$

of class D^1. We suppose that the corresponding curve γ joins the endpoints of g in X. We then term γ *admissible* relative to the *base curve* g.

The Integral **J**. We shall compare the Riemann integral

$$\mathbf{J}(\gamma) = \int_{a^1}^{a^2} f(x,\gamma(x),\gamma'(x))\, dx \tag{1.4}$$

"along" the curve γ with the integral $\mathbf{J}(g)$ "along" g and prove the following theorem. In $\mathbf{J}(\gamma)$, γ is understood to be either the curve γ or the mapping γ.

THEOREM 1.1. *In order that* g *(of class* D^1*) afford a minimum to* **J** *relative to neighboring "admissible" curves* γ,† *it is necessary that there exist constants* $c_1, \ldots, c_m$ *such that*

$$f_{p_i}(x,g(x),g'(x)) = \int_{a^1}^{x} f_{y_i}(a,g(a),g'(a))\, da + c_i \qquad (i = 1, \ldots, m) \tag{1.5}$$

for each $x \in [a^1,a^2]$ *at which* $g'(x)$ *is defined.*

* We term f a *preintegrand* because of its use in defining the integral **J**.

† "Relative to neighboring admissible curves γ" means relative to admissible curves γ in some neighborhood of g in X.

For each i on the range $1, \ldots, m$ let $x \to \eta_i(x)$ be a mapping of $[a^1,a^2]$ into R of class D^1 such that $\eta_i(a^1) = \eta_i(a^2) = 0$. A curve on which

$$y_i = g_i(x) + e\eta_i(x) \qquad (x \in [a^1,a^2],\ i = 1, \ldots, m) \tag{1.6}$$

joins g's endpoints and is in X for $|e|$ sufficiently small. Let $J(e)$ be the value of **J** along this curve. Observe that

$$J'(0) = \int_{a^1}^{a^2} (\eta_i'(x) f_{p_i}^x + \eta_i(x) f_{y_i}^x)\, dx, \tag{1.7}^*$$

where the superscript x means evaluation for $(x,y,p) = (x,g(x),g'(x))$ when $g'(x)$ exists, with p the left or right limit of $g'(x)$ at other points of $[a^1,a^2]$. The right member of (1.7) is called the *first variation* of **J** *along* g induced by the *variation* η. Since $J(e) \geq J(0)$ by hypothesis, we conclude that $J'(0) = 0$.

Upon integrating each product $\eta_i(x) f_{y_i}^x$ in (1.7) by parts, one finds that

$$J'(0) = \int_{a^1}^{a^2} \eta_i'(x) \left(f_{p_i}^x - \int_{a^1}^{x} f_{y_i}^a\, da \right) dx = 0. \tag{1.8}$$

Theorem 1.1 will follow from (1.8) and the modified DuBois Reymond Lemma 1.1.

LEMMA 1.1. *If* $\mathrm{x} \to \varphi(\mathrm{x}) : [\mathrm{a}^1,\mathrm{a}^2] \to \mathrm{R}$ *is a mapping which is of class* D^0 *and if*

$$0 = \int_{\mathrm{a}^1}^{\mathrm{a}^2} \zeta'(\mathrm{x})\varphi(\mathrm{x})\, \mathrm{dx} \tag{1.9}$$

for each mapping $\mathrm{x} \to \zeta(\mathrm{x}) : [\mathrm{a}^1,\mathrm{a}^2] \to \mathrm{R}$ *of class* D^1 *which vanishes at* a^1 *and* a^2, *then* φ *has the same value* c *at each point* $\mathrm{x} \in [\mathrm{a}^1,\mathrm{a}^2]$ *at which* φ *is continuous*.

Let c be a constant such that $\int_{a^1}^{a^2} (\varphi(x) - c)\, dx = 0$. The mapping

$$x \to \zeta(x) = \int_{a^1}^{x} (\varphi(a) - c)\, da$$

of $[a^1,a^2]$ into R is "admissible" as ζ in Lemma 1.1. For this ζ, (1.9) takes the form

$$0 = \int_{a^1}^{a^2} (\varphi(x) - c)\varphi(x)\, dx = \int_{a^1}^{a^2} (\varphi(x) - c)^2\, dx = 0$$

from which it follows that $\varphi(x) = c$ at points of continuity of φ.

To complete the proof of Theorem 1.1 we prefer one of the integers, say k,

* We follow the convention that an index i repeated in two factors of a monomial implies the summation of the monomial over the range of i.

on the range $1, \ldots, m$. For $i \neq k$ we take η_i identically zero on $[a^1,a^2]$. The mapping

$$x \rightarrow \varphi(x) = f^x_{p_k} - \int_{a^1}^{x} f^a_{y_k}\, da \tag{1.10}$$

of $[a^1,a^2]$ into R satisfies the conditions on φ in Lemma 1.1, so that (1.5) holds for $i = k$ by virtue of Lemma 1.1. Since k can be prescribed on the range $1, \ldots, m$, Theorem 1.1 follows.

COROLLARY 1.1. *Under the hypotheses of Theorem* 1.1 *the following is true:*

(i) *Each subarc of* g *which is of class** C^1 *satisfies the Euler equations*

$$\frac{d}{dx} f_{p_i}(x,y,y') = f_{y_i}(x,y,y') \qquad (i = 1, \ldots, m). \tag{1.11}$$

(ii) *At a corner of* g *with* x *coordinate* x_0

$$\lim_{x \uparrow x_0} f^x_{p_i} = \lim_{x \downarrow x_0} f^x_{p_i} \qquad (i = 1, \ldots, m). \tag{1.12}$$†

(iii) *If* g *is of class* C^1 *and if the* m*-square determinant*

$$|f^x_{p_i p_j}| \neq 0 \qquad (x \in [a^1,a^2]) \tag{1.13}$$

then g *is of at least class* C^2.

Statements (i) and (ii) are immediate consequences of Theorem 1.1. Mason and Bliss [1] prove (iii) as follows. Given g, the m conditions,

$$f_{p_i}(x,g(x),z) = \int_{a^1}^{x} f_{y_i}(a,g(a),g'(a))\, da + c_i \tag{1.14}$$

on the m-tuple z for $x \in [a^1,a^2]$, have a solution $z(x) = g'(x)$ for $x \in [a^1,a^2]$. By virtue of (1.13) the classical implicit function theorems imply that the solution $x \rightarrow z(x) = g'(x)$ is of class C^1.

At the end of the next section we shall see that g is of class C^3 for $x \in [a^1,a^2]$ when (1.13) holds.

Definition 1.1. *An Extremal.* A C^2-mapping

$$x \rightarrow (y_1(x), \ldots, y_m(x)) : [a^1,a^2] \rightarrow R^m \tag{1.15}$$

* We understand that an x-parameterized curve γ is of *class* C^r, $r > 0$ on a *closed* interval I of the x-axis, if γ admits an extension as an x-parameterized curve of class C^r on an *open* extension of I.

† Conditions (1.12) are called "Weierstrass-Erdmann corner conditions."

satisfying the Euler equations (1.11) defines an x-parameterized curve $y = y(x)$ called an *extremal*. The extremal is the "*graph*" of the *solution* (1.15). We shall call the extremal (1.15) *primary* to distinguish it from *secondary* extremals of Section 4.

Exercise

1.1. Suppose that g is an extremal and that the variation η, introduced in (1.6), has arbitrary end values $\eta(a^2)$ and $\eta(a^1)$. Show that the corresponding "*first variation*" of **J** along g, as given by (1.7), has the value

$$J'(0) = \left[f^x_{p_i} \eta_i(x) \right]^{x=a^2}_{x=a^1}. \tag{1.16}$$

2. The Existence of Extremals Near g

There is given an x-parameterized extremal g in X of class C^2 for $x \in [a^1,a^2]$. *We shall assume that* (1.13) *holds along* g.

Recall that $f(x,y,y')$ was defined for triples $(x,y,y') \in X \times R^m$. We shall seek solutions of the Euler equations conditioning triples (x,y,y') in an open neighborhood N_g in $X \times R^m$ of the set of triples (x,y,y') on g. Triples (x,y,y') in N_g provide a point (x,y) and a set of slopes y'_i. They are called *ordinary triples* to distinguish them from *canonical triples* (x,y,v) which we shall now define.

Canonical triples (x,y,v). We introduce triples

$$(x,y,v) = (x, y_1, \ldots, y_m, v_1, \ldots, v_m) \tag{2.1}$$

obtained from ordinary triples $(x,y,y') \in N_g$ by setting

$$v_i = f_{p_i}(x,y,y') \qquad (i = 1, \ldots, m). \tag{2.2}$$

We shall write $v = f_p(x,y,y')$ when (2.2) holds and term (x,y,v) a *canonical triple*.

Under the assumption that (1.13) holds along the extremal g, classical implicit function theorems imply the following.

LEMMA 2.1. *If* $\mathrm{N_g}$ *is a sufficiently small open neighborhood in* $\mathrm{X \times R^m}$ *of the subset of ordinary triples* $(\mathrm{x,y,y'})$ *on* g, *the determinant*

$$|\mathrm{f}_{\mathrm{p_i p_j}}(\mathrm{x,y,y'})| \neq 0 \qquad ((\mathrm{x,y,y'}) \in \mathrm{N_g}) \tag{2.3}$$

*and there exists a diff**

$$\mathrm{T} : (\mathrm{x,y,y'}) \to (\mathrm{x,y,v}) : \mathrm{N_g} \to \mathrm{T(N_g)} \tag{2.4}$$

* A diff is by definition biunique and surjective and has an inverse which is of the same differentiability class.

of class C^2 *of* N_g *onto an open neighborhood* $T(N_g)$ *in* $X \times R^m$ *of the subset of canonical triples* (x,y,v) *on* g *such that when* (x,y,y') *goes into* (x,y,v), (2.2) *holds and, equivalently,*

$$y_i' = P_i(x,y,v) \qquad (i = 1, \ldots, m) \tag{2.5$'$}$$

where for each i, P_i *is a mapping*

$$(x,y,v) \to P_i(x,y,v) : T(N_g) \to R \tag{2.5$''$}$$

of class C^2.

The Euler equations transformed. For N_g such that the diff T of Lemma 2.1 is well-defined we introduce the differential equations

$$\frac{dy_i}{dx} = P_i(x,y,v) \tag{2.6$'$}$$

$$\frac{dv_i}{dx} = f_{y_i}(x,y,P(x,y,v)) \qquad (i = 1, \ldots, m) \tag{2.6$''$}$$

restricting the triples (x,y,v) to $T(N_g)$. Given a solution $x \to y(x) = (y_1(x), \ldots, y_m(x))$ of the Euler equations such that $(x,y(x),y'(x)) \in N_g$, one obtains a solution $x \to (y(x),v(x))$ of the differential equations (2.6) by setting $y = y(x)$ and

$$v_i(x) = f_{p_i}(x,y(x),y'(x)) \qquad (x \in [a^1,a^2], (i = 1, \ldots, m)). \tag{2.7}$$

Conversely a solution $x \to (y(x),v(x))$ of the equations (2.6) is such that (2.7) holds when (2.6)$'$ holds and, by virtue of (2.6)$''$,

$$\frac{d}{dx} f_{p_i}(x,y(x),y'(x)) = f_{y_i}(x,y(x),y'(x)) \qquad (x \in [a^1,a^2]) \tag{2.8}$$

with $(x,y(x),y'(x))$ in N_g.

We summarize as follows.

LEMMA 2.2. *If the Euler equations are restricted by the condition that triples* (x,y,y') *be in the domain* N_g *of the diff* T, *and the equations* (2.6) *by the condition that triples* (x,y,v) *be in* $T(N_g)$, *then the set of ordinary triples* (x,y,y') *on a solution* $x \to y(x)$ *of the restricted Euler equations is mapped by* T *biuniquely onto the set of canonical triples on a solution* $x \to (y(x),v(x))$ *of the restricted differential equations* (2.6).

The mapping of solutions of the Euler equations, restricted as above, into solutions of the differential equations (2.6), *restricted as above, is biunique and surjective.*

The dependence of extremals upon initial parameters. We shall make use of classical theorems of ordinary differential equation theory to show how extremals near the extremal g depend upon appropriate initial points and slopes. To do this we first show how solutions of (2.6) depend upon initial canonical triples near the initial canonical triple on g. We begin with notation. We first extend the interval $[a^1,a^2]$.

The interval $\aleph \supset [a^1,a^2]$. If (x,y,y') is an ordinary triple we shall set $\pi(x,y,y') = x$. We refer to the domain N_g of the diff T introduced in Lemma 2.1. Under conditions to be defined let

$$\aleph \supset [a^1,a^2] \tag{2.9}$$

be an open interval of the x-axis containing the interval $[a^1,a^2]$. Let U be an open neighborhood in N_g of the initial triple $(\bar{x},\bar{y},\bar{y}')$ on g. We suppose U so small that $\pi(U) \subset \aleph$. Set $V = T(U)$. The set V is an open neighborhood in $T(N_g)$ of the initial canonical triple $(\bar{x},\bar{y},\bar{v})$ on g. Canonical triples in V will be denoted by (x^0,y^0,v^0). Classical theorems in ordinary differential equation theory imply the following.

A. If *the interval* $\aleph$ differs sufficiently little from $[a^1,a^2]$ and if the above neighborhood V of $(\bar{x},\bar{y},\bar{v})$ is sufficiently small with $\pi(U) \subset \aleph$, there then exist mappings

$$(x : x^0,y^0,v^0) \to \mathscr{Y}_i(x : x^0,y^0,v^0) : \aleph \times V \to R \qquad (i = 1, \ldots, m) \tag{2.10$'$}$$

$$(x : x^0,y^0,v^0) \to \mathscr{V}_i(x : x^0,y^0,v^0) : \aleph \times V \to R \tag{2.10$''$}$$

of class C^2 such that, for (x^0,y^0,v^0) fixed in V, the partial mappings of $\aleph$ into R induced by the mappings (2.10) afford solutions of the differential equations (2.6) such that

$$\begin{cases} y_i^0 = \mathscr{Y}_i(x^0 : x^0,y^0,v^0) & (i = 1, \ldots, m) \quad (2.11)' \\ v_i^0 = \mathscr{V}_i(x^0 : x^0,y^0,v^0) & (2.11)'' \end{cases}$$

Extremals near g. The family (2.10) of solutions of the differential equations (2.6) leads to a family of solutions of the Euler equations as follows. Triples $(x,y,y') \in U$ will be denoted by (c,a,b) or more explicitly, by

$$(c, a_1, \ldots, a_m, b_1, \ldots, b_m). \tag{2.12}$$

For $(x : c,a,b) \in \aleph \times U$ and $i = 1, \ldots, m$, we set

$$\mathscr{Y}_i(x : c,a,f_p(c,a,b)) = Y_i(x : c,a,b) \tag{2.13}$$

and state the following theorem.

THEOREM 2.1. *If* $\aleph$ *and* U *are conditioned as in* **A**, *the mappings*

$$(x : c,a,b) \to Y_i(x : c,a,b) : \aleph \times U \to R \qquad (i = 1, \ldots, m) \tag{2.14}$$

are of class C^2 *and for fixed* $(c,a,b) \in U$ *induce solutions*

$$x \rightarrow Y_i(x : c,a,b) : \aleph \rightarrow R \qquad (i = 1, \ldots, m) \tag{2.15}$$

of the Euler differential equations such that

$$\begin{cases} a_i = Y_i(c : c,a,b) \\ b_i = Y_{ix}(c : c,a,b) \end{cases} \tag{2.16}$$

and the resultant triples (x,Y,Y_x) *are in* N_g.

We have seen that (2.8) is satisfied by a mapping $x \rightarrow y(x)$ whenever (2.6) is satisfied by a pair of mappings $x \rightarrow y(x)$, $x \rightarrow v(x)$. Similarly, for fixed (c,a,b) the partial mapping (2.15) satisfies (2.8), since for fixed (x^0,y^0,v^0), the partial mappings

$$\begin{cases} x \rightarrow \mathscr{Y}(x : x^0,y^0,v^0) \\ x \rightarrow \mathscr{V}(x : x^0,y^0,v^0) \end{cases} \qquad (x \in \aleph) \tag{2.17}$$

satisfy (2.6) and the identity

$$\mathscr{Y}(x : x^0,y^0,v^0) \equiv Y(x : c,a,b) \qquad (x \in \aleph)$$

in x is implied by (2.13) if $(x^0,y^0) = (c,a)$ and if $v^0 = f_p(c,a,b)$.

That (2.16) holds is verified as follows. A canonical triple $(x^0,y^0,v^0) \in V$ is on the solution (2.10) of (2.6) when $x = x^0$. If then (c,a,b) is an ordinary triple such that $(x^0,y^0,v^0) = T(c,a,b)$, Lemma 2.2 implies that (c,a,b) is a triple (x,y,y') on the solution (2.15) of the Euler equation when $x = c$. That is, (2.16) holds.

Note. The solutions $x \rightarrow y(x)$, $x \rightarrow v(x)$ of the differential equations (2.6) given by (2.10) are of at least class C^2, and since the right hand members of the differential equations (2.6) are of class C^2, the solutions $x \rightarrow y(x)$, $x \rightarrow v(x)$ are of at least class C^3. The solutions $x \rightarrow y(x)$ of the Euler equations affirmed to exist in Theorem 2.1 are accordingly of class C^3 by virtue of their definition in (2.13). These solutions have been derived under the assumption that (1.13) holds along each of their graphs.

The mappings Y_i of Theorem 2.1 are affirmed to be of class C^2. It follows that *the mappings* Y_{ix} *are also of class* C^2.

The mappings Y_i have been derived under the assumption that (1.13) holds along the graph of the solutions (2.15) of the Euler equations. That the mappings Y_{ix} are of class C^2 now follows from the fact that the right hand members of the differential equations (2.6) are of class C^2.

Definition 2.1. *The extension* $g^{\aleph}$ *of* g. In particular the extremal g with its x-domain $[a^1,a^2]$ admits an extension $g^{\aleph}$, for $x \in \aleph$, as an extremal of class

C^3. We suppose $\aleph$ so small that along $g^{\aleph}$ the determinant,

$$|f_{p_i p_j}(x,y,y')| \neq 0. \tag{2.18}$$

There are three conditions on $\aleph$ in Section 2, the condition (2.9), the condition **A** of Theorem 2.1 and the condition that (2.18) hold for $x \in \aleph$.

On the other hand, the interval $\aleph$ of Appendix I will be taken as an arbitrary open interval.

3. The Necessary Conditions of Weierstrass and Legendre

The Weierstrass E-function is defined by its values

$$E(x,y,p,q) = f(x,y,q) - f(x,y,p) - (q_i - p_i)f_{p_i}(x,y,p) \tag{3.0}$$

for $(x,y) \in X$ and p and q m-tuples in R^m.

We shall prove the following theorem.

THEOREM 3.1. (*Weierstrass*) *If an arc* g *of form* (1.2), *of class* C¹, *affords a minimum to* **J** *relative to neighboring* x-*parameterized curves* γ *of class* D¹, *joining the endpoints of* g, *then*

$$E(x,y,y',q) \geq 0 \tag{3.1}$$

for each triple (x,y,y') *on* g *and for any* m-*tuple* $q \in R^m$.

Let (x^1,y^1) be a point of g. We treat the case in which $x^1 > a^1$. The case in which $x^1 = a^1$ requires at most obvious changes in which a^2 plays the role of a^1.

The Arc k. Let k be a short open x-parametrized arc in X meeting the point (x^1,y^1) on g. We suppose that k has a representation $y_i = k_i(x)$, $i = 1, \ldots, m$, where the mappings $x \to k_i(x)$ are of class C^1 on a small open interval for x containing x^1. We suppose moreover that the slopes $k_i'(x^1)$ equal q_i, where q is prescribed in the Theorem.

We shall compare **J** along g with **J** along an x-parameterized curve γ_α of class D^1 joining the endpoints of g and consisting of three successive arcs each of class* C^1, the first of which, λ_α, will now be defined.†

The Arc λ_α. Let a parameter α be restricted to an arbitrarily small subinterval $[x^1 - e, x^1]$ of the interval $(a^1,x^1]$. The coordinates y_i of points (x,y) on λ_α shall have the values

$$y_i(x,\alpha) = \frac{(x - a^1)}{(\alpha - a^1)}(k_i(\alpha) - g_i(\alpha)) + g_i(x) \qquad (i = 1, \ldots, m) \tag{3.2}$$

* See first footnote to Corollary 1.1.

† The reader should graph γ_α.

for $x \in [a^1,\alpha]$. The arc λ_α joins the initial point of g to the point P_α on k at which $x = \alpha \leq x^1$. When the parameter $\alpha = x^1$, $k_i(\alpha) = g_i(\alpha)$, and the arc λ_α reduces to the subarc of g on which $a^1 \leq x \leq x^1$.

The Curve γ_α. The curve γ_α shall consist of three successive arcs:

(1) the arc λ_α terminating at the point P_α of k
(2) the subarc of k on which $\alpha \leq x \leq x^1$
(3) the subarc of g on which $x^1 \leq x \leq a^2$

If $\alpha = x^1$ the second subarc of γ_α reduces to the point (x^1,y^1) on g and γ_α to g.

Set $J(\alpha) = \mathbf{J}(\gamma_\alpha)$. One finds that $J(\alpha)$ equals

$$(3.3) \quad \int_{a^1}^{\alpha} f(x,y(x,\alpha),y_x(x,\alpha))\,dx + \int_{\alpha}^{x^1} f(x,k(x),k'(x))\,dx + \int_{x^1}^{a^2} f(x,g(x),g'(x))\,dx$$

If g is a minimizing arc the left derivative $J_-'(x^1) \leq 0$. Upon setting $p_i = g_i'(x^1)$ for $i = 1, \ldots, m$, we find that $J_-'(x^1)$ equals

$$\int_{a^1}^{x^1} (y_{i\alpha}(x,x^1)f_{y_i}^x + y_{i\alpha x}(x,x^1)f_{p_i}^x)\,dx - f(x^1,y^1,q) + f(x^1,y^1,p)$$

where the superscript x means evaluation for $(x,y,p) = (x,g(x),g'(x))$. If one integrates each term $y_{i\alpha x}(x,x^1)f_{p_i}^x$ by parts one finds that

$$(3.4) \qquad J_-'(x^1) = f(x^1,y^1,p) - f(x^1,y^1,q) + \Big[y_{i\alpha}(x,x^1)f_{p_i}^x\Big]_{x=a^1}^{x=x^1}.$$

From the identities $y_i(a^1,\alpha) = g_i(a^1)$ and $y_i(\alpha,\alpha) = k_i(\alpha)$ in α for α near x^1, it follows, upon differentiating as to α and setting $\alpha = x^1$, that

$$y_{i\alpha}(a^1,x^1) = 0, \qquad p_i + y_{i\alpha}(x^1,x^1) = q_i.$$

Using these results in (3.4) we find that

$$J_-'(x^1) = f(x^1,y^1,p) - f(x^1,y^1,q) + (q_i - p_i)f_{p_i}(x^1,y^1,p).$$

Since $J_-'(x^1) \leq 0$ and $p = g'(x^1)$, (3.1) follows.

When g is of class C^1 we distinguish between weak and strong minima afforded to $\mathbf{J}$ by g as follows.

Weak and Strong Minima. A minimum afforded by g to $\mathbf{J}$ is called a *strong* minimum if there are no conditions on the slopes of the curves γ compared with g other than that γ be of class D^1. Suppose that the curve γ is the graph of a mapping $x \to \gamma(x)$ of class D^1. The curve g will be said to afford a *weak minimum* to $\mathbf{J}$, if for each sufficiently small positive constant e

and for each admissible curve γ such that

$$\|\gamma(x) - g(x)\| + \|\gamma'(x) - g'(x)\| < e \qquad (a^1 \le x \le a^2) \tag{3.5}$$

when $\gamma'(x)$ exists, the relation $\mathbf{J}(g) \le \mathbf{J}(\gamma)$ is true.

The proof of Theorem 1.1 shows that the following theorem is true. See footnote to Corollary 1.1 for a convention.

THEOREM 3.2. *If an* x*-parameterized curve* g *of class* C^1 *on an interval* $[a^1,a^2]$ *affords a weak minimum to* $\mathbf{J}$ *then for* i *on the range* $1, \ldots, m$

$$\frac{d}{dx} f_{p_i}(x,g(x),g'(x)) - f_{y_i}((x,g(x),g'(x))) = 0$$

for x *on* $[a^1,a^2]$.

For g to afford a weak minimum to $\mathbf{J}$ it is necessary that condition (3.1) hold in its "weak form," that is for sets $(x,y,y',q) = (x,g(x),g'(x),q)$ such that $|q - g'(x)| < \delta$ for some sufficiently small positive constant δ. This weak Weierstrass condition implies a condition due to Legendre when $m = 1$.

THEOREM 3.3. *If* g *gives a weak minimum to* $\mathbf{J}$ *it is necessary that*

$$f_{p_ip_j}(x,g(x),g'(x))z_iz_j \ge 0 \qquad (x \in [a^1,a^2]) \tag{3.6}$$

for a summation with respect to i, j *on the range* $1, \ldots, m$ *and for arbitrary* m*-tuples* $(z_1, \ldots, z_m)$.

To establish (3.6) one fixes the m-tuple z and introduces the mapping

$$e \to \varphi(e) = E(x,g(x),g'(x),g'(x) + ez)$$

for arbitrary e. One finds that $\varphi'(0) = 0$ and that $\varphi''(0)$ equals the left member of (3.6). The weak Weierstrass condition implies that $\varphi(e) \ge \varphi(0)$ if $|e|$ is sufficiently small. Hence $\varphi''(0) \ge 0$, implying (3.6).

4. The Jacobi Condition

In Section 4 let g be an x-parameterized extremal of class C^2 defined by mappings of class C^2 of form (1.2). Let γ^e be an x-parameterized curve which joins the endpoints of g and on which (1.6) holds. Let $J(e) = \mathbf{J}(\gamma^e)$. Given g, for each $x \in [a^1,a^2]$ we introduce the quadratic form

$$2\Omega(x,\eta,\zeta) = f^x_{p_ip_j}\zeta_i\zeta_j + 2f^x_{p_iy_j}\zeta_i\eta_j + f^x_{y_iy_j}\eta_i\eta_j \tag{4.0}$$

in the variables $\eta_1, \ldots, \eta_m, \zeta_1, \ldots, \zeta_m$. As previously, the superscript x means evaluation for $(x,y,p) = (x,g(x),g'(x))$.

In the representation (1.6) of γ^e we suppose that the mapping $x \to \eta(x)$ is of class D^1. The mapping $x \to \eta'(x)$ is well-defined for $x \in [a^1,a^2]$ except at a finite number of points and can be extended so as to be of class D^0 over $[a^1,a^2]$. Because of this the Riemann integral,

$$\mathscr{I}_0(\eta) = \int_{a^1}^{a^2} 2\Omega(x,\eta(x),\eta'(x))\, dx \tag{4.1)$'$(}$$

is well-defined. Differentiation of the integral $J(e) = \mathbf{J}(\gamma^e)$ under the integral sign leads to the equations

$$J'(0) = 0, \qquad J''(0) = \mathscr{I}_0(\eta). \tag{4.1)$''$(}$$

One terms $\mathscr{I}_0(\eta)$ the value at η of the second variation $\mathscr{I}_0$ of **J** *along* g.

The Jacobi Differential Equations. If g affords a weak minimum to **J** it is necessary that $J'(0) = 0$ and that $J''(0) \geq 0$. Zero is thus a minimum value of the second variation $\mathscr{I}_0(\eta)$ for η conditioned as above. A mapping $x \to u(x) = (u_1(x), \ldots, u_m(x))$ of $[a^1,a^2]$ into R^m of class C^1, such that $u(a^1) = u(a^2) = \mathbf{0}$ and $\mathscr{I}_0(\eta) \geq \mathscr{I}_0(u)$, for all admissible η of class D^1, must satisfy the Euler equations for the integral (4.1)$'$, that is the equations

$$\frac{d}{dx}\Omega_{\zeta_i}(x,u,u') - \Omega_{\eta_i}(x,u,u') = 0 \qquad (i = 1, \ldots, m). \tag{4.2}$$

The equations (4.2) are called the *Jacobi differential equations* (abbreviated JDE) based on g.

To verify that the JDE are satisfied by the mapping $x \to u(x)$ under the above conditions on u, one must review the proof of Corollary 1.1 when 2Ω replaces f. The conditions that f_{y_i}, f_{p_i}, and f be of class C^2 on the domain of f are not in general satisfied since the coefficients in the quadratic form $2\Omega(x,\eta,\zeta)$ are not in general of class C^2. However a review of the proof of Corollary 1.1 reveals the following.

LEMMA 4.1. *Corollary* 1.1 *holds in analogous form when the integral* (1.4) *is replaced by the integral* (4.1)$'$ *and* g *is replaced by the graph* γ *of a mapping* $x \to v(x)$ *of* $[a^1,a^2]$ *into* E_m *of class* D^1 *such that* $\mathbf{0} = v(a^1) = v(a^2)$ *and* $\mathscr{I}_0(\gamma) \leq \mathscr{I}_0(\eta)$ *for each* η *conditioned as in* (4.1)$'$.

Lemma 4.1 here implies the following:

(i) *Each subarc* $x \to v(x)$ *of* γ *of class* C^1 *satisfies the* JDE (4.2).

(ii) *At a corner of* γ *with* x *coordinate* α

$$\lim_{x \downarrow \alpha} \Omega_{\zeta_i}(x,v(x),v'(x)) = \lim_{x \uparrow \alpha} \Omega_{\zeta_i}(x,v(x),v'(x)) \qquad (i = 1, \ldots, m).$$

(iii) *If* γ *is of class* C^1 *and if the determinant* (1.13) *does not vanish for* $x \in [a^1,a^2]$ *then* γ *is of at least class* C^2.

A point $x = c$ at which (1.13) holds will be called a *nonsingular point* of the JDE. If each point $x \in [a^1,a^2]$ is a nonsingular point of the JDE, the left members of the equations (4.2) can be given a form, linear and homogeneous in the variables u_i, u_i', u_i'', $i = 1, \ldots, m$, with coefficients which are of class C^1 on $[a^1,a^2]$. The determinant of the coefficients of the variables $u_1'', \ldots, u_m''$ is $|f^x_{p_i p_j}|$.

In a sufficiently small open neighborhood w on $[a^1,a^2]$ of a nonsingular point $x = c$ of the JDE, a solution of the JDE, if known to be of class C^1, is necessarily of class C^2. On the interval w all solutions are linearly dependent on $2m$ such solutions. A solution $x \to u(x)$ such that $u(\alpha) = u'(\alpha) = \mathbf{0}$ at some point $x = \alpha \in w$, vanishes identically for $x \in w$. Cf. Appendix I. In (7.12) and in Appendix I the "derived" quadratic form 2Ω : (4.0) is replaced by a "free" quadratic form 2ω whose coefficients R_{ij}, Q_{ij}, P_{ij} are required to be no more than continuous. When Ω is replaced by ω, the resultant JDE have solutions which, in general, are of class C^1 (not class C^2).

Definition 4.1 The *graph* in the euclidean space of coordinates $(x, u_1, \ldots, u_m)$ of a solution $x \to u(x)$ of class C^2 of the JDE will be called a *secondary extremal u.**

Lemma 4.1 is an aid in proving Theorem 4.1 below. A second lemma useful in proving Theorem 4.1 follows.

LEMMA 4.2. *A solution* $x \to u(x)$ *of the* JDE* *of class* C^1 *on a subinterval* $[a^1,\alpha]$ *of* $[a^1,a^2]$ *which vanishes when* $x = a^1$ *and* $x = \alpha$ *is such that*

$$\int_{a^1}^{\alpha} \Omega(x,u(x),u'(x))\,dx = 0. \tag{4.3}$$

There is no assumption that points $x \in [a^1,\alpha]$ are nonsingular points of the JDE.

We begin the proof by noting that

$$2\Omega^x = u_i(x)\Omega^x_{\eta_i} + u_i'(x)\Omega^x_{\zeta_i} \tag{4.4}$$

where the superscript x means evaluation for $(x,\eta,\zeta) = (x,u(x),u'(x))$. If the terms $u_i'(x)\Omega^x_{\zeta_i}$ are integrated by parts we find that

$$\int_{a^1}^{\alpha} 2\Omega^x\,dx = \Big[u_i(x)\Omega^x_{\zeta_i}\Big]_{x=a^1}^{x=\alpha} + \int_{a^1}^{\alpha} u_i(x)\Big[\Omega^x_{\eta_i} - \frac{d}{dx}\,\Omega^x_{\zeta_i}\Big]\,dx \tag{4.5}$$

and the relation (4.3) follows.

* Relative to g.

Definition 4.2. *Conjugate points.* Let $x \to u(x)$ be a solution of the JDE of class C^1 on a closed subinterval $H = [c,\alpha]$ or $[\alpha,c]$ of $[a^1,a^2]$ such that $u(\alpha) = u(c) = \mathbf{0}$. If u does not vanish identically on H arbitrarily near $x = \alpha$ then the point $x = \alpha$ on the x-axis will be termed *conjugate* to the point $x = c$ on the x-axis. The point on g at which $x = \alpha$ will then be said to be *conjugate* to the point on g at which $x = c$.

Note. This definition of a conjugate point reduces in Section 5 to a simpler form with a classical geometric interpretation under the hypothesis that (2.3) holds at each point of g, that is, that each $x \in [a^1,a^2]$ is a nonsingular point of the JDE.

The following theorem gives the Jacobi necessary condition.

THEOREM 4.1. *If* g (*of class* C²) *affords a weak minimum to* **J**, *there can be no point* $x = \alpha$ *on the interval* (a^1,a^2) *which is both conjugate to the point* $x = a^1$ *and a nonsingular point of the* JDE *based on* g.

Although our definition of a conjugate point differs somewhat from that of Bliss [1], nevertheless we can still apply his method of proof.

Suppose the theorem false in that there exists a solution $x \to u(x)$ of the JDE, of class C^1 on the interval $[a^1,\alpha]$, which vanishes when $x = a^1$ and $x = \alpha$ but does not vanish identically for $x \in (a^1,\alpha)$ near $x = \alpha$. Suppose, moreover, that $x = \alpha$ is a "nonsingular" point of the JDE. For i on the range $1, \ldots, m$, let $x \to v_i(x)$ be a mapping of $[a^1,a^2]$ into R such that $v_i(x) = u_i(x)$ for $x \in [a^1,\alpha]$ and $v_i(x) = 0$ for $x \in [\alpha,a^2]$. By virtue of Lemma 4.2, $\mathscr{I}_0(v) = 0$. Moreover, if g (of class C^2) affords a weak minimum* to **J**, as we are supposing, $\mathscr{I}_0(\eta) \geq 0$ for η of class D^1 on $[a^1,a^2]$, with $\eta(a^1) = \eta(a^2) = \mathbf{0}$.

It follows from Lemma 4.1 that the mapping $x \to v(x)$, minimizing as it does the second variation $\mathscr{I}_0$, must satisfy the Weierstrass-Erdmann corner conditions at $x = \alpha$. That is, for $i = 1, \ldots, m$,

$$\lim_{x \downarrow \alpha} \Omega_{\zeta_i}(x,v(x),v'(x)) = \lim_{x \uparrow \alpha} \Omega_{\zeta_i}(x,v(x),v'(x)). \tag{4.6}$$

The left member of (4.6) vanishes and hence the right. Moreover, $v(\alpha) = \mathbf{0}$. If $x = \alpha$ is a nonsingular point of the JDE, (4.6) is possible only if $v'(\alpha)$ exists and vanishes. It follows that $u'(\alpha) = u(\alpha) = \mathbf{0}$, so that $u(x)$ must vanish identically near $x = \alpha$. This is contrary to the hypothesis that $x = \alpha$ is conjugate to $x = a^1$.

Thus Theorem 4.1 is true

The Quadratic Form Ω. The integral **J** has the preintegrand f, the integral $\mathscr{I}_0/2$ the preintegrand Ω. If one replaces the extremal g by a secondary

* See the definition of a weak minimum in Section 3.

extremal u one finds that a form (4.0), *based* on Ω and u, is identical with the form (4.0), based on f and g. Conjugate points on the x-axis based on Ω and u are accordingly conjugate points based on f and g. If a point $x = \alpha$ on the x-axis is conjugate to a point $x = c$ on the x-axis the point on u at which $x = \alpha$ will be termed *conjugate* to the point on u at which $x = c$.

5. Conjugate Points in the Nonsingular Case

Conjugate points and their extensions, focal points, play a fundamental role in variational theory and differential topology. They are needed, in particular, in formulating sufficient conditions that the extremal g give a relative minimum to **J**.

From this point on we shall suppose, as at the end of Section 2, that g *is an* x*-parameterized extremal of class* C^3 *for* x *in the interval* $[a^1,a^2]$ *and that* g *admits an extension* $g^{\aleph}$ *for* x *in an open interval* $\aleph \supset [a^1,a^2]$ *such that along* $g^{\aleph}$ *the determinant*

$$|f_{p_i p_j}(x,y,y')| \neq 0. \tag{5.0}$$

This case is called the nonsingular case.

The extension $g^{\aleph}$ is a member of the family (2.15) for

$$(c,a,b) = (a^1,g(a^1),g'(a^1)).$$

In I, II, and III, below, we shall represent conjugate points on $\aleph$ of a prescribed point $x = c \in \aleph$ in three different ways.

We understand that $2\Omega(x,\eta,\zeta)$, as defined in (4.0), with x restricted to the interval $[a^1,a^2]$, is now extended in definition, so as to be "based" on the extended $g^{\aleph}$. The JDE (4.2) are extended correspondingly. If the JDE are expanded formally, the terms in $u_1'', \ldots, u_m''$, have a nonsingular matrix of coefficients for each $x \in \aleph$; for (5.0) holds, by hypothesis, for $[a^1,a^2]$ replaced by its extension $\aleph$. We now extend Definition 4.2 of a point $x = \alpha$ *conjugate* to a point $x = c$ to the case in which $[c,\alpha]$ or $[\alpha,c]$ is an arbitrary subinterval of $\aleph$.

I. **The** n**-square determinant** $D(x,c)$. For $x \in \aleph$ let $\|v_i^j(x,c)\|$ be an m-square matrix of elements $v_i^j(x,c)$ such that for fixed j and fixed $c \in \aleph$, the mapping (Appendix I, Lemma 3.1),

$$x \to (v_1^j(x,c), \ldots, v_m^j(x,c)) : \aleph \to R^m \tag{5.1}$$

is a solution of the extended JDE for which

$$v_i^j(c,c) = 0, \; v_{ix}^j(c,c) = \delta_i^j \qquad (i = 1, \ldots, m). \tag{5.1$'$}$$

The m solutions (5.1) are linearly independent by virtue of (5.1)′. We introduce the m-square determinant

$$D(x,c) = |v_i^j(x,c)|. \tag{5.2}$$

The determinant $D(x,c)$ vanishes when $x = c$ in a way we shall now make clear. One can set

$$v_i^j(x,c) = (x - c)a_i^j(x,c) \qquad (i,j = 1, \ldots, m)$$

where (see Jordan [1], p. 251)

$$a_i^j(x,c) = \int_0^1 v_{ix}^j(c + t(x - c),c)\, dt$$

We see that a_i^j is continuous for $(x,c) \in \aleph \times \aleph$, and that $a_i^j(c,c) = v_{ix}^j(c,c) = \delta_i^j$. If $A(x,c)$ is the determinant of the elements $a_i^j(x,c)$, then $D(x,c)$ has a representation

$$D(x,c) = (x - c)^m A(x,c) \qquad (A(c,c) = 1) \tag{5.3}$$

for $(x,c) \in \aleph \times \aleph$. Conjugate points can then be determined as follows.

LEMMA 5.1. *The conjugate points of a point* $x = c$ *in* $\aleph$ *are the points* $x \neq c$ *at which* $D(x,c) = 0$.

Let $u(x) = (u_1(x), \ldots, u_m(x))$ be a nontrivial linear combination of the columns of $D(x,c)$ with constant coefficients, that is real numbers independent of $x \in \aleph$. The mapping $x \to u(x)$ satisfies the JDE and vanishes when $x = c$. If $D(\alpha,c) = 0$ with $\alpha \in \aleph$ and $\alpha \neq c$, we can suppose that coefficients, not all zero, have been chosen so that $u(\alpha) = \mathbf{0}$. The solution u will be identically zero in no open subinterval of $\aleph$. By the definition of conjugate points, $x = \alpha$ is then a conjugate point of $x = c$.

Conversely, if $x = \alpha$ is a conjugate point of $x = c$ there must be a nontrivial solution $x \to u(x)$ of the JDE which is defined for $x \in \aleph$ and vanishes when $x = c$ and $x = \alpha$. A nontrivial solution $x \to u(x)$ which vanishes when $x = c$ has a value $u(x) = (u_1(x), \ldots, u_m(x))$ which is a nontrivial linear combination of the columns of $D(x,c)$. Since $u(\alpha) = \mathbf{0}$, $D(\alpha,c) = 0$.

Thus Lemma 5.1 is true.

The continuity of the mapping $(x,c) \to A(x,c)$ implies the following.

Note A. If there is no point in $(a^1,a^2]$ conjugate to the initial point of $[a^1,a^2]$ then if $a^1 - c$ is sufficiently small and positive, there is no conjugate point on the interval $(c,a^2]$ of the point $x = c$.

II. A principle of Jacobi. Let there be given a 1-parameter family of extremals G_μ in X such that $G_\mu = g$ when the parameter $\mu = \mu_0$. On G_μ we suppose that

$$y_i = h_i(x,\mu) \qquad (x \in \aleph;\ i = 1, \ldots, m) \tag{5.4}$$

and that the mappings $(x,\mu) \to h_i(x,\mu)$, $i = 1, \ldots, m$ are of class C^2 for $x \in \aleph$ and $|\mu - \mu_0| < \varepsilon$, a sufficiently small positive constant. We set $\eta_i(x) = h_{i\mu}(x,\mu_0)$ for $x \in \aleph$ and $i = 1, \ldots, m$.

According to Jacobi the mapping $x \to \eta(x)$ *satisfies the* JDE *for* $x \in \aleph$.

Proof. By hypothesis the identities in x

$$\frac{\partial}{\partial x} f_{p_i}(x,h(x,\mu),h_x(x,\mu)) \equiv f_{y_i}(x,h(x,\mu),h_x(x,\mu)) \tag{5.5}$$

are valid for $i = 1, \ldots, m$, for $x \in \aleph$ and for $|\mu - \mu_0| < \varepsilon$. If we differentiate the members of (5.5) with respect to μ, interchange the order of differentiation with respect to x and μ and set $\mu = \mu_0$, we find that

$$\frac{d}{dx}\Omega_{\zeta_i}(x, \eta(x), \eta'(x)) = \Omega_{\eta_i}(x, \eta(x), \eta\ (x)) \qquad (i = 1, \ldots, m) \tag{5.6}$$

for $x \in \aleph$, thereby establishing the principle of Jacobi.

An Application of the Principle of Jacobi. In (2.15) there is given a family of extremals $E(c,a,b)$ on which

$$y_i = Y_i(x : c,a,b) \qquad (i = 1, \ldots, m) \tag{5.7}$$

for $x \in \aleph$. If e and δ are sufficiently small positive constants, then for triples (c,a,b) such that

$$|c - a^1| < \delta, \qquad \|a - g(c)\| + \|b - g'(c)\| < e \tag{5.8}$$

and for $x \in \aleph$, the mappings Y_i are of class C^2 and the relations (2.16) hold.

For $x \in \aleph$, $|c - a^1| < \delta$ and $i, j = 1, 2, \ldots, m$, set

$$\begin{cases} Y_{ia_j}(x : c,g(c),g'(c)) = w_i^j(x,c) & (5.9)\\ Y_{ib_j}(x : c,g(c),g'(c)) = z_i^j(x,c). & (5.10)\end{cases}$$

By the above "principle of Jacobi" for fixed j and c the mapping

$$x \to (w_1^j(x,c), \ldots, w_m^j(x,c)) \tag{5.11}$$

is a solution of the JDE, as is the mapping

$$x \to (z_1^j(x,c), \ldots, z_m^j(x,c)). \tag{5.12}$$

The relations (2.16) imply that

$$\begin{cases} w_i^j(c,c) = \delta_i^j, & z_i^j(c,c) = 0 & (5.13)' \\ w_{ix}^j(c,c) = 0, & z_{ix}^j(c,c) = \delta_i^j. & (5.13)'' \end{cases}$$

As j takes on the values $1, \ldots, m$ the mappings (5.11) afford m solutions of the JDE. Similarly the mappings (5.12) afford m additional solutions of the JDE. Together, the mappings (5.11) and (5.12) define $2m$ solutions of the JDE. These $2m$ solutions are linearly independent by virtue of the conditions (5.13). Moreover, the relations (5.13) imply that each solution of the JDE for x on the interval $\aleph$, is linearly dependent on the $2m$ solutions afforded by (5.11) and (5.12).

The Kneser Representation of Conjugate Points. We shall consider an m-parameter family of extremals E_b which meet a fixed point $(x,y) = (c,g(c))$ of $g^{\aleph}$ and have slopes $(b_1, \ldots, b_m) = b$ at this point. We shall restrict c and the slope m-tuple b by the conditions

$$|c - a^1| < \delta, \qquad \|b - g'(c)\| < e. \tag{5.14}$$

If δ and e are sufficiently small positive constants, an extremal E_b with parameters conditioned by (5.14), admits a representation

$$y_i = \varphi_i(x,b) = Y_i(x : c,g(c),b) \qquad (i = 1, \ldots, m) \tag{5.15}$$

for fixed c and for $x \in \aleph$. See Theorem 2.1. We introduce the Jacobian

$$\frac{D(\varphi_1, \ldots, \varphi_m)}{D(b_1, \ldots, b_m)}(x,b) \qquad (x \in \aleph) \tag{5.16}$$

and affirm the following.

LEMMA 5.2. *The Jacobian* (5.16), *evaluated along* g *by setting* b = g′(c), *vanishes at* x = c, *at the conjugate points of* x = c *on* $\aleph$, *and at no other points* x ≠ c *on* $\aleph$.

Proof. Let $d(x,c)$ denote the Jacobian (5.16), evaluated along g as in the lemma. That the j-th columns of the determinants $d(x,c)$ and $D(x,c)$ are identical is seen as follows. These j-th columns give the m-coordinates $u_i^j(x)$ at $x \in \aleph$ of a solution of the JDE which vanishes at $x = c$ and for fixed j has the slopes $\lambda_i^j(c) = \delta_i^j$ when $x = c$. It follows that when $|c - a^1| < \delta$ and $x \in \aleph$

$$d(x,c) = D(x,c). \tag{5.17}$$

Lemma 5.2 accordingly follows from Lemma 5.1.

III. *A Mayer determinant.* As in I, for arbitrary x and $c \in \aleph$ let $\|v_i^j(x,c)\|$ be an m-square matrix whose columns (indexed by j) define solutions (5.1) of the JDE such that (5.1)′ holds. Similarly for x and $c \in \aleph$ let $\|u_i^j(x,c)\|$ be an m-square matrix such that for fixed j and c the mapping

$$x \to (u_1^j(x,c), \ldots, u_m^j(x,c)) : \aleph \to R^m \tag{5.18}$$

is a solution of the JDE such that

$$u_i^j(c,c) = \delta_i^j, \qquad u_{ix}^j(c,c) = 0. \tag{5.19}$$

With i on the range $1, \ldots, m$, p on the range $1, \ldots, 2m$ and x arbitrary in $\aleph$, let $\|\eta_i^p(x)\|$ be a matrix whose $2m$ columns (indexed by p) define $2m$ independent solutions of the JDE. The $2m$-square determinant

$$\Delta(x,c) = \begin{vmatrix} \eta_i^p(c) \\ \eta_i^p(x) \end{vmatrix} \qquad \begin{cases} i = 1, \ldots, m \\ p = 1, \ldots, 2m \end{cases} \tag{5.20}$$

is called a *Mayer determinant*. We note that the $2m$-square determinant

$$W(c) = \begin{vmatrix} \eta_i^p(c) \\ \eta_i^{p\prime}(c) \end{vmatrix} \neq 0$$

since the $2m$ columns of $\|\eta_i^p(x)\|$ are linearly independent. We shall establish the basic relation

$$\boxed{\Delta(x,c) = D(x,c)W(c)} \qquad ((x,c) \in \aleph \times \aleph). \tag{5.21}$$

Proof of (5.21). The relation (5.21) is a consequence of the matrix identity

$$\left\| \begin{matrix} \eta_i^p(c) \\ \eta_i^p(x) \end{matrix} \right\| = \left\| \begin{matrix} u_i^j(c,c), & v_i^j(c,c) \\ u_i^j(x,c), & v_i^j(x,c) \end{matrix} \right\| \left\| \begin{matrix} \eta_i^p(c) \\ \eta_i^{p\prime}(c) \end{matrix} \right\|. \tag{5.22}$$

The matrix identity (5.22) is valid for x and c in $\aleph$, for p on the range $1, \ldots, 2m$ and i on the range $1, \ldots, m$, since

$$\eta_i^p(x) = u_i^j(x,c)\eta_j^p(c) + v_i^j(x,c)\eta_j^{p\prime}(c). \tag{5.23}$$

Terms on the right of (5.23) are summed for j on the range $1, \ldots, m$. To verify (5.23) we note that for fixed p both members of (5.23) define solutions of the JDE. These two solutions are identical, since (5.23) holds when $x = c$, and the equations obtained by differentiating the members of (5.23) with respect to x also hold when $x = c$. Thus (5.23) is an identity in x implying (5.22). Hence (5.21) is true.

Since $W(c) \neq 0$, (5.21) implies the following.

LEMMA 5.3. *The conjugate points on* $\aleph$ *of a point* $x = c$ *in* $\aleph$ *are the points* $x \neq c$ *at which a Mayer determinant* $\Delta(x,c)$ *vanishes.*

With the aid of Lemma 5.3 a fundamental theorem can now be proved. We begin with notation.

The extremal family F_e. Let the family of extremals (2.15) be restricted by setting $c = a^1$, and requiring that

$$\|a - g(c)\| + \|b - g'(c)\| < e \qquad \text{(cf. (5.8))} \tag{5.24}$$

with e so small a positive constant that triples (a^1,a,b) conditioned by (5.24) are in U of Theorem 2.1. The family of extremals (2.15) so restricted will be denoted by F_e.

THEOREM 5.1. *Suppose that the initial point* Q_1 *of* g *is not conjugate to the terminal point* Q_2 *of* g. *Corresponding to a sufficiently small positive constant* e *there exist open neighborhoods* N_1^e *of* Q_1 *and* N_2^e *of* Q_2 *so small that the following is true.*

A point $(x^1,y^1) \in N_1^e$ *and a point* $(x^2,y^2) \in N_2^e$ *are met by a unique extremal in the family* F_e *with a representation of the form*

$$y_i = \psi_i(x : x^1,y^1 : x^2,y^2) \qquad (i = 1, \ldots, m) \tag{5.25}$$

where the mappings ψ_i *and* ψ_{ix} *are of class* C^2 *on their domains* $\aleph \times N_1^e \times N^e$.

The family of extremals F_e is a subfamily of the family of extremals (2.15) and is accordingly representable in the form

$$y_i = Y_i(x : c,a,b) \qquad (i = 1, \ldots, m, x \in \aleph, c = a^1) \tag{5.26}$$

subject to the condition (5.24) on (a,b). To satisfy the theorem we fix c as a^1, and seek solutions of the equations

$$\begin{cases} y_i^1 = Y_i(x^1 : a^1,a,b) \\ y_i^2 = Y_i(x^2 : a^1,a,b) \end{cases} \qquad (i = 1, \ldots, m) \tag{5.27}$$

for (a,b) in terms of $(x^1,y^1 : x^2,y^2)$ in accord with the classical implicit function theorems.

The equations (5.27) have initial solutions

$$(a,b) = (g(a^1),g'(a^1)); \qquad (x^1,y^1 : x^2,y^2) = (a^1,g(a^1) : a^2,g(a^2)). \tag{5.28}$$

When (5.27) holds the Jacobian of the $2m$ mappings defined by the right members of (5.27) for fixed x^1, x^2, a^1, with respect to the $2m$ variables $a_1, \ldots, a_m, b_1, \ldots, b_m$, evaluated for the initial values listed in (5.28), is a

Mayer determinant

$$\Delta(a^2,a^1) = \begin{vmatrix} \eta_i^p(a^1) \\ \eta_i^p(a^2) \end{vmatrix} \tag{5.29}$$

provided the columns of the m by $2m$ matrix $\| \eta_i^p(x) \|$ are the successive columns of the m-square matrices,

$$w_i^j(x,c), \; z_i^j(x,c) \qquad (\text{with } c = a^1) \tag{5.30}$$

introduced in (5.9) and (5.10), respectively. The Mayer determinant $\Delta(x,c)$ so defined with $c = a^1$, and evaluated for $x = a^2$, does not vanish by virtue of Lemma 5.3, since the point $x = a^1$ on g is by hypothesis not conjugate to the point $x = a^2$.

Corresponding to a prescribed positive constant e let H_e be the open subspace of R^{2m} of sets

$$(a,b) = (a_1, \ldots, a_m : b_1, \ldots, b_m)$$

such that (5.24) holds when $c = a^1$. The classical implicit function theorems imply the following. Corresponding to the subspace H_e of R^{2m} defined by (5.24) for a sufficiently small positive e, there exist open neighborhoods N_1^e of Q_1 and N_2^e of Q_2 and a C^2-mapping

$$(x^1,y^1 : x^2,y^2) \to (a(x^1,y^1 : x^2,y^2), b(x^1,y^1 : x^2,y^2)) : N_1^e \times N_2^e \to H_e \tag{5.31}$$

which defines solutions of equations (5.27) which are unique among mappings of $N_1^e \times N_2^e$ into H_e.

Mappings ψ_i which are defined for $c = a^1$ and $i = 1, \ldots, m$ by setting

$$\psi_i(x : x^1,y^1 : x^2,y^2) = Y_i(x : c, a(x^1,y^1 : x^2,y^2), b(x^1,y^1 : x^2,y^2))$$

for $(x : x^1,y^1 : x^2,y^2) \in \aleph \times N_1^e \times N_2^e$ will satisfy the theorem. See Note following Theorem 2.1.

The following theorem is needed in the study of broken extremals in Chapter 2. We are concerned with the extremal g with its endpoints Q_1 and Q_2 at which $x = a^1$ and $x = a^2$ respectively, and along which (5.0) holds. We refer to the extremal $E(x^1,y^1 : x^2,y^2)$ of Theorem 5.1 with endpoints (x^1,y^1) and (x^2,y^2). We suppose that the neighborhoods N_1^e of Q_1 and N_2^e of Q_2 are so small that the nonsingularity condition (5.0) holds on each extremal $E(x^1,y^1 : x^2,y^2)$. As a consequence "conjugate points" on $E(x^1,y^1 : x^2,y^2)$ of (x^1,y^1) are well-defined or fail to exist.

THEOREM 5.1 SUPPLEMENTED. *Suppose that there is no point on* g *conjugate to the initial point* Q_1 *of* g. *If the neighborhoods* N_1^e *and* N_2^e *of the initial and terminal points* Q_1 *and* Q_2 *of* g *are sufficiently small, there is no point on the extremal* $\mathrm{E}(\mathrm{x}^1,\mathrm{y}^1 : \mathrm{x}^2,\mathrm{y}^2)$ *joining* $(\mathrm{x}^1,\mathrm{y}^1) \in \mathrm{N}_1^e$ *to* $(\mathrm{x}^2,\mathrm{y}^2) \in \mathrm{N}_2^e$, *which is conjugate to* $(\mathrm{x}^1,\mathrm{y}^1)$.

We shall make use of Theorem 2.1 in which the family of extremals (5.26) is presented. The initial triple $(x,y,y') = (a^1,g(a^1),g'(a^1))$ on g will be denoted by (c_0,a_0,b_0). In the space of triples (c,a,b) of Theorem 2.1 let U be the neighborhood of (c_0,a_0,b_0) introduced in Theorem 2.1. For $(c,a,b) \in U$ and $c \in \aleph$ of Theorem 2.1 the Jacobian

$$G(x : c,a,b) = \frac{D(Y_1, \ldots, Y_m)}{D(b_1, \ldots, b_m)}(x : c,a,b) \tag{5.32}$$

is well-defined. This is the Jacobian (5.16) introduced by Kneser in defining the conjugate points of the point $x = c$ on the extremal with initial point $(x,y) = (c,a)$ and initial slope m-tuple b.

By virtue of our hypothesis on g we infer from Lemma 5.2 that

$$G(x : c_0,a_0,b_0) \neq 0 \qquad (a^1 < x \leq a^2). \tag{5.33}$$

We shall show that if $N \subset U$ is a sufficiently small neighborhood of (c_0,a_0,b_0) then

$$G(x : c,a,b) \neq 0 \qquad (c < x \leq a^2;\ (c,a,b) \in N). \tag{5.34}$$

One sees easily that (5.34) implies Theorem 5.1, as supplemented.

Proof of (5.34). Let $E(c,a,b)$ be the extremal with initial point $(x^1,y^1) = (c,a)$ and initial slope m-tuple b.

The condition (5.34) will follow from (5.33) and the continuity of G provided we prove that if N is a sufficiently small neighborhood of (c_0,a_0,b_0) and e a sufficiently small positive constant, then

$$G(x : c,a,b) \neq 0 \qquad (0 < x - c < e;\ (c,a,b) \in N). \tag{5.35}$$

Proof of (5.35). Let $v_i^j(x : c,a,b)$ be the element in the Jacobian (5.32) in the i-th row and j-th column. It follows from the identities (2.16) that for $i,j = 1, \ldots, m$,

$$v_i^j(c : c,a,b) = 0, \qquad \frac{\partial}{\partial x} v_i^j(c : c,a,b) = \delta_i^j. \tag{5.36}$$

Hence by a proof similar to the proof of (5.3)

$$G(x : c,a,b) = (x - c)^m A(x : c,a,b) \tag{5.37}$$

where A is continuous, $A(c : c,a,b) = 1$ and hence

$$A(x : c,a,b) > 0 \qquad (0 < x - c < e;\ (c,a,b) \subset N) \tag{5.38}$$

for e and N sufficiently small.

The conditions (5.35) follow and together with (5.33), imply (5.34). The supplement of Theorem 5.1 follows.

6. The Hilbert Integral

The Hilbert integral is a line integral which serves as an aid in establishing sufficient conditions for a relative minimum of **J**. It is best introduced by studying m-parameter families of extremals of the type now to be defined.

The Family F_Π. m-Tuples $(\beta_1, \ldots, \beta_m) = \beta$ will serve as parameters with β restricted to an open connected subspace B of R^m. Let I be an open interval of R and set $I \times B = \Pi$. We shall refer to $(m + 1)$-tuples $(x,\beta) \in I \times B$. There is given a C^2-mapping

$$(6.0) \quad (x,\beta) \to \psi(x,\beta) : \Pi \to R^m \qquad (\psi(x,\beta) = (\psi_1(x,\beta), \ldots, \psi_m(x,\beta)))$$

where the partial mapping

$$(6.1)' \qquad x \to \psi(x,\beta) : I \to R^m \qquad (\beta \in B)$$

is a solution of the Euler equations whose graph is an extremal E_β in X. Points $(x, y_1, \ldots, y_m) = (x,y)$ on E_β are such that

$$(6.1)'' \qquad y = \psi(x,\beta) \qquad ((x,\beta) \in \Pi).$$

The resultant family of extremals will be denoted by F_Π.

The family F_Π will "induce" a line integral

$$(6.2) \qquad \int_\Gamma C(x,\beta)\,dx + D_h(x,\beta)\,d\beta_h \qquad (\text{range } h = 1, \ldots, m)$$

said to be *over the family* F_Π and defined by setting

$$\begin{aligned} &(6.3)' && C(x,\beta) = f(x,\psi(x,\beta),\psi_x(x,\beta)) \\ &(6.3)'' && D_h(x,\beta) = \psi_{i\beta_h}(x,\beta)\, f_{p_i}(x,\psi(x,\beta),\psi_x(x,\beta)) \end{aligned}$$

for $(x,\beta) \in \Pi$, where $\psi(x,\beta)$ and $\psi_x(x,\beta)$ are the m-tuples:

$$\begin{aligned} &(6.4)' && \psi(x,\beta) = (\psi_1(x,\beta), \ldots, \psi_m(x,\beta)) \\ &(6.4)'' && \psi_x(x,\beta) = (\psi_{1x}(x,\beta), \ldots, \psi_{mx}(x,\beta)). \end{aligned}$$

If Π is simply connected, it is a classical theorem that the line integral (6.2) is independent of admissible* paths Γ in Π joining two prescribed points in

* By a regular path in Π we here mean a path along which the coordinates $x, \beta_1, \ldots, \beta_m$ are given by C^1-mappings

$$t \to x(t), \quad t \to \beta_1(t), \ldots, t \to \beta_m(t)$$

of an interval $[t_1,t_2]$ into R with

$$(x'(t), \beta_1'(t), \ldots, \beta_m'(t)) \neq \mathbf{0} \qquad (t_1 \leq t \leq t_2).$$

We here "admit" *piecewise regular paths* in Π, that is continuous paths composed of a finite sequence of regular paths.

Π if and only if the integrability conditions

$$(6.5)' \qquad \frac{\partial C}{\partial \beta_h} = \frac{\partial D_h}{\partial x} \qquad (h = 1, \ldots, m)$$

$$(6.5)'' \qquad \frac{\partial D_k}{\partial \beta_h} = \frac{\partial D_h}{\partial \beta_k} \qquad (h, k = 1, \ldots, m)$$

are satisfied. If one sets

$$(6.6)' \qquad v_i(x,\beta) = f_{p_i}(x, \psi(x,\beta), \psi_x(x,\beta)) \qquad (i = 1, \ldots, m)$$

then

$$(6.6)'' \qquad D_h(x, \beta) = v_i(x,\beta)\frac{\partial \psi_i}{\partial \beta_h} \qquad (h = 1, \ldots, m)$$

and (6.5)′ and (6.5)″ take the respective forms

$$(6.7)' \qquad \frac{\partial C}{\partial \beta_h} = \frac{\partial}{\partial x}\left(v_i \frac{\partial \psi_i}{\partial \beta_h}\right) \qquad (h = 1, \ldots, m)$$

$$(6.7)'' \qquad \frac{\partial}{\partial \beta_h}\left(v_i \frac{\partial \psi_i}{\partial \beta_k}\right) = \frac{\partial}{\partial \beta_k}\left(v_i \frac{\partial \psi_i}{\partial \beta_h}\right) \qquad (h, k = 1, \ldots, m)$$

The following lemma concerns the satisfaction of these conditions.

LEMMA 6.1. (i) *The conditions* (6.5)′ *are identically satisfied on evaluation of the members at arbitrary points* $(x,\beta) \in \Pi$.

(ii) *Subject to evaluation at an arbitrary point* $(x,\beta) \in \Pi$

$$(6.8) \qquad \frac{\partial}{\partial \beta_k}\left(v_i \frac{\partial \psi_i}{\partial \beta_h}\right)^{(x,\beta)} - \frac{\partial}{\partial \beta_h}\left(v_i \frac{\partial \psi_i}{\partial \beta_k}\right)^{(x,\beta)} = d_{hk}(\beta) \qquad (h, k = 1 \ldots, m)$$

where $d_{hk}(\beta)$ *is independent of* x *in the pair* (x,β).

Proof of (i). To verify (6.5)′ we introduce mappings

$$(x,\beta) \rightarrow f_{y_i}(x,\psi(x,\beta),\psi_x(x,\beta)) = w_i(x,\beta) : \Pi \rightarrow R \qquad (i = 1, \ldots, m).$$

Conditions (6.5)′ then take the form

$$(6.9) \qquad w_i \frac{\partial \psi_i}{\partial \beta_h} + v_i \frac{\partial^2 \psi_i}{\partial \beta_h \, \partial x} = v_i \frac{\partial^2 \psi_i}{\partial x \, \partial \beta_h} + \frac{\partial v_i}{\partial x}\frac{\partial \psi_i}{\partial \beta_h} \qquad (h = 1, \ldots, m)$$

and hold identically after evaluation at $(x,\beta) \in \Pi$ since

$$w_i(x, \beta) = \frac{\partial}{\partial x} v_i(x,\beta) \qquad (i = 1, \ldots, m)$$

by virtue of the Euler conditions on the extremal E_β of the family F_{Π} : (6.0)

Proof of (ii). If $m > 1$, examples show that the left member of (6.8) need not vanish identically. To establish (6.8) when $m > 1$ we introduce the integral

$$J(x,\beta) = \int_{x0}^{x} f\big(x,\psi(x,\beta),\psi_x(x,\beta)\big)\,dx$$

along an extremal E_β of the family (6.0) between x_0 and x in the range of x on E_β. If one differentiates the integral $J(x,\beta)$ under the integral sign and integrates the usual term by parts one finds that for h, k on the range $1, \ldots, m$

$$\text{(6.10)} \qquad \frac{\partial J}{\partial \beta_h}(x,\beta) = v_i \frac{\partial \psi_i}{\partial \beta_h}\Bigg]_{(x^0,\beta)}^{(x,\beta)}; \qquad \frac{\partial J}{\partial \beta_k}(x,\beta) = v_i \frac{\partial \psi_i}{\partial \beta_k}\Bigg]_{(x^0,\beta)}^{(x,\beta)}.$$

Upon differentiating the members of the first identity with respect to β_k and the second with respect to β_h and equating the results, we infer that

$$\text{(6.11)} \quad \left[\frac{\partial}{\partial \beta_k}\left(v_i \frac{\partial \psi_i}{\partial \beta_h}\right) - \frac{\partial}{\partial \beta_h}\left(v_i \frac{\partial \psi_i}{\partial \beta_k}\right)\right]^{(x,\beta)} = \left[\frac{\partial}{\partial \beta_k}\left(v_i \frac{\partial \psi_i}{\partial \beta_h}\right) - \frac{\partial}{\partial \beta_h}\left(v_i \frac{\partial \psi_i}{\partial \beta_k}\right)\right]^{(x^0,\beta)}$$

from which Lemma 6.1 (ii) follows.

Note. It is clear from Lemma 6.1 that the conditions (6.5)′ are always satisfied, and that the conditions (6.5)″ are satisfied if and only if $d_{hk}(\beta) = 0$ in (6.8) for each m-tuple β and for $h, k = 1, \ldots, m$.

A field F of extremals. If the condition

$$\text{(6.12)}' \qquad \frac{D(\psi_1, \ldots, \psi_m)}{D(\beta_1, \ldots, \beta_m)}(x,\beta) \neq 0 \qquad ((x,\beta) \in \Pi)$$

is satisfied and if the union of the point set carriers of the extremals E_β of F_Π is an open subset X_0 of X through each point of which passes one and only one extremal E_β, the family F_Π will be called a *field* F *over* X_0. When F_Π is a field over X_0 the mapping

$$\text{(6.12)}'' \qquad (x,\beta) \to (x,\psi(x,\beta)) = (x,y) : \Pi \to X_0$$

is a C^2-diff of Π onto X_0. We term X_0 the *carrier* of the field F.

Definition 6.1. *A field* F_Π *with exterior pole* (x^0,y^0). If the extremals E_β of a field F_Π over X_0 can be extended† to form a family F^* of extremals meeting a point $(x^0,y^0) \in X - X_0$, then the field F_Π will be said to have an *exterior pole* (x^0,y^0). It is understood that the extended family of extremals is defined by an extended family of solutions of the Euler equations, conditioned as is the family (6.0).

We digress to state a special theorem of importance.

† Extended in the sense of decreasing x.

THEOREM 6.1. *Let there be given a family* F_{Π} *of extremals which form a field over a simply connected subspace* X_0 *of* X. *If the family* F_{Π} *has an "exterior pole"* (x^0,y^0), *the line integral* (6.2) *induced by* F_{Π} *is independent of admissible paths joining any two points in* Π.

Suppose that the mappings $x \to \psi(x,\beta)$ of (6.1)′ have been extended for each $\beta \in B$ as solutions of the Euler equations whose extremal graphs meet the pole (x^0,y^0), while the extended mapping $(x,\beta) \to \psi(x,\beta)$ remains of class C^2. For the extended mappings each partial derivative $\psi_{i\beta_h}$ vanishes for $x = x^0$ and $\beta \in B$. Hence $d_{hk}(\beta)$ in (6.8) vanishes for $\beta \in B$. The relations (6.5)″ are then true.

Hence Theorem 6.1 is true.

Note on Proof of Theorem 6.1. In the proof of Lemma 6.1 and of Theorem 6.1 no use has been made of the existence of the partial derivatives f_x. This fact has important applications when $f(x,y,y')$ is replaced by $\omega(x,\eta,\eta')$ of Appendix I, since ω_x may not exist. Moreover when the mappings ψ_i of (6.0) are linear in $\beta_1, \ldots, \beta_m$ it is sufficient for the proof to assume that the coefficient of each β_j is of class C^1 in x.

To prepare for the definition of the Hilbert integral we assume that the family (6.1)″ *of extremals is a field** F_{Π} *over* X_0 *and denote* X_0 *by* X_F.

Field Slopes $p_i(x,y)$. Given the field F_{Π} and a point $(x,y) \in X_F$, let $\beta(x,y)$ be the m-tuple of parameters $(\beta_1, \ldots, \beta_m)$ of the extremal E_β meeting (x,y). By virtue of (6.12), the mapping

$$(6.13) \qquad (x,y) \to \beta(x,y) : X_F \to R^m$$

is of class C^2 and the mapping

$$(6.14) \qquad \Theta : (x,y) \to (x,\beta(x,y)) : X_F \to \Pi$$

a C^2-diff onto Π, with the inverse diff

$$(6.15) \qquad (x,\beta) \to (x,\psi(x,\beta)) : \Pi \to X_F$$

of Π onto X_F. For $(x,y) \in X_F$ set

$$(6.16) \qquad p_i(x,y) = \psi_{ix}(x,\beta(x,y)) \qquad (i = 1, \ldots, m).$$

We term $p_i(x,y)$ the i-th *slope* of the field F_{Π} at the point (x,y) and

$$(6.17) \qquad (p_1(x,y), \ldots, p_m(x,y)) = p(x,y)$$

the slope m-tuple of the field F_{Π} at (x,y).

* We do not assume that the field has an "exterior pole" although this case is not excluded.

Hilbert integrals and Mayer fields. By definition, the Hilbert integral *induced* by the above field F_Π is a line integral of the form

$$I_F(\gamma) = \int_\gamma A(x,y)\,dx + B_i(x,y)\,dy_i \qquad ((x,y) \in X_F) \tag{6.18}$$

with γ an admissible path in X_F and

$$A(x,y) = f(x,y,p(x,y)) - p_i(x,y) f_{p_i}(x,y,p(x,y))$$
$$B_i(x,y) = f_{p_i}(x,y,p(x,y)) \qquad (i = 1, \dots, m)$$

Fields of extremals for which the Hilbert integral I_F is independent of the path γ joining two points of X_F are called *Mayer fields* and are of basic importance. We state an existence theorem for Mayer fields.

THEOREM 6.2. *If the family* F_Π *of extremals is a field over a subspace* X_F *of* X, *the Hilbert integral* $\mathrm{I}_\mathrm{F}(\gamma)$ *induced by the field* F_Π *is independent of admissible paths in* X_F *joining two points of* X_F, *if and only if the line integral* (6.2) *induced by* F_Π *is independent of admissible paths in* Π *joining two points in* Π.

We shall represent $I_F(\gamma)$ as a line integral (6.2). By virtue of the diff Θ, (6.14), of X_F onto Π, the Hilbert integral over an admissible arc γ in X_F equals a line integral of the form (6.2) over the arc $\Gamma = \Theta(\gamma)$ in Π, provided one sets $y = \psi(x,\beta)$ in the coefficients $A(x,y)$ and $B_i(x,y)$ in (6.18) and sets

$$dy_i = \psi_{ix}(x,\beta)\,dx + \psi_{i\beta_h}(x,\beta)\,d\beta_h \qquad (i = 1, \dots, m). \tag{6.19}$$

Subject to the diff (6.14) and the relations (6.19)

$$\int_\gamma A(x,y)\,dx + B_i(x,y)\,dy_i = \int_{\Theta(\gamma)} C(x,\beta)\,dx + D_h(x,\beta)\,d\beta_h \tag{6.20}$$

provided $C(x,\beta)$ and $D_h(x,\beta)$ are given by (6.3) for $(x,\beta) \in \Pi$, and γ is an admissible arc in X_F.

Theorem 6.2 is an immediate consequence of the relation (6.20).

The relation (6.20) has another important consequence.

LEMMA 6.2. *If* γ *is a subarc of an extremal* E_β *of the field* F_Π, *then* $\mathrm{I}_\mathrm{F}(\gamma) = \mathrm{J}(\gamma)$.

In the coordinates $x, \beta_1, \dots, \beta_m$ the arc γ of Lemma 6.2 has a representation $y = \psi(x,\beta)$ with β constant. For this arc γ one must set each $d\beta_h = 0$ in the right member of (6.20). If $x \to \gamma(x) = y$ is an x-parameterization of γ with $[\alpha^1,\alpha^2]$ the range of x on γ, then (6.20) shows that

$$I_F(\gamma) = \int_{\alpha^1}^{\alpha^2} C(x,\beta)\,dx. \tag{6.21}$$

By virtue of (6.3)′ it follows that

$$I_F(\gamma) = \int_{a^1}^{a^2} f(x,\gamma(x),\gamma'(x))\, dx = \mathbf{J}(\gamma). \tag{6.22}$$

Thus Lemma 6.2 is true.

We return to the basic extremal g studied in Sections 2 to 5. Recall that the x domain of g is a closed interval $[a^1,a^2]$ with an open extension $\aleph$. According to Theorem 2.1, $\aleph$ can be taken as so small an extension of $[a^1,a^2]$ that g admits an extension $g^{\aleph}$ as an extremal of class C^3, with x-domain $\aleph$ and with (2.18) holding along $g^{\aleph}$.

The following lemma is needed to prove the second theorem of Section 7.

LEMMA 6.3. *If there is no point on the extremal* g *conjugate to the initial point of* g, *then* g *is a subarc of an extremal of a Mayer field* F *covering an open simply connected neighborhood of* g.

Reasoning as in Section 5 we see that if δ and e are sufficiently small positive constants there exists an m-parameter family $\Phi^e_{\delta,c}$ of extremals E_b near g with representations

$$y_i = \varphi_i(x,b) = Y_i(x : c,g(c),b) \qquad (i = 1, \ldots, m) \tag{6.23}$$

for $x \in \aleph$, where c and b are subject to the conditions

$$0 < a^1 - c < \delta; \qquad \|b - g'(c)\| < e. \tag{6.24}$$

(Cf. (5.14) and Theorem 2.1.) The extremal E_b meets $g^{\aleph}$ when $x = c$ and there has the slope m-tuple b. According to Note A of Section 5, if there is no point on g conjugate to the initial point of g and if $a^1 - c$ is sufficiently small and positive, there is no point on the interval $(c,a^2]$ conjugate to the point $x = c$ on $g^{\aleph}$. We suppose c so chosen.* If then ε is a sufficiently small positive constant, the interval

$$\aleph_\varepsilon = (a^1 - \varepsilon,\, a^2 + \varepsilon) \tag{6.25}$$

will have a closure included in $\aleph$ which contains neither c nor any value conjugate to c. We shall now define a *special field* F.

Let $\Phi^{e,\varepsilon}_{\delta,c}$ be the family of extremals obtained from the family $\Phi^e_{\delta,c}$ by replacing the extremal E_b of $\Phi^e_{\delta,c}$ by the subarc of E_b with x-domain $\aleph_\varepsilon$. By virtue of the choice of c and ε

$$\frac{D(\varphi_1, \ldots, \varphi_m)}{D(b_1, \ldots, b_m)}(x,b) \neq 0 \qquad (x \in \mathrm{Cl}\, \aleph_\varepsilon) \tag{6.26}$$

* This value of c is fixed for the rest of Section 6.

for $b = g'(c)$. It follows that if e is sufficiently small and (6.24) holds, then the family $\Phi_{\delta,c}^{e,\varepsilon}$ will be a field F whose "carrier" is an open, simply-connected neighborhood M_g of g.

The extremals of F are defined by (6.23) with $x \in \aleph_\varepsilon$, but have extensions defined by (6.23) for $x \in \aleph$ and so meet the point $(c,g(c))$. The field F thus has an "*exterior pole*" $(c,g(c))$. Let B_e denote the open ball of m-tuples b such that $\|b - g'(c)\| < e$. We see that the mapping

$$(x,b) \to (x,\varphi(x,b)) : \aleph_\varepsilon \times B_e \to E_{m+1} \tag{6.27}$$

is a C^2-diff into the carrier M_g of the field F. The diff (6.27) *defines* the field F in that the graphs of its partial mappings

$$x \to \varphi(x,b) : \aleph_\varepsilon \to E_m \qquad (b \in B_e) \tag{6.27$'$}$$

are the extremal arcs of F.

Set $\hat{\pi} = \aleph_\varepsilon \times B_e$ and denote the above field F by $F_{\hat{\pi}}$. According to Theorem 6.1 the line integral (6.2) induced by the field $F_{\hat{\pi}}$ is independent of admissible paths in $\hat{\pi}$ joining two points of $\hat{\pi}$. It follows then from Theorem 6.2 that $F_{\hat{\pi}}$ is a Mayer field. Since g is a subarc of an extremal of this field, Lemma 6.3 is true.

Definition 6.2. *Inner fields.* We term a field F "defined" by a diff of form (6.27) an *inner field* if the diff (6.27) is *extendable* by a diff of the same character "defined" by a family of extremals with the same pole $(c,g(c))$, but with ε and e replaced by larger constants ε' and e'.

The proof of Lemma 6.3 implies the truth of the following lemma.

LEMMA 6.4. *If ε and e are sufficiently small positive constants, a diff Θ of form (6.27) exists which is "extendable" in the sense of Definition 6.2 and so defines an "inner" field of extremals whose carrier is an open neighborhood M_g of g. As a diff Θ is such that*

$$\frac{D(\varphi_1, \ldots, \varphi_m)}{D(b_1, \ldots, b_m)}(x, b) \neq 0 \tag{6.28}$$

for $x \in \aleph_\varepsilon$ *and* $b \in B_e$.

An extension of Lemma 6.3. We shall need an extension of Lemma 6.3 which provides fields for extremals γ near g such that the carriers of these fields include a common open neighborhood of g. The following lemma states the desired conclusion.

LEMMA 6.5. *The extremal g is assumed free of conjugate points. If then N is a sufficiently small open neighborhood of g and N_1 and N_2 are sufficiently small open neighborhoods of the respective end points of g, the following is true.*

(1) *Each extremal* γ *in* N *which joins a point in* N_1 *to a point in* N_2 *is a proper subarc of an extremal in a Mayer field* F_γ.

(2) *The fields* F_γ *can be so defined that their carriers include a common open neighborhood of* g.

Notation for Proof. The x-domain of g is $[a^1,a^2]$. As in the proof of Lemma 6.3 let $\aleph \supset [a^1,a^2]$ be an open interval such that g admits an extremal extension $g^\aleph$ free of conjugate points. The m-tuples a and b appearing in $Y(x : c,a,b)$ of Theorem 2.1 are restricted to neighborhoods A_μ and B_e, respectively, of $g(c)$ and $g'(c)$, defined by inequalities

$$\|a - g(c)\| < \mu, \qquad \|b - g'(c)\| < e \tag{6.29}$$

while $x \in \aleph_\varepsilon$ is conditioned as in (6.25) and $c < a^1$ remains fixed.

If μ, e, and ε are sufficiently small, $\varphi^a(x,b)$ is well-defined by setting

$$\varphi^a(x,b) = Y(x : c,a,b) \tag{6.30}$$

for

$$x \in \aleph_\varepsilon, \qquad a \in A_\mu, \qquad b \in B_e, \tag{6.31}$$

and for each m-tuple $a \in A_\mu$, the mapping

$$(x,b) \to (x,\varphi^a(x,b)) : \aleph_\varepsilon \times B_e \to E_{m+1} \tag{6.32}$$

is a C^2-mapping, with

$$\frac{D(\varphi_1^a, \ldots, \varphi_m^a)}{D(b_1, \ldots, b_m)}(x, b) \neq 0. \tag{6.33}$$

Note that as $a \to g(c)$ the mapping (6.32) converges pointwise to the mapping (6.27). This convergence is *uniform* if the mapping (6.32) is restricted to any compact subset of its domain $\aleph_\varepsilon \times B_e$.

We suppose the mapping (6.27) is "extendable" in accord with Lemma 6.4. Because (6.27) is extendable and because (6.33) holds, the following is true.

(i) *If* μ *is sufficiently small and if* $\mathrm{a} \in \mathrm{A}_\mu$, *the* C^2*-mapping* (6.32) *is a diff* Θ_a *which maps* $\aleph_\varepsilon \times \mathrm{B}_\mathrm{e}$ *onto a simply connected subset* X_a *of* $\mathrm{E}_{\mathrm{m}+1}$. *Moreover* Θ_a *converges pointwise as* $\mathrm{a} \to \mathrm{g(c)}$, *to the diff* (6.27) *of* $\aleph_\varepsilon \times \mathrm{B}_\mathrm{e}$ *onto the neighborhood* $\mathrm{M_g}$ *of* $\mathrm{g} \subset \mathrm{E}_{\mathrm{m}+1}$.

Completion of Proof of Lemma 6.5. If N, N_1 and N_2 of Lemma 6.5 are sufficiently small each extremal γ admitted in Lemma 6.3 can be extended as an extremal in the sense of decreasing x to meet the m-plane $x = c$ in a point $(x,y) = (c,a)$ such that $a \in A_\mu$, and have a slope m-tuple b at $x = c$ such that $b \in B_\mathrm{e}$, where μ and e are conditioned as above. The diff Θ_a : (6.32) corresponding to our earlier choice of c will define a field to be denoted by F_γ. The extremal E_b of F_γ will carry γ as a proper subarc if N_1 and N_2 are sufficiently small.

Each field F_γ is a Mayer field as a consequence of Theorem 6.2. That the carriers of the fields include a common open neighborhood of g if the neighborhoods N_1 and N_2 are sufficiently small, follows from the fact that, as $a \to g(c)$, the diff Θ_a converges uniformly on any compact subset W of $\aleph_\varepsilon \times B_e$ to the diff (6.27), restricted to W.

7. Sufficient Conditions

The following theorem is due to Weierstrass, at least when $m = 1$.

THEOREM 7.1. *Suppose that* g *is a closed subarc of an extremal of a Mayer field* F *with slope* m*-tuple* p(x,y) *for* (x,y) $\in$ X$_F$. *If*

$$\text{(7.1)} \qquad E(x,y,p(x,y),q) > 0 \qquad ((x,y) \in X_F,\ q \neq p(x,y))$$

for arbitrary m*-tuples* q, *then* g *affords a proper strong minimum to* **J** *relative to* x*-parameterized curves* γ *of class* D^1 *which join the endpoints of* g *in* X$_F$.

A curve γ of Theorem 7.1 is defined by D^1-mappings

$$\text{(7.2)} \qquad x \to y_i(x) : [a^1,a^2] \to R \qquad (i = 1, \ldots, m)$$

Let I_F be the Hilbert integral associated with the field F. The integrals $\mathbf{J}(\gamma)$ and $\mathbf{J}(g)$ are defined in (1.4). According to Lemma 6.2, $\mathbf{J}(g) = I_F(g)$. Since F is a Mayer field, $I_F(g) = I_F(\gamma)$ so that $\mathbf{J}(g) = I_F(\gamma)$. Thus $\mathbf{J}(g)$ is the line integral (6.18) taken along γ. Explicitly

$$\text{(7.3)} \qquad \mathbf{J}(g) = \int_{a^1}^{a^2} (f^F - p_i(x,y(x))f^F_{p_i} + y_i'(x)f^F_{p_i})\, dx$$

where the superscript F means evaluation with

$$\text{(7.4)} \qquad (x,y,p) = (x,y(x),p(x,y(x))).$$

Since

$$\text{(7.5)} \qquad \mathbf{J}(\gamma) = \int_{a^1}^{a^2} f(x,y(x),y'(x))\, dx$$

we are led to the Weierstrass formula

$$\text{(7.6)} \qquad \mathbf{J}(\gamma) - \mathbf{J}(g) = \int_{a^1}^{a^2} E(x,y(x),p(x,y(x)),y'(x))\, dx.$$

By virtue of hypothesis (7.1),

$$\text{(7.7)} \qquad \mathbf{J}(\gamma) - \mathbf{J}(g) > 0$$

unless the mappings $x \to y_i(x)$ defining γ satisfy the differential equations

$$\frac{dy_i}{dx} = p_i(x, y) \qquad (i = 1, \ldots, m, (x, y) \in X_F) \tag{7.8}$$

between corners. The uniqueness of solutions of the differential equations (7.8) meeting a prescribed point of X_F implies that when $\mathbf{J}(\gamma) = \mathbf{J}(g)$, the m mappings $x \to y_i(x)$ afford a solution of (7.8) and accordingly define the extremal of the field F meeting the initial point of g, that is, the extremal g. Thus (7.7) holds if $\gamma \neq g$.

This establishes Theorem 7.1.

We shall continue this chapter by enumerating certain conditions which are fundamental in the nonparametric theory. In this enumeration S abbreviates *sufficient*. An extremal g of class C^2 is given with $[a^1,a^2]$ the range of x on g. The conditions are on f and g.

Under the *Jacobi* S-*condition* there is no point on g conjugate to the initial point of g.

Under the *Legendre* S-*condition* the quadratic form

$$f_{p_ip_j}(x,g(x),g'(x))z_iz_j \tag{7.9}$$

in the variables $z_1, \ldots, z_m$ is positive definite for each $x \in [a^1,a^2]$.

Under the *Weierstrass* S-*condition* $E(x,y,p,q) > 0$ for triples (x,y,p) sufficiently near triples (x,y,y') on g and for any m-tuple $q \neq p$.

Definition 7.1. The preintegrand f is *positive regular* on a subset X of the space E_{m+1} of points (x,y) if for each $(x,y) \in X$ and unrestricted m-tuples p the quadratic form

$$f_{p_ip_j}(x,y,p)z_iz_j \tag{7.10}$$

is positive definite.

We come to a principal theorem.

THEOREM 7.2. *In order that the extremal* g *afford a proper, strong minimum to* **J** *relative to* x-*parameterized curves of class* D¹ *which join the endpoints of* g *in a sufficiently small open neighborhood of* g, *it is sufficient that the Weierstrass, Legendre and Jacobi Conditions hold relative to* f *and* g.

A particular consequence of the Legendre S-condition, all that we use here, is that the determinant $|f_{p_ip_j}| \neq 0$ along g. With this condition satisfied, the results on conjugate points in Section 5 are valid. According to Lemma 6.3 if there is no point on g conjugate to the initial point of g, there exists a Mayer field of extremals covering a simply connected open neighborhood of g. Theorem 7.2 then follows from Theorem 7.1.

We state the following lemma.

LEMMA 7.1. *If* f *satisfies the condition of positive regularity for* (x,y) *in some neighborhood* N *of* g *in* X, *then the Legendre and Weierstrass* S-*conditions hold for* g *and* f.

That the Legendre S-condition holds along g under the hypothesis of Lemma 7.1 is trivial.

To show that the Weierstrass S-condition holds for g we note the following. For (x,y) in N and p and q arbitrary m-tuples, Taylor's formula implies that

$$f(x,y,q) - f(x,y,p) - (q_i - p_i)f_{p_i}(x,y,p) = \tfrac{1}{2}(q_i - p_i)(q_j - p_j)f_{p_i p_j}(x,y,p^*)$$

where $p_i^* = p_i + \theta(q_i - p_i)$ with $0 < \theta < 1$. Thus $E(x,y,p,q) > 0$ for $(x,y) \in N$ and $p \neq q$.

This completes the proof of Lemma 7.1.

The following extension of Theorem 7.2 is a consequence of Theorem 7.1 and Lemma 6.5.

THEOREM 7.3. *Let* g *be an* x-*parameterized extremal relative to which the Weierstrass, Legendre and Jacobi* S-*Conditions hold. If* N *is a sufficiently small open neighborhood of* g *and* N′ *and* N″ *sufficiently small disjoint open neighborhoods of the endpoints* A^1 *and* A^2 *of* g, *then any* x-*parameterized extremal* E(p′,p″) *which joins* p′ ∈ N′ *and* p″ ∈ N″ *in* N *affords a proper strong minimum to* **J** *relative to* x-*parameterized curves of class* D^1 *which join* p′ *and* p″ *in* N.

A positive definite quadratic functional. The second variation

$$\mathscr{I}_0(\eta) = \int_{a^1}^{a^2} 2\Omega(x,\eta(x),\eta'(x))\, dx \tag{7.11}$$

of **J** along g, was introduced in (4.1)′. As defined in (4.0), for x in the open interval $\aleph$ of Section 2, Ω is a quadratic form in the m-tuples η and ζ whose coefficients are C^1-mappings of $\aleph$ into R. For an initial study we prefer the more general quadratic form

$$2\omega(x,\eta,\zeta) = R_{ij}(x)\zeta_i\zeta_j + 2Q_{ij}(x)\zeta_i\eta_j + P_{ij}(x)\eta_i\eta_j, \tag{7.12}$$

conditioned as in Section 1 of Appendix I, because the coefficients in the form (7.12) are free of any dependence on the preintegrand f and extremal g and are not required to be more than continuous for $x \in \aleph$. By hypothesis of Appendix I

$$R_{ij}(x)\zeta_i\zeta_j > 0 \qquad (\zeta \neq \mathbf{0},\ x \in \aleph). \tag{7.13}$$

Theorem 7.4 below gives sufficient conditions that

$$\int_{a^1}^{a^2} \omega(x,\eta(x),\eta'(x))\,dx > 0 \tag{7.14}$$

for each nonnull D^1-mapping η of $[a^1,a^2]$ into R^m which vanishes when $x = a^1$ and $x = a^2$. Our Theorem 7.4 below is classical except for the fact that we do not require that the coefficients R_{ij}, Q_{ij}, and P_{ij} be more than continuous in x.

The principal condition in Theorem 7.4 follows.

A NEW JACOBI CONDITION H. *Condition* H *requires that a point* a^1 *be conjugate relative to the JDE*

$$\frac{d}{dx}\,\omega_{\zeta_i}(x,\eta,\eta') = \omega_{\eta_i}(x,\eta,\eta') \qquad (i = 1, \ldots, m) \tag{7.15}$$

to no point in the interval $(a^1,a^2]$. See Section 3 of Appendix I.

Some Consequences of Condition H. The conjugate points of a point $c \in \aleph$ are the values x at which the m-square determinant

$$D(x, c) = |v_i^j(x,c)| \qquad \text{(cf. (5.2), or (3.6) of Appendix I)} \tag{7.16$'$}$$

vanishes. Recall that

$$D(x,c) = (x - c)^m A(x,c) \qquad \big(A(c,c) = 1\big) \tag{7.16$''$}$$

where $A(x,c)$ is continuous for $(x,c) \in (\aleph \times \aleph)$. The Condition H implies that $A(x,a^1) \neq 0$ for $a^1 \leq x \leq a^2$. Hence if $a^1 - c$ is sufficiently small and positive, $D(x,c) \neq 0$ for $a^1 \leq x \leq a^2$. We suppose c so chosen.

For j on the range $1, \ldots, m$, let $v^j(x)$ be the m-tuple which is the j-th column of the determinant $|v_i^j(x,c)|$. The "canonical associate" of v^j, as introduced in (1.3) of Appendix I, has component values,

$$z_i^j(x) = \omega_{\zeta_i}(x,v^j(x),\dot{v}^j(x)) \qquad (i = 1 \ldots, m) \tag{7.17}$$

where the superdot* indicates derivation as to x.

We shall verify the identity

$$v_i^h(x)z_i^k(x) - v_i^k(x)z_i^h(x) \equiv 0 \qquad (x \in \aleph) \tag{7.18}$$

for each h and k on the range $1, \ldots, m$.

Proof of (7.18). The derivative, as to x, of the left member of (7.18) vanishes for $x \in \aleph$, because v^h and v^k are C^1-solutions of the JDE (7.15) with "canonical associates" which are z^h and z^k, respectively. Hence the left

* The superdot is introduced for notational simplicity in the proof of Theorem 7.4.

member of (7.18) is a constant K for $x \in \aleph$. Since $v^h(c) = v^k(c) = \mathbf{0}$ by hypothesis, $K = 0$, and (7.18) follows.

von Escherich Reduction Formulas. Formulas useful in transforming the second variation in the Lagrange problem were devised by Clebsch and greatly simplified by von Escherich. See Bolza [1], pages 630–631. On deleting the side conditions of Lagrange, these formulas are of the type we need except for the excessive conditions of differentiability on the coefficients R_{ij}, Q_{ij}, P_{ij}. In 1931, Bliss and Schoenberg derived a related simple reduction formula.* Our derivation† of reduction formula (7.21) makes it clear that the coefficients R_{ij}, Q_{ij}, P_{ij} need be no more than *continuous*.

Let η be a D¹-mapping of $[a^1,a^2]$ into R^m. A key step is to use Cramer's rule to solve the m-equations

$$\eta_i(x) = v_i^j(x)\sigma_j(x) \qquad (i = 1, \ldots, m;\ x \in [a^1,a^2]) \tag{7.19}$$

for the values $\sigma_j(x)$, obtaining thereby m D^1-mappings $\sigma_1, \ldots, \sigma_m$ of $[a^1,a^2]$ into R. This is possible since the m-square determinant $|v_i^j(x)| \neq 0$ for $x \in [a^1,a^2]$ and the mappings v^j are of class C^1.

Our variant (7.36) of a Clebsch formula is a corollary of the following lemma. In this lemma we set

$$w_i(x) = v_i^j(x)\dot{\sigma}_j(x) \qquad (i = 1, \ldots, m). \tag{7.20}$$

LEMMA 7.2. *Under the Condition* H, *and with* η *a* D¹-*mapping of* $[a^1,a^2]$ *into* R^m

$$2\omega(x,\eta(x),\eta'(x)) = \frac{d}{dx}(\eta_i(x)z_i^j(x)\sigma_j(x)) + R_{ij}(x)w_i(x)w_j(x) \tag{7.21}$$

for each $x \in [a^1,a^2]$ *at which* $\eta'(x)$ *exists.*

Notation for the Proof of Lemma 7.2. Recall the Euler formula,

$$2\omega(x,\eta,\zeta) = \eta_i\omega_{\eta_i}(x,\eta,\zeta) + \zeta_i\omega_{\zeta_i}(x,\eta,\zeta), \tag{7.22}$$

valid for $x \in \aleph$, and arbitrary m-tuples η and ζ. To prove (7.21), for $x \in [a^1,a^2]$ we shall set

$$\gamma_i(x) = \dot{v}_i^j(x)\sigma_j(x) \qquad (i = 1, \ldots, m). \tag{7.23}$$

It follows from (7.19) and (7.20) that the m-tuple

$$\eta'(x) \equiv \gamma(x) + w(x) \qquad (x \in [a^1,a^2]). \tag{7.24}$$

* See Bliss and Schoenberg for the basic formula (1.8) on page 784, and a hypothesis on line 11, page 783, that R_{ij}, Q_{ij}, P_{ij} be of class C^1.

† Cf. Philip Hartman [1], Ch. XI, Th. 10.3 for a related exposition.

Proof of Lemma 7.2.* Note that for $x \in [a^1,a^2]$

$$2\omega(x,\eta(x),\eta'(x)) = 2\omega(x,\eta(x),\gamma(x) + w(x)) \tag{7.25}$$

and that, as a matter of elementary algebra, this value equals

$$2\omega(x,\eta(x),\gamma(x)) + 2w_i(x)\omega_{\zeta_i}(x,\eta(x),\gamma(x)) + R_{ij}(x)w_i(x)w_j(x). \tag{7.26}$$

To prove Lemma 7.2 it is necessary and sufficient to prove that for $x \in [a^1,a^2]$,

$$2\omega(x,\eta(x),\gamma(x)) + 2w_i(x)\omega_{\zeta_i}(x,\eta(x),\gamma(x)) = \frac{d}{dx}(\eta_i(x)z_i^j(x)\sigma_j(x)). \tag{7.27}$$

Proof of (7.27). We shall represent the two terms on the left of (7.27) separately.

Representation of $2\omega(x,\eta(x),\gamma(x))$. First note that by the Euler formula

$$2\omega(x,\eta(x),\gamma(x)) = \eta_i(x)\omega_{\eta_i}(x,\eta(x),\gamma(x)) + \gamma_i(x)\omega_{\zeta_i}(x,\eta(x),\gamma(x)). \tag{7.28}$$

By virtue of the linearity of $\omega_{\eta_i}(x,\eta,\zeta)$ in the components of η and ζ we find from (7.19) and (7.23) that

$$\omega_{\eta_i}(x,\eta(x),\gamma(x)) = \sigma_j(x)\omega_{\eta_i}(x,v^j(x),\dot{v}^j(x)) \tag{7.29}$$

$$= \sigma_j(x)\frac{d}{dx}\,\omega_{\zeta_i}(x,v^j(x),\dot{v}^j(x)) = \sigma_j(x)\dot{z}_i^j(x) \tag{7.30}$$

where the first equality sign in (7.30) is valid since v^j is a solution of the JDE (7.15) and the final equality follows from the definition of $z_i^j(x)$ in (7.17). Similarly for i on the range $1, \ldots, m$,

$$\omega_{\zeta_i}(x,\eta(x),\gamma(x)) = \sigma_j(x)\omega_{\zeta_i}(x,v^j(x),\dot{v}^j(x)) = \sigma_j(x)z_i^j(x). \tag{7.31}$$

It follows from (7.28), (7.29), (7.30), and (7.31) that

$$2\omega(x,\eta(x),\gamma(x)) = \eta_i(x)(\sigma_j(x)\dot{z}_i^j(x)) + \gamma_i(x)(\sigma_j(x)z_i^j(x)). \tag{7.32}$$

Representation of the Second Term in (7.27). From (7.31) we find that

$$w_i(x)\omega_{\zeta_i}(x,\eta(x),\gamma(x)) = w_i(x)(\sigma_k(x)z_i^k(x)) \tag{7.33$'$}$$

where k has the range $1, \ldots, m$. By virtue of the definition (7.20) of $w_i(x)$ and the relations (7.18), the sum (7.33)′ also equals

$$(v_i^j(x)\dot{\sigma}_j(x))(\sigma_k(x)z_i^k(x)) = (v_i^k(x)\sigma_k(x))(z_i^j(x)\dot{\sigma}_j(x)) = \eta_i(x)(z_i^j(x)\dot{\sigma}_j(x)). \tag{7.33$''$}$$

From (7.32) and (7.33) we conclude that the left member of (7.27) equals

$$\eta_i(x)(\sigma_j(x)\dot{z}_i^j(x)) + \gamma_i(x)(\sigma_j(x)z_i^j(x)) + w_i(x)(\sigma_k(x)z_i^k(x)) + \eta_i(x)(z_i^j(x)\dot{\sigma}_j(x)). \tag{7.34}$$

* Strictly the values of x at which $\eta'(x)$ fails to exist should be deleted from $[a^1,a^2]$ throughout this proof.

Since $\eta'(x) = \gamma(x) + w(x)$, this sum clearly equals the x-derivative of $\eta_i(x)z_i^j(x)\sigma_j(x)$. The relation (7.27) follows and, together with (7.26), implies Lemma 7.2.

Theorem 7.4 concerns a quadratic functional positive definite on its domain of definition.

THEOREM 7.4. *Under the Condition* H

$$\int_{a^1}^{a^2} 2\omega(x,\eta(x),\eta'(x))\,dx > 0 \tag{7.35}$$

for each D^1*-mapping* η *of* $[a^1,a^2]$ *into* R^m *which vanishes when* $x = a^1$ *and* a^2 *but does not vanish identically for* $x \in [a^1,a^2]$.

According to (7.21) of Lemma 7.2

$$\int_{a^1}^{a^2} 2\omega(x,\eta(x),\eta'(x))\,dx = \int_{a^1}^{a^2} R_{ij}(x)w_i(x)w_j(x)\,dx \tag{7.36}$$

since $\eta(a^1) = \eta(a^2) = \mathbf{0}$ by hypothesis. Moreover the m-tuple $w(x) \not\equiv \mathbf{0}$. Otherwise (7.20) implies that the m-tuple $\dot{\sigma}(x) \equiv \mathbf{0}$. In such a case $\sigma(x) \equiv \mathbf{0}$ since $\sigma(a^1) = \mathbf{0}$, and $\eta(x) \equiv \mathbf{0}$ in accord with (7.19), contrary to hypothesis.

Thus Theorem 7.4 is true.

We state a corollary of Theorem 7.4.

COROLLARY 7.1. *Let* $x \to u(x)$ *be a* C^1*-solution of the* JDE (7.15) *for* $x \in [a^1,a^2]$. *If* $x \to y(x)$ *is a* D^1*-mapping of* $[a^1,a^2]$ *into* R^m *such that* $u(a^1) = y(a^1)$ *and* $u(a^2) = y(a^2)$, *then under the Jacobi Condition* H,

$$\int_{a^1}^{a^2} \omega(x,y(x),y'(x))\,dx \geq \int_{a^1}^{a^2} \omega(x,u(x),u'(x))\,dx, \tag{7.37}$$

with the equality sign excluded when $y(x) \not\equiv u(x)$.

Consider the case in which $y(x) \not\equiv u(x)$. Set

$$\eta_i(x) = y_i(x) - u_i(x) \qquad (i = 1, \ldots, m). \tag{7.38}$$

Then $x \to \eta(x)$ is a D^1-mapping of $[a^1,a^2]$ into R^m which vanishes when $x = a^1$ and a^2, but which does not vanish identically. The interval $[a^1,a^2]$ can be broken up into a finite set of subintervals $[\alpha,\beta]$ on each of which η is of class C^1. Let the partial derivatives ω_{η_i} and ω_{ζ_i}, evaluated when $(x,\eta,\zeta) = (x,u(x),u'(x))$, be denoted by $\omega_{\eta_i}^x$ and $\omega_{\zeta_i}^x$, respectively.

For x in $[\alpha,\beta]$

$$\omega(x,y(x),y'(x)) = \omega(x,u(x) + \eta(x),u'(x) + \eta'(x)). \tag{7.39}$$

For x fixed we shall expand the right member of (7.39) as a Taylor's series in the $2m$ variables

$$(7.40) \qquad \eta_1(x), \ldots, \eta_m(x); \eta_1'(x), \ldots, \eta_m'(x).$$

The value (7.39) thereby equals

$$(7.41) \qquad \omega(x,u(x),u'(x)) + \{\eta_i(x)\omega_{\eta_i}^x + \eta_i'(x)\omega_{\zeta_i}^x\} + \omega(x,\eta(x),\eta'(x)).$$

The integral with respect to x of the brace in (7.41), taken over each subinterval $[\alpha,\beta]$, reduces, by integration by parts, to

$$(7.42) \qquad \Big[\eta_i(x)\omega_{\zeta_i}^x\Big]_{x=\alpha}^{x=\beta},$$

and taken over $[a^1,a^2]$, accordingly equals zero, since $\omega_{\zeta_i}^x$ is continuous for $x \in [a^1,a^2]$ and $\eta(a^1) = \eta(a^2) = \mathbf{0}$. We infer from the representation (7.41) of $\omega(x,y(x),y'(x))$ that

$$(7.43) \qquad \int_{a^1}^{a^2} \omega(x,y(x),y'(x))\,dx = \int_{a^1}^{a^2} \omega(x,u(x),u'(x))\,dx + \int_{a^1}^{a^2} \omega(x,\eta(x),\eta'(x))\,dx.$$

Corollary 7.1 follows from (7.43) and Theorem 7.4.

The following theorem supplements Theorem 7.4. It is a consequnce of Exercises 15.1 and 15.2.

THEOREM 7.5. *Let* $I_0(\eta)$ *and* $I_0(\hat{\eta})$ *be values of the integral* (7.35) *for* D^1*-mappings* η *and* $\hat{\eta}$ *of* $[a^1,a^2]$ *into* R^m *which vanish when* $x = a^1$ *and* a^2. *If for fixed* $\hat{\eta}$ *and arbitrary admissible* η

$$I_0(\eta) \geq I_0(\hat{\eta})$$

then $\hat{\eta}$ *is a* C^1*-solution of the* DE (7.15) *and* $I_0(\hat{\eta}) = 0$.

The following theorem should be bracketed with Theorems 7.4 and 7.5.

THEOREM 7.6. *If* a^2 *is the first conjugate point of* $x = a^1$, *following* $x = a^1$, *relative to the* DE (7.15), *then*

$$(7.44) \qquad \int_{a^1}^{a^2} \omega(x,\eta(x),\eta'(x))\,dx \geq 0$$

for each D^1*-mapping of* $[a^1,a^2]$ *into* R^m *which vanishes when* $x = a^1$ *and* a^2. *Moreover the equality holds in* (7.44) *if and only if* η *is a* C^1*-solution of the* DE (7.15) *which vanishes at* $x = a^1$ *and* $x = a^2$.

Theorem 7.6 follows from Theorem 19.6.

Theorems 7.4, 7.5, 7.6 are classical if the underlying quadratic form ω has coefficients of class C^1 in x. The proofs which we have given are under the assumption that the coefficients are merely *continuous*. In this case Theorem 7.6 should be supplemented by the following theorem.

THEOREM 7.7. *If the relation* (7.44) *holds for each* D^1*-mapping of* $[a^1,a^2]$ *into* R^m *which vanishes when* $x = a^1$ *and* a^2, *then* $x = a^1$ *has no conjugate point in* (a^1,a^2).

If on the contrary $\alpha \in (a^1,a^2)$ is a conjugate point of $x = a^1$, then by definition (Section 3 of Appendix I) there exists a nonnull solution u of the free DE (7.15) which vanishes at $x = a^1$ and $x = \alpha$. As in the proof of Lemma 4.2 one finds that

$$\int_{a^1}^{\alpha} \omega(x,u(x),u'(x))\, dx = 0. \tag{7.45}$$

Let $\bar{u}$ be a D^1-mapping of $[a^1,a^2]$ into R^m which extends u with $\bar{u}(x) = 0$ for $\alpha \leq x \leq a^2$. The equality (7.44) holds if η is replaced by $\bar{u}$. Hence by Theorem 7.6, $\bar{u}$ is a C^1-solution of (7.15). This, however, is contrary to the fact that $\bar{u}$ has a corner at $x = \alpha$.

Hence Theorem 7.7 is true.

An Alternative Proof of Theorem 7.4. Theorem 7.4 can be proved essentially as Theorem 7.2 was proved, making use of an appropriately defined Hilbert integral and Weierstrass E-function, provided the background integrability conditions are properly reformulated, verified, and applied. This is done in a paper [8] by Morse on "Singular quadratic functionals" in Section 6 and Section 7 where the Weierstrass E-function plays a special role.

An Alternative Proof of Theorem 7.6. The proof of Theorem 19.6 and thereby the proof of Theorem 7.6 has been postponed until the techniques associated with broken extremals have been developed. It is, however, possible to give a proof of Theorem 7.6 with the techniques developed preceding Theorem 7.6, as the reader can verify.

CHAPTER 2

The Algebraic Index of a Critical Extremal g

An alternative title to Chapter 2 could be "general endpoint conditions, locally defined." A systematic study of end conditions must precede the definition of the indices of critical extremals.

An analogy. Let φ be a C^{∞}-differentiable function defined in the neighborhood of the origin in a Euclidean space E_n of coordinates $(x_1, \ldots, x_n)$. Suppose that the origin is a critical point of φ. The algebraic index of this critical point can be found as follows. If $(u_1, \ldots, u_n)$ is an arbitrary n-tuple, the second derivative as to e of $\varphi(u_1 e, \ldots, u_n e)$ evaluated when $e = 0$, is a quadratic form $a_{ij}u_i u_j$. The algebraic index of the origin as a critical point of φ is then the number of negative characteristic roots of the quadratic form

$$a_{ij}u_i u_j - \lambda u_i u_i. \tag{8.0}$$

Cf. Theorem 3.4 of MC[1].

In nonparametric variational theory one studies curve integrals whose domains of definition are classes of admissible x-parameterized curves γ satisfying prescribed endpoint conditions. This study is *local* if the endpoint conditions prescribe endpoints (x^1, y^1) and (x^2, y^2) of γ near the endpoints A^1 and A^2 of a given x-parameterized curve g, and if γ itself is required to remain in a neighborhood of g. The curve g is termed an *extremal* if it is of class C^2 and satisfies the Euler equations, and a *critical extremal** if, in addition, g satisfies the "tranversality conditions" associated with prescribed endpoint conditions. Cf. Section 9. When the prescribed endpoint conditions require that the curves γ all pass through A^1 and A^2, there are no transversality conditions and the given base extremal g is by convention a *critical extremal* of **J**. We shall treat this simple special case in Section 12.

In Chapter 2 we shall make more precise the analogies between a critical point of φ and a "critical extremal" by defining the algebraic *index* of a

* Termed *primary* to distinguish it from the critical secondary extremals of Sections 11 and 12.

"critical extremal" as the number of negative characteristic roots λ of an integral analogue of the form (8.0). So defined, the index of a critical extremal will depend in part upon the way the conditions on the endpoint (x^1,y^1) and (x^2,y^2) are prescribed.

8. General End Conditions C_r, Locally Defined

We continue with the preintegrand f of Chapter 1 with domain $X \times R^m$, with the curve integral **J** of (1.4) and an extremal

$$g : x \to g(x) : [a^1,a^2] \to E_m \tag{8.1}$$

of class C^3 admitting an extension $g^{\aleph}$, again of class C^3, for x on an interval

$$\aleph \supset [a^1,a^2]. \tag{8.2}$$

We suppose that the Legendre S-condition (7.9) holds along $g^{\aleph}$. The nonsingularity condition (5.0) is thereby implied and is assumed to hold unless the contrary is stated.

The endpoints of g will be denoted by A^1 and A^2, respectively, and points near A^1 and A^2 by (x^1,y^1) and (x^2,y^2), respectively.

The simplest endpoint conditions fix (x^1,y^1) and (x^2,y^2) at A^1 and A^2, respectively. In Section 8 we shall set up endpoint conditions in three other important cases and in the general case of variable endpoints. These cases are named as follows:

Case I. One endpoint manifold.
Case II. Two endpoint manifolds.
Case III. The condition of periodicity.
Case IV. General variable endpoint conditions.

Case I. *One endpoint manifold.* In this case one endpoint, say (x^2,y^2), is fixed, while the other endpoint (x^1,y^1) is required to be on a regular, connected, manifold M^1 in X of class C^2 and dimension ρ, where $1 \leq \rho \leq m$. The manifold M^1 shall meet the curve $g^{\aleph}$ only in A^1 and be nontangent to $g^{\aleph}$.

The manifold M^1 shall be "presented" as follows. Let E_ρ be a Euclidean space of points $\alpha = (\alpha_1, \ldots, \alpha_\rho)$. Let N^1 be an open neighborhood of the point $\alpha = \mathbf{0} \in E_\rho$. The end manifold M^1 will be supposed "presented" by a regular biunique C^2-mapping

$$\alpha \to (\hat{x}^1(\alpha),\hat{y}^1(\alpha)) : N^1 \to X \tag{8.3}$$

of N^1 into X, onto M^1. We suppose that

$$(\hat{x}^1(\mathbf{0}),\hat{y}^1(\mathbf{0})) = A^1. \tag{8.4}$$

We are requiring that to each point (x^1,y^1) in M^1 there corresponds a unique value $\alpha \in N^1$ such that

$$(x^1,y^1) = (\hat{x}^1(\alpha),\hat{y}^1(\alpha)). \tag{8.5}$$

Case II. *Two endpoint manifolds.* Let p and q be positive integers at most m. In Case II there are given two disjoint regular, connected manifolds M^1 and M^2 of dimensions p and q, respectively, of class C^2, meeting g in A^1 and A^2, respectively, and not tangent to $g^\aleph$.

The manifolds M^1 and M^2 shall be presented as follows. Let E_p and E_q be Euclidean spaces of points $\mu = (\mu_1, \ldots, \mu_p)$ and $\nu = (\nu_1, \ldots, \nu_q)$, respectively. Let N^1 and N^2 be open neighborhoods of the origin in E_p and E_q, respectively. The manifolds M^1 and M^2 will be supposed "presented" by regular, biunique C^2-mappings,

$$\begin{cases} \mu \to (\hat{x}^1(\mu),\hat{y}^1(\mu)) : N^1 \to X \\ \nu \to (\hat{x}^2(\nu),\hat{y}^2(\nu)) \ : N^2 \to X \end{cases} \tag{8.6}$$

of N^1 and N^2 into X, onto M^1 and M^2, respectively. We suppose that

$$(\hat{x}^1(\mathbf{0}_p),\hat{y}^1(\mathbf{0}_p)) = A^1, \qquad (\hat{x}^2(\mathbf{0}_q),\hat{y}^2(\mathbf{0}_q)) = A^2 \tag{8.7}$$

where $\mathbf{0}_p$ is the origin in E_p, $\mathbf{0}_q$ the origin in E_q. To be admissible, end points

$$\begin{cases} (x^1,y^1) = (\hat{x}^1(\mu),\hat{y}^1(\mu)) \\ (x^2,y^2) = (\hat{x}^2(\nu),\hat{y}^2(\nu)) \end{cases} \tag{8.8}$$

for some point $\mu \in N^1$ and point $\nu \in N^2$. The conditions (8.8) supplement rather than replace the conditions preceding (8.8).

Case III. *The condition of periodicity.* We suppose that the preintegrand f and the given extremal g have the *period* τ in x. That is, we suppose that when (x,y) is a point in X, $(x + \tau, y)$ is in X and that

$$\begin{cases} f(x + \tau, y, p) \equiv f(x,y,p) & ((x,y;p) \in X \times R^m) \\ \quad g(x + \tau) \equiv g(x) & (x \in R). \end{cases} \tag{8.9}$$

For simplicity, we suppose that A^1 is the origin $(0,\mathbf{0}_m) \in X$ and that A^2 is the point $(x,y) = (\tau,\mathbf{0}_m) \in X$.

Let N be an open neighborhood of the origin in R^m and let $\alpha = (\alpha_1,\ldots,\alpha_m)$ be a point in N. Our endpoint conditions in Case III require that

$$(x^1,y^1) = (0,\alpha) \qquad (x^2,y^2) = (\tau,\alpha) \tag{8.10}$$

for some value of α in N.

Case IV. *The general variable endpoint conditions.* The endpoint conditions in the above three cases can be given a common form as follows. Let r be an integer which equals ρ in Case I, $p + q$ in Case II, and m in Case III. In each case $1 \leq r \leq 2m$. If $\alpha = (\alpha_1, \ldots, \alpha_r)$ is a point in E_r in a suitably chosen open neighborhood $\mathbf{A}_r$ of the point $\alpha = \mathbf{0}$, conditions of Cases I, II, and III on (x^1,y^1) and (x^2,y^2) require the following:

*Conditions C_r

$$(8.11)' \quad \begin{aligned} (x^1,y^1) &= (X^1(\alpha), Y^1(\alpha)) \\ (x^2,y^2) &= (X^2(\alpha), Y^2(\alpha)) \end{aligned} \quad (\alpha \in \mathbf{A}_r)$$

$$(8.11)'' \quad \begin{aligned} (X^1(\mathbf{0}), Y^1(\mathbf{0})) &= A^1 \\ (X^2(\mathbf{0}), Y^2(\mathbf{0})) &= A^2 \end{aligned}, \quad (r > 0)$$

$$(8.11)''' \quad X^1(\alpha) < X^2(\alpha) \quad (\alpha \in \mathbf{A}_r)$$

for proper definition of $X^s(\alpha)$, $Y^s(\alpha)$, $s = 1, 2$, and of the point $\mathbf{0} \in \mathbf{A}_r$ as we now verify.

Verification in Case I. In this case one identifies X^1 with $\hat{x}^1$, Y^1 with $\hat{y}^1$, $\mathbf{A}_r$ with N^1 and sets $(X^2(\alpha), Y^2(\alpha)) \equiv A^2$ for $\alpha \in \mathbf{A}_r$. The conditions C_r then hold in a special way.

Verification in Case II. Here $r = p + q$. One sets $E_r = E_p \times E_q$, and $\mathbf{A}_r = N^1 \times N^2$. Given a point $\mu \in N^1$ and a point $\nu \in N^2$ set

$$(8.12) \qquad (\alpha_1, \ldots, \alpha_r) = (\mu_1, \ldots, \mu_p : \nu_1, \ldots, \nu_q),$$

$$(8.13) \qquad \begin{aligned} (X^1(\alpha), Y^1(\alpha)) &= (\hat{x}^1(\mu), \hat{y}^1(\mu)) \\ (X^2(\alpha), Y^2(\alpha)) &= (\hat{x}^2(\nu), \hat{y}^2(\nu)) \end{aligned} \qquad (\alpha \in \mathbf{A}_r).$$

Conditions (8.8) then take the form (8.11)′, and Conditions (8.11)″ and (8.11)‴ hold.

Verification in Case III. In Case III $r = m$. For $\alpha \in \mathbf{A}_m$ one sets

$$(8.14) \qquad \begin{cases} X^1(\alpha) \equiv 0, & Y^1(\alpha) \equiv \alpha \\ X^2(\alpha) \equiv \tau, & Y^2(\alpha) \equiv \alpha \end{cases}.$$

Conditions (8.14) then take the form (8.11)′ and the other conditions of (8.11) hold.

In Case I, II, and III the function pairs (X^1, Y^1) and (X^2, Y^2) are well-defined, of class C^2 for $\alpha \in \mathbf{A}_r$ and have values which are points in X. For i on the range $1, \ldots, m$, the functions

$$(8.15) \qquad X^1,\ Y^1_i,\ X^2,\ Y^2_i,$$

* Not to be confused with the differentiability conditions C^r.

are in number $2m + 2$. If h has the range $1, \ldots, r$ the Jacobian matrix of the $2m + 2$ functions (8.15) with respect to $\alpha_1, \ldots, \alpha_r$ has the form

$$\left\| \frac{\partial X^1}{\partial \alpha^h}, \frac{\partial Y_i^1}{\partial \alpha^h}, \frac{\partial X^2}{\partial \alpha^h}, \frac{\partial Y_i^2}{\partial \alpha^h} \right\| \qquad (\alpha \in \mathbf{A}_r) \tag{8.16}$$

with rows, r in number, indexed by h, and with $2m + 2$ columns. In Cases I, II, III, the matrix (8.16) has the rank r for each $\alpha \in \mathbf{A}_r$.

With this background we introduce the following definition.

Definition 8.1. **General variable endpoint conditions, locally defined.** For a positive* integer r at most $2m$ and for $\mathbf{A}_r$ an open neighborhood of the origin in the Euclidean space E_r of points $(\alpha_1, \ldots, \alpha_r)$, conditions on (x^1,y^1) and (x^2,y^2) of the form C_r are understood as induced by a biunique C^2-mapping

$$\alpha \to [X^1(\alpha), Y^1(\alpha); X^2(\alpha), Y^2(\alpha)] : \mathbf{A}_r \to X \times X \tag{8.17}$$

with a functional matrix (8.16) of rank r.

Definition 8.2. C_r*-admissible pairs* (χ,α). Let α be an r-tuple in $\mathbf{A}_r$. If the graph in X of a D^1-mapping

$$x \to \chi(x) : [X^1(\alpha), X^2(\alpha)] \to R^m \tag{8.17$'$}$$

has C_r-admissible endpoints $[X^s(\alpha), Y^s(\alpha)]$, then χ will be called a C_r*-admissible curve* and (χ,α) a C_r*-admissible pair*.

Values $J(\chi,\alpha)$. Corresponding to endpoint conditions C_r, (8.11), let a mapping

$$\alpha \to \Theta(\alpha) : \mathbf{A}_r \to R \tag{8.18}$$

of class C^2 be prescribed. We term Θ an *endpoint function* associated with the end conditions C_r. Given a C_r-"admissible" pair (χ,α), set

$$\boxed{J(\chi,\alpha) = \Theta(\alpha) + \int_{X^1(\alpha)}^{X^2(\alpha)} f(x,\chi(x),\chi'(x))\, dx.} \tag{8.19}$$

The "Functional" (J,C_r). Given endpoint conditions C_r, $r > 0$, and an endpoint function Θ associated with C_r, the mapping

$$(\chi,\alpha) \to J(\chi,\alpha) \tag{8.20}$$

of C_r-admissible pairs (χ,α) into R will be denoted by (J,C_r).

The addition of values $\Theta(\alpha)$ in (8.19) to the integral in defining $J(\chi,\alpha)$ is essential to the proper development of the variable endpoint theory. This will

* The integer $r = 0$ is associated with fixed endpoint conditions, as we shall see in Section 12. We denote these conditions by C_0 and set $(J,C_0) = \mathbf{J}$.

become clearer when it is seen that the second variation associated with the "functional" (J,C_r) *necessarily* has a form (10.23) similar to that of $J(\chi,\alpha)$.

The presentation of endpoint conditions *in parametric form* originated with Morse in M[1]. See Chapter 2. The study of functionals such as (J,C_r), locally defined, is a necessary prelude to the global variational theory. In this approach the nonparametric variational theory becomes a local aspect of the global theory.

In our study of endpoint and boundary conditions the following convention is needed.

Definition 8.3. *A differentiability convention.* Let E_μ be a euclidian space of rectangular coordinates $(v_1, \ldots, v_\mu)$. Let Z be a connected subset of E_μ included in the closure of its interior. A continuous mapping $v \to \varphi(v) : Z \to R$ will be said to be of class C^s, $s > 0$, if φ admits an extension ψ of class C^s on some open subset W of E_μ such that $W \supset Z$. The partial derivatives up to a total order s of *two different extensions* ψ of φ will have the *same* values on the point set boundary βZ of Z and will be taken as the values of the corresponding partial derivatives of φ at points on βZ.

9. C_r-Transversality Conditions on g

The transversality conditions to be established under the conditions of Theorem 9.1, make use of the notation of the endpoint conditions C_r: (8.11), $r > 0$ and of an "endpoint function" Θ associated with C_r when $r > 0$. The conditions (9.1) will be called *C_r-transversality conditions* on g.* They exist only when $r > 0$, and have the form

$$d\Theta + \Big[(f - p_i f_{p_i})\, dX^s + f_{p_i}\, dY_i^s\Big]_{(s=1)}^{(s=2)} \equiv 0 \qquad (\text{at } \alpha = \mathbf{0}) \tag{9.1}$$

the notation of which we will now clarify.

The range of i is $1, \ldots, m$. The differentials $d\Theta$, dX^s, and dY_i^s are with respect to the parameters $\alpha_1, \ldots, \alpha_r$. They are evaluated when $\alpha = \mathbf{0}$ and are linear forms in $d\alpha_1, \ldots, d\alpha_r$. The bracket in (9.1), evaluated for $s = 1$, is to be subtracted from the bracket evaluated for $s = 2$. To evaluate the bracket indexed by s, the functions f, f_{p_i} and p_i are to be given their values when

$$(x,y,p) = (a^s, g(a^s), g'(a^s)).$$

The identity (9.1) implies that the coefficients of the differentials $d\alpha_1, \ldots, d\alpha_r$ are zero. Exercise 9.1 at the end of this section will clarify the meaning of the transversality conditions.

* These transversality conditions are termed *primary* to distinguish them from the *secondary* transversality conditions of Section 11.

A Weak Minimum Defined. Given boundary conditions C_r : (8.11), $r > 0$, a C_r-admissible pair is afforded by $(g,\mathbf{0})$. Recall that g can be extended as an extremal for an open interval $\aleph \supset [a^1,a^2]$. $J(g,\mathbf{0})$ will be said to afford a *weak minimum* to the functional (J,C_r) if for some sufficiently small constant $\varepsilon > 0$ and interval $(a^1 - e, a^2 + e) \subset \aleph$ the following is true. For all C_r-admissible pairs (χ,α) such that χ is of class D^1, $J(g,\mathbf{0}) \leq J(\chi,\alpha)$, provided

$$a^1 - e < X^1(\alpha) < X^2(\alpha) < a^2 + e \tag{9.2)$'$}$$

and for g extended as an extremal,

$$\|\chi(x) - g(x)\| + \|\chi'(x) - g'(x)\| < \varepsilon \qquad (X^1(\alpha) \leq x \leq X^2(\alpha)) \tag{9.2)$''$}$$

for points x at which $\chi'(x)$ exists.

THEOREM 9.1. *A necessary condition that* J(g,**0**) *afford a weak minimum to* (J,C$_r$) *when* r > 0 *is that the* C$_r$*-transversality conditions* (9.1) *on* g *be satisfied.*

Points (x^s,y^s) near the endpoints of g can be joined by an x-parameterized curve on which $y = y(x)$, where

$$y(x) = g(x) + (y^1 - g(x^1)) + [(y^2 - g(x^2)) - (y^1 - g(x^1))]\left(\frac{x - x^1}{x^2 - x^1}\right) \tag{9.3)$'$}$$

as one readily verifies. One sets $(x^s,y^s) = (X^s(\alpha), Y^s(\alpha))$ in the right member of (9.3)$'$ and denotes the resultant right member by $\varphi(x,\alpha)$. For fixed $\alpha \in \mathbf{A}_r$ the curve on which $y = \varphi(x,\alpha)$ joins the endpoints $(X^s(\alpha), Y^s(\alpha))$. For $\alpha = \mathbf{0}$, this curve reduces to g.

Let a C^1-mapping $e \to A(e)$ into $\mathbf{A}_r$ be prescribed with $A(0) = \mathbf{0}$ and $|e| < d$. If d is sufficiently small, $\varphi(x,A(e))$ is defined. Set

$$\varphi(x,A(e)) = \gamma(x,e) \qquad (X^1(A(e)) \leq x \leq X^2(A(e))) \tag{9.3)$''$}$$

for $|e| < d$. A 1-parameter family of C_r-admissible pairs (Definition 8.2) $(\gamma^e,A(e))$* is thereby defined. Hence, for $s = 1$ and 2

$$Y^s(A(e)) \equiv \gamma(X^s(A(e)),e) \qquad (|e| < d). \tag{9.4}$$

Reference to (8.19) shows that if one sets $X^s(A(e)) = x^s(e)$,

$$J(\gamma^e,A(e)) = \Theta(A(e)) + \int_{x^1(e)}^{x^2(e)} f(x,\gamma(x,e),\gamma_x(x,e))\,dx. \tag{9.5}$$

The left member of (9.5) reduces to $J(g,\mathbf{0})$ when $e = 0$ and, by virtue of the hypotheses of Theorem 9.1, has a null derivative when $e = 0$. This will imply the transversality Conditions (9.1).

* For fixed e the partial mapping $x \to \gamma(x,e)$, restricted to the interval $[X^1(A(e)), X^2(A(e))]$, will be denoted by γ^e.

Notation. When $\alpha = \mathbf{0}$, a superscript x on f_{y_i} and f_{p_i} means evaluation for $(x,y,p) = (x,g(x),g'(x))$. An evaluation $[\ ]_1^2$ shall have the meaning of $[\ ]_{(s=1)}^{(s=2)}$ in (9.1). The superscript 0 means evaluation for $e = 0$ after differentiation.

From (9.5) we find a so-called *first variation* of (J,C_r)

$$(9.6)\quad \frac{d}{de} J(\gamma^e,A(e))^0 = \frac{d\Theta^0}{de} + \int_{a^1}^{a^2} (f_{y_i}^x \gamma_{ie}(x,0) + f_{p_i}^x \gamma_{iex}(x,0))\, dx + \left[f \frac{dX^{s,0}}{de}\right]_1^2$$

$$(9.7)\quad \frac{d}{de} J(\gamma^e,A(e))^0 = \frac{d\Theta^0}{de} + \left[f \frac{dX^{s,0}}{de} + f_{p_i}\gamma_{ie}(a^s,0)\right]_1^2,$$

where the integral in (9.6) has been reduced by an integration by parts of the terms with factors $f_{p_i}^x$, making use of the fact that g satisfies the Euler equations. If the members of the identity (9.4) are differentiated as to e, and evaluated when $e = 0$, we find that

$$(9.8)\quad \frac{dY_i^{s,0}}{de} = \gamma_{ix}(a^s,0)\frac{dX^{s,0}}{de} + \gamma_{ie}(a^s,0) \qquad (s = 1,2).$$

If we now eliminate $\gamma_{ie}(a^s,0)$ from (9.7) by means of (9.8), we find that

$$(9.9)^*\quad \frac{d}{de} J(\gamma^e,A(e))^0 = \frac{d\Theta^0}{de} + \left[(f - p_i f_{p_i})\frac{dX^{s,0}}{de} + f_{p_i}\frac{dY_i^{s,0}}{de}\right]_1^2 = 0.$$

The mapping $e \to A(e)$ can be chosen so as to have prescribed differentials at the point $\alpha = \mathbf{0}$ in E_r. The identities (9.1) follow from (9.9).

Definition 9.1. *A critical extremal* g *of* (J,C_r). An extremal g, conditioned as in the first paragraph of Section 8, satisfying the conditions C_r : (8.11) with $\alpha = \mathbf{0}$, and the C_r-transversality conditions (9.1), when $\alpha = \mathbf{0}$, will be called a *critical extremal* of (J,C_r). Here $r > 0$.

For g as a critical extremal of (J,C_0) see Section 12.

Transversality Conditions (9.1) *in Another Form.* Conditions (9.1) on g take the form of an identity in the differentials $d\alpha_1, \ldots, d\alpha_m$ of α, with coefficients of these differentials evaluated as in (9.1) when α is the r-tuple $\mathbf{0}$. Hence the transversality conditions (9.1) on g are equivalent to the following m conditions,

$$(9.10)\quad \left[\Theta_{\alpha_h} + (f - p_i f_{p_i})X_{\alpha_h}^s + f_{p_i}Y_{i\alpha_h}^s\right]_{s=1}^{s=2} = 0 \qquad (h = 1, \ldots, m)$$

* When $e = 0$, γ^e reduces to g and $\gamma_{ix}(a^s,0) = g_i'(a^s)$ gives the value assigned to p_i when $x = a^s$.

where the evaluation at $\alpha = \mathbf{0}$ is as in (9.1). A particular case of importance arises when $X^s(\alpha) \equiv a^s$. In this case the transversality conditions take the form

$$\left[\Theta_{\alpha_h} + f_{p_i} Y^s_{i\alpha_h}\right]_{s=1}^{s=2} = 0 \qquad (h = 1, \ldots, m). \tag{9.11}$$

As we shall see, this is a form which the secondary transversality conditions will take when 2Ω replaces f.

Exercise

9.1. With i on the range $1, \ldots, m$, let

$$f(x,y,p) = (1 + p_i p_i)^{1/2} \qquad ((x,y;p) \in E_{m+1} \times R^m).$$

Let g be an x-parameterized straight line in E_{m+1} with endpoints A^1 and A^2. Let endpoint conditions C_r: (8.11) require that (x^2,y^2) be fixed and (x^1,y^1) be conditioned as in Case I of Section 8. If $\Theta(\alpha) \equiv 0$, show that the transversality conditions (9.1) imply that g is orthogonal to the end manifold M^1 at A^1.

10. The Second Variation of (J,C_r), $r > 0$

In Chapter 1 the endpoints of admissible curves are fixed and a curve integral **J** introduced. The second variation $\mathscr{I}_0$ of **J** was introduced. Given conditions C_r : (8.11) and an associated endpoint function Θ, together with a critical extremal g of (J,C_r), we shall define a quadratic functional $\mathscr{I}_r$ with values $\mathscr{I}_r(\eta)$ extending the *second variation** of (J,C_r). The form which $\mathscr{I}_r$ should take is determined by a study of the 1-parameter family of C_r-admissible pairs which we now define. In Section 10 and Section 11, $r > 0$. In Section 12, $r = 0$.

Definition 10.0. *A 1-parameter family* $((\gamma,A))$ *of* C_r*-admissible pairs* (Cf. Definition 8.2.). Let (X^s, Y^s) define the end conditions of C_r as in (8.11). If u is prescribed as an r-tuple, if e is a scalar and if $d > 0$ is sufficiently small, the mapping

$$e \to A(e) = eu : (-d,d) \to \mathbf{A}_r \tag{10.0)'$$

is into the domain $\mathbf{A}_r$ of the mappings X^s and Y^s, where $\mathbf{A}_r$ is an open neighborhood in R^r of $\mathbf{0}$. For $|e| < d$ set

$$X^s(eu) = x^s(e), \qquad Y^s(eu) = y^s(e) \qquad (s = 1, 2). \tag{10.0)''$$

Let $[a',a'']$ be an interval included in the open interval $\aleph$ of Theorem 2.1 and

* It is recommended that in a first reading the reader stop with Theorem 10.1. The proof of Theorem 10.1 is formal. The meaning of the comparison matrix $\|b_{hk}\|$ will become clear in the following section.

such that $[a^1,a^2] \subset (a',a'')$. We suppose that d is so small that $a' < x^1(e) < x^2(e) < a''$. Let

$$(x,e) \to \gamma(x,e) : (a',a'') \times (-d,d) \to R^m \tag{10.1)'$$

be a C^2-mapping. Let γ^e be the partial mapping

$$x \to \gamma(x,e) : [x^1(e),x^2(e)] \to R^m. \tag{10.1)''$$

We suppose that $\gamma^0 = g$ and that d is so small that for $|e| < d$ the graph of γ^e is in X. We require also that (γ^e,eu) be a C_r-admissible pair for each $e \in (-d,d)$ (Definition 8.2). That is suppose that, subject to (10.0)″,

$$\gamma(x^s(e),e) \equiv y^s(e) \qquad (|e| < d : s = 1, 2) \tag{10.2)'$$

or equivalently

$$\gamma(X^s(eu),e) \equiv Y^s(eu). \tag{10.2)''$$

We denote the family (10.1)′ *of C_r-admissible pairs $(\gamma^e,A(e))$ by $((\gamma,A))$.*

A value of the functional (J,C_r) : (8.19) is defined for $|e| < d$ by setting

$$J(\gamma^e,eu) = \Theta(eu) + \int_{x^1(e)}^{x^2(e)} f(x,\gamma(x,e),\gamma_x(x,e))\, dx. \tag{10.3}$$

The first derivative as to e of $J(\gamma^e,eu)$ vanishes when $e = 0$, by virtue of the C_r-transversality condition (9.1) on g. The second derivative

$$\frac{d^2}{de^2} J(\gamma^e,eu)\Big|^{e=0} \tag{10.4}$$

is a value of the second variation of (J,C_r) in accord with definitions to be given. In addition to its general dependence on the preintegrand f and Θ, the value (10.4) depends only on the "variation"* $\eta = \gamma_e\big|^{e=0}$ and upon the r-tuple $(u_1, \ldots, u_r)$. This section is devoted to disclosing the dependence of the derivative (10.4) upon η and u.

The Secondary End Conditions C_r''. The identities (10.2)″ imply that when $r > 0$ the "end values"

$$\eta^s = (\eta_1(a^s), \ldots, \eta_m(a^s)) \qquad (s = 1, 2) \tag{10.5}$$

of the variation $\eta = \gamma_e\big|^{e=0}$ depend linearly on the r-tuple u. We shall make this dependence explicit.

Notation. It will simplify the many derivations required in this section if one indicates partial derivatives of X^s, Y^s, and Θ with respect to α_h and α_k by adding subscripts h and k, respectively, to these symbols. The range of h and k

* Restricting the variation $\eta = \gamma_e\big|^{e=0}$ to the interval $[a^1,a^2]$.

is $1, \ldots, r$. If one differentiates the members of (10.2)″ with respect to e, sets $e = 0$ and makes use of the relations $\gamma\big|^{e=0} = g$, $\gamma_e\big|^{e=0} = \eta$, one finds *secondary end conditions*,

$$C_r'' : \eta_i^s = c_{ih}^s u_h \qquad (i = 1, \ldots, m;\ s = 1, 2) \tag{10.6}$$

on the end values of η, where for $h = 1, \ldots, r$, $i = 1, \ldots, m$ and $s = 1$ or 2

$$c_{ih}^s = Y_{ih}^s(\mathbf{0}) - g_i'(a^s)X_h^s(\mathbf{0}). \tag{10.7}$$

Under C_r'' we understand that $x^s = a^s$ for $s = 1, 2$.

The matrix $\|c_{ih}^s\|$ is understood to have $2m$ rows of which the first m rows are indexed by i with $s = 1$, and the second m rows are indexed by i with $s = 2$. This matrix has r columns indexed by h.

We shall prove the following lemma.

LEMMA 10.1. *In the Cases* I, II, III *of Section* 8, *the matrix* $\|c_{ih}^s\|$ *has a rank* $r > 0$ *equal to the number of its columns.*

Verification in Case I. In this case the matrix reduces to $\|c_{ih}^1\|$. Were this matrix of rank less than r there would exist r constants $d_1, \ldots, d_r$, not all zero such that

$$d_h(Y_{ih}^1(\mathbf{0}) - g_i'(a^1)X_h^1(\mathbf{0})) = 0 \qquad (i = 1, \ldots, m). \tag{10.8}$$

There would then exist a constant c such that

$$\begin{cases} d_h X_h^1(\mathbf{0}) = c \\ d_h Y_{ih}^1(\mathbf{0}) = c g_i'(a^1) \qquad (i = 1, \ldots, m). \end{cases} \tag{10.9}$$

It is impossible that $c = 0$, since the rank of the matrix of coefficients of $d_1, \ldots, d_r$ on the left of (10.9) is r by hypothesis. It is also impossible that $c \neq 0$. For the $m + 1$ numbers $1, g_1'(a^1), \ldots, g_m'(a^1)$ define a direction in E_{m+1} tangent to g at A^1, and this direction is not tangent to the manifold M^1 at A^1.

Thus Lemma 10.1 is true.

Verification in Case II. This follows from the verification in Case I, since the endpoint conditions in Case II on (x^1,y^1) and (x^2,y^2) are independent and similar to the conditions on (x^1,y^1) in Case I.

Verification in Case III. In this case $X^1(\alpha) \equiv 0$ and $X^2(\alpha) \equiv \tau$, a constant. Reference to (8.14) shows that $\|c_{ih}^1\|$ is a m-square diagonal matrix with a diagonal of 1's.

Thus Lemma 10.1 is true.

Definition 10.1. *The nontangency condition.* We shall assume that the matrix $\|c_{ih}^s\|$ has a rank equal to the number $r > 0$ of its columns.

This hypothesis is known as the *nontangency condition* and was first introduced by Morse on page 525 of Morse [4]. It is satisfied in Cases I, II, III of Section 8 and distinguishes a general case from exceptional cases of little significance. In our present treatment its importance depends upon the obvious truth of the following lemma.

LEMMA 10.2. *Under the condition that the matrix* $\|c^s_{ih}\|$ *have a rank equal to the number* r *of its columns, the secondary end conditions*

$$(10.10) \qquad C''_r : \eta^s_i = c^s_{ih} u_h \qquad (i = 1, \ldots, m; s = 1, 2)$$

give a biunique representation of the end values $(\eta^1_1, \ldots, \eta^1_m; \eta^2_1, \ldots, \eta^2_m)$ *in terms of the* r*-tuples,* $(u_1, \ldots, u_r)$.

The fundamental theorem. In preparation for the fundamental Theorem 10.1 we differentiate the members of (10.3) as to e. In (10.3) f is evaluated by setting $(x,y,p) = (x,\gamma(x,e),\gamma_x(x,e))$. To indicate such an evaluation of f or its partial derivatives, the superscript x will be added. From (10.3) we find that for $|e| < d$

$$(10.11) \quad \frac{d}{de} J(\gamma^e,eu) \equiv \Theta_h(eu)u_h + \int_{x^1(e)}^{x^2(e)} (f^x_{y_i}\gamma_{ie}(x,e) + f^x_{p_i}\gamma_{iex}(x,e))\, dx$$
$$+ u_h\Big[f\,(x^s(e),\gamma(x^s(e),e),\gamma_x(x^s(e),e))X^s_h(eu)\Big]^{s=2}_{s=1}.$$

In evaluating the second derivative as to e of $J(\gamma^e,eu)$ when $e = 0$, we shall be led to an r-square symmetric matrix $\|b_{hk}\|$ of real constants with values,

$$(10.12)' \qquad b_{hk} = b^2_{hk} - b^1_{hk} + \Theta_{hk}(\mathbf{0})$$

where for $s = 1, 2$ and $h, k = 1, \ldots, r$,

$$(10.12)'' \quad b^s_{hk} = (f - p_i f_{p_i})X^s_{hk} + (f_x - p_i f_{y_i})X^s_h X^s_k$$
$$+ f_{y_i}(X^s_h Y^s_{ik} + X^s_k Y^s_{ih}) + f_{p_i} Y^s_{ihk}$$

and the right member is to be evaluated by setting $(x,y,p) = (a^s,g(a^s),g'(a^s))$ and setting $\alpha = \mathbf{0}$ in each of the partial derivatives of X^s and of the m-tuple Y^s.

THEOREM 10.1. *If* g *is an* x*-parameterized extremal of class* C^2, *satisfying the* C_r*-transversality conditions* (9.1), *and if* $((\gamma,A))$ *is the family of Definition* 10.0 *of* C_r*-admissible pairs* $(\gamma^e,A(e))$ *with* $\gamma^0 = g$, *then*

$$(10.13)^* \qquad \frac{d^2}{de^2} J(\gamma^e,eu)^{e=0} = b_{hk}u_h u_k + \int_{a^1}^{a^2} 2\Omega(x,\eta(x),\eta'(x))\, dx,$$

* The quadratic form $Q(u) = b_{hk}u_h u_k$ will define a "secondary endpoint function" in contrast with the primary endpoint function Θ of Section 8. The matrix $\|b_{hk}\|$ will be called the *comparison matrix* associated with g as a critical extremal of (J,C_r).

where η *is the variation* $\gamma_e\big|^{e=0}$ *and* u *the* r-*tuple* $A'(0)$. *Moreover, the end values* η^s *of* η *and the* r-*tuple* u *satisfy the secondary endpoint conditions,*

$$C_r'': \eta_i^s = c_{ih}^s u_h \qquad (i = 1, \ldots, m;\ s = 1, 2). \tag{10.14}$$

The following is especially noteworthy. While $((\gamma,A))$ is a family of C_r-admissible pairs (γ^e,eu) near g, when $r > 0$, nevertheless the constants, b_{hk}, as given by (10.12)′, do not involve the mappings γ or u in any way. These constants b_{hk} enter in a fundamental way into our generalizations of the Sturm separation, comparison and oscillation theorems, when $m > 1$. See Chapter 4.

Proof of Theorem 10.1. The proof of Theorem 10.1 consists of three parts. Part I is preparatory. It presents a set of consequences of the identities (10.2). Part II contains the differentiation of the right member of (10.11). The resulting terms include the integral $\int_{a^1}^{a^2} 2\Omega\, dx$ and terms not a priori quadratic (as desired) in the variables $(u_1, \ldots, u_r)$. In Part III we use the relations derived in Part I to reduce the sum of the terms obtained in Part II to the right member of (10.13).

Part I *of Proof.* A differentiation as to e of the identity (10.2)″ gives the identity in e

$$\gamma_{ie} \equiv (Y_{ih}^s(eu) - \gamma_{ix}X_h^s(eu))u_h \qquad (i = 1, \ldots, m) \tag{10.15}$$

where γ_{ie} and γ_{ix} are evaluated after differentiation, as is γ in (10.2)″. Upon differentiating the members of the identity (10.15) as to e and setting $e = 0$ and $x^s(0) = a^s$, we find that (with $\gamma_{ix}(x,0) = g_i'(x)$)

$$\gamma_{iee} + 2\gamma_{iex}X_h^s u_h = (Y_{ihk}^s - g_i'X_{hk}^s - g_i''X_h^sX_k^s)u_hu_k. \tag{10.16}$$

Part II. Setting $e = 0$ and $x = a^s$ in γ or its partial derivatives will be indicated by a superscript $\circ$. When $e = 0$ a superscript x on f, f_x, f_{y_i}, f_{p_i} means evaluation with

$$(x,y,p) = (x,\gamma(x,e),\gamma_x(x,e))^{e=0} = (x,g(x),g'(x)). \tag{10.17}$$

The derivative as to e of the right member of (10.11), evaluated at $e = 0$, is a sum $S = A + B + C + D + E$ where

$$A = \Theta_{hk}^\circ u_h u_k$$

$$B = \Big[(f_x^x + g_i'f_{y_i}^x + g_i''f_{p_i}^x)X_h^sX_k^s + (f^xX_{hk}^s)\Big]_1^2 u_hu_k$$

$$C = \Big[(f_{y_i}^x\gamma_{ie}^\circ + f_{p_i}^x\gamma_{ixe}^\circ)X_k^s\Big]_1^2 u_k$$

$$D = \Big[(f_{y_i}^x\gamma_{ie}^\circ + f_{p_i}^x\gamma_{iex}^\circ)X_h^s\Big]_1^2 u_h$$

$$E = \int_{a^1}^{a^2} 2\Omega(x,\eta(x),\eta'(x))\, dx + \Big[\gamma_{iee}^\circ f_{p_i}^x\Big]_1^2.$$

The evaluation $]_1^2$ is for $s = 2$ and 1, $x = a^2$ and a^1, and $\alpha = \mathbf{0}$. As indicated previously, $\eta = \gamma_e|^{e=0}$.

A comes from differentiating the term in (10.11) involving Θ.

B comes from differentiating the last term in (10.11), first setting $e = 0$, except in $x^s(e)$ and $X_h^s(eu)$.

C is complementary to B. It comes from differentiating the last term in (10.11), first setting $e = 0$ in $x^s(e)$ and $X_h^s(eu)$.

D comes from differentiating the integral in (10.11), after first setting $e = 0$ in the integrand.

E has two terms. The first term, the integral, is obtained by differentiating the integral in (10.11) under the integral sign, after first setting the limits equal to a^1 and a^2 and setting $e = 0$ in γ_{ie} and γ_{iex}. The last term in E is obtained by first setting $e = 0$ in $f_{y_i}^x$ and $f_{p_i}^x$ in the integrand of (10.11), then integrating by parts to get the term,

$$\Big[\gamma_{ie}(x,e) f_{pi}^x\Big]_1^2 \qquad \text{(with } e = 0 \text{ in } f_{pi}^x)$$

and finally differentiating as to e and setting $e = 0$.

Part III. To reduce the sum S we replace $\overset{\circ}{\gamma}_{ie}$ in C and D by its value,

$$\text{(10.18)} \qquad \overset{\circ}{\gamma}_{ie} = (Y_{ih}^s(\mathbf{0}) - \overset{\circ}{\gamma}_{ix} X_h^s(\mathbf{0}))u_h \qquad (i = 1, \ldots, m;\ s = 1, 2)$$

obtained from (10.15). The terms in C and D thereby modified, have as replacement the quadratic form

$$\text{(10.19)} \qquad G = \Big[f_{yi}^x(X_k^s Y_{ih}^s + X_h^s Y_{ik}^s - 2g_i' X_h^s X_k^s)\Big]_1^2 u_h u_k$$

with coefficients evaluated as in (9.1).

Apart from the integral in E, the terms in the modified S which are linear in the variables $u_1, \ldots, u_r$ have the sum

$$\text{(10.20)} \qquad K = \Big[f_{pi}^x(\overset{\circ}{\gamma}_{iee} + 2\overset{\circ}{\gamma}_{iex} X_h^s u_h)\Big]_1^2.$$

If the parenthesis on the right of (10.20) is replaced by the right member of (10.16) we find that

$$\text{(10.21)} \qquad K = \Big[f_{pi}^x(Y_{ihk}^s - g_i'' X_h^s X_k^s - g_i' X_{hk}^s\Big]_1^2 u_h u_k.$$

We conclude that

$$\text{(10.22)} \qquad S = \overset{\circ}{\Theta}_{hk} u_h u_k + B + G + K + \int_{a^1}^{a^2} 2\Omega(x,\eta(x),\eta'(x))\, dx.$$

The terms $\Theta^\circ_{hk}u_h u_k + B + G + K$ reduce to $b_{hk}u_h u_k$ in accord with (10.12). Theorem 10.1 follows.

The right member of (10.13) is a special value of a functional $\mathscr{I}_r$ whose general domain we now define.

Definition 10.2. *C_r''-admissible mappings.* A mapping $x \to \eta(x) : [a^1,a^2] \to R^m$ of class D^1, is termed *C_r''-admissible* when $r > 0$, if its end values $\eta^s = \eta(a^s)$, $s = 1, 2$ satisfy the conditions C_r'' : (10.14) with some r-tuple u. When $r = 0$ conditions $\eta(a^1) = \eta(a^2) = \mathbf{0}$ are to be satisfied.

With this understood $\mathscr{I}_r$ can now be defined.

Definition 10.3. The quadratic functional $\mathscr{I}_r$ is based on g as a critical extremal of (J,C_r). $\mathscr{I}_r(\eta)$ is defined for each C_r''-admissible mapping η and when $r > 0$ has the value

$$\mathscr{I}_r(\eta) = b_{hk}u_h u_k + \int_{a^1}^{a^2} 2\Omega(x,\eta(x),\eta'(x))\,dx \tag{10.23}$$

where the coefficients b_{hk} are determined as in (10.12)′ by g, C_r, and the external function Θ associated with (J,C_r) and where η and the r-tuple u satisfy the conditions C_r'' : (10.10). When $r = 0$, all reference to u is deleted. When $r > 0$ our definition of the quadratic functional $\mathscr{I}_r$ is logically independent of any concept of the "second variation."

The following definition enables us to formulate an essential extension of Theorem 10.1.

Definition 10.4. *Piecewise C_x^2-mappings.* The mapping

$$(x,e) \to \gamma(x,e) : (a',a'') \times (-d,d) \to R^m \tag{10.24}$$

was given in (10.1)′, as of class C^2. Let $c_1 < \cdots < c_p$ be a set of values in the interval (a',a'') dividing (a',a'') into the set of intervals

$$(a',c_1],\ [c_1,c_2],\ \ldots,\ [c_{p-1},c_p],\ [c_p,a''). \tag{10.25}$$

A mapping (10.24) will be said to be *piecewise* of class C_x^2* if it is continuous and if for each of the intervals H in the set (10.25), the restriction $\gamma \mid (H \times (-d,d))$ is of class C^2 in the sense of Definition 8.3.

If γ : (10.24) is piecewise of class C_x^2, one sees that the partial derivatives γ_e and γ_{ee} exist and are continuous on the domain of γ. A review of the proof of Theorem 10.1 shows that the following is true.

THEOREM 10.2. *If a mapping γ of form* (10.24) *is conditioned as was the mapping γ of* (10.1)′, *except that γ* : (10.24) *is piecewise of class C_x^2 rather than*

* Piecewise of class C_x^2 relative to the points of division $c_1 < \cdots < c_p$ of (a',a'').

of class C^2, *the conclusion of Theorem* 10.1 *still holds for the variation* $\eta = \gamma_e\big|^{e=0}$ *of* γ. *That is, if* u *is an* r*-tuple satisfying* (10.4) *with* η, *then when* $r > 0$

$$\text{(10.26)} \qquad \frac{d^2}{de^2} J(\gamma^e, eu)\bigg|^{e=0} = b_{hk}u_h u_k + \int_{a^1}^{a^2} 2\Omega(x,\eta(x),\eta'(x))\,dx.$$

In terms of the quadratic functional $\mathscr{I}_r$, introduced in Definition 10.3, (10.26) can be written in the form

$$\text{(10.27)} \qquad \mathscr{I}_r(\eta) = \frac{d^2}{de^2} J(\gamma^e, eu)\bigg|^{e=0},$$

subject to the conditions of Theorem 10.2 on the variation $\eta = \gamma_e\big|^{e=0}$. When (10.27) holds we say that $\mathscr{I}_r(\eta)$ gives a value of the *second variation* (J,C_r) based on the critical extremal g of (J,C_r) and induced by the *variation* $\eta = \gamma_e\big|^{e=0}$.

For us the most useful variations η will be derived in Section 14 from 1-parameter families of C_r-admissible broken extremals. A variation η, so derived along g, will be a C_r''-admissible D^1-mapping of $[a^1,a^2]$ into R^m whose graph is a broken secondary extremal.

Conversely, let $\bar{\eta}$ be a D^1-mapping of $[a^1,a^2]$ into R^m whose graph is a prescribed C_r''-admissible broken, secondary extremal. Our constructions in Sections 13 and 14 of families of broken extremals will make it clear that there exists a mapping γ of form (10.24) which is piecewise of class C_x^2 and which is such that $\gamma_e\big|^{e=0} = \bar{\eta}$ for $x \in [a^1,a^2]$. For this γ and $\bar{\eta}$, Theorem 10.2 holds.

A Summary. In Section 10 the main emphasis has been on setting up a quadratic functional $\mathscr{I}_r$ based on a critical extremal g of the given primary functional (J,C_r). In the next two sections the emphasis will shift to setting up a system of selfadjoint differential equations and boundary conditions whose characteristic roots and solutions will serve eventually to characterize the index and nullity of the functional $\mathscr{I}_r$.

Exercises

In the following exercises suppose that a preintegrand f is given with values $f(p) = (1 + p_1^2 + \cdots + p_m^2)^{1/2}$, $\Theta(\alpha) \equiv 0$ and that g is an x-parameterized extremal g. Suppose that endpoint conditions C_r require that (x^2,y^2) be fixed and that (x^1,y^1) rest on a manifold M_m, meeting g at A^1 and orthogonal to g at A^1. Near A^1, M_m shall have a presentation $\alpha \to (X^1(\alpha), Y^1(\alpha))$ of class C^∞ with $\alpha \in \mathbf{A}_m$, as in Case I of Section 8. Verify the following.

10.1. The $(m + 1)$-tuple

$$\text{(10.28)} \qquad f - p_i f_{p_i}, f_{p_1}, \ldots, f_{p_m},$$

evaluated by setting the m-tuple $p = g'(a^1)$, is a set of direction cosines in E_{m+1} of g at A^1.

10.2. An evaluation of b^1_{hk} in (10.12)′ in which $p = g'(a^1)$ and $\alpha = \mathbf{0}_r$, gives the formula at A^1

$$(f - p_i f_{p_i})X^1_{hk} + f_{p_i} Y^1_{ihk} = -b_{hk}. \tag{10.29}$$

10.3. Let Π_m be the m-plane tangent in E_{m+1} to M_m at A^1. Let normals to Π_m be sensed as is g. If $d(\alpha)$ is the algebraic distance along the normal to Π_m from $(X^1(\alpha), Y^1(\alpha))$, then for sufficiently small values of $\|\alpha\|$, $2d(\alpha) \equiv b_{hk}\alpha_h\alpha_k$ + (terms of higher order).

Comment. It follows that $b_{hk}\, d\alpha_h\, d\alpha_k$ is the *second fundamental form* of M_m at A^1. Cf. Eisenhart [1], p. 115.

11. Critical Secondary Extremals: Endpoints Variable

A nonnull secondary extremal η (cf. Definition 4.1) which satisfies the secondary endpoint conditions C_r'' : (10.6) $r > 0$, with an r-tuple u will be said to be a *C_r''-admissible secondary extremal.* It will be called a *critical,* secondary extremal of the quadratic functional $\mathscr{I}_r$ of (J,C_r), if it satisfies the C_r''-transversality conditions relative to $\mathscr{I}_r$, to be defined. Cf. Definition 10.3.

The definition of C_r''-transversality conditions calls for a notational innovation.

The canonical m-*tuple* $\xi = (\xi_1, \ldots, \xi_m)$. For $x \in \aleph$ and arbitrary m-tuples η and ζ we shall set

$$\xi_i = \Omega_{\zeta_i}(x,\eta,\zeta) \qquad (i = 1, \ldots, m;\ \text{see (4.0) for } \zeta). \tag{11.1}$$

For x and η fixed, (11.1) defines a nonsingular homogeneous linear transformation from the m-tuple ζ to the "canonical" m-tuple ξ. When a secondary extremal η is given together with η', we shall define the m-tuples ξ^s, $s = 1, 2$, by setting

$$\xi_i(x) = \Omega_{\zeta_i}(x,\eta(x),\eta'(x)); \qquad \xi_i^s = \xi_i(a^s) \qquad (i = 1, \ldots, m;\ s = 1, 2). \tag{11.2}$$

The first variation of $\mathscr{I}_r$. When $r > 0$ the quadratic functional $\mathscr{I}_r$, based on a critical extremal g of (J,C_r), has values $\mathscr{I}_r(\eta)$ given by (10.23) for C_r''-admissible mappings η of $[a^1,a^2]$ into R^m. In particular let η be a C_r''-admissible secondary extremal which satisfies the conditions

$$C_r'' : \eta_i^s = c_{ih}^s u_h \qquad (s = 1, 2;\ h = 1, \ldots, r) \tag{11.3$'$}$$

with an r-tuple u. Let $\bar\eta$ be an arbitrary C_r''-admissible mapping of $[a^1,a^2]$ into R^m which satisfies the conditions C_r'' with an r-tuple $\bar u$. For an arbitrary scalar

e, the mapping $\eta + e\bar{\eta}$ is C_r''-admissible and satisfies the conditions C_r'' with an r-tuple $u + e\bar{u}$. By Definition 10.3

$$(11.3)'' \quad \mathscr{I}_r(\eta + e\bar{\eta}) = b_{hk}(u_h + e\bar{u}_h)(u_k + e\bar{u}_k) + 2\int_{a^1}^{a^2} \Omega(x, \eta + e\bar{\eta}, \eta' + e\bar{\eta}')\, dx.$$

When η is a secondary extremal one finds that

$$(11.4)' \qquad \frac{d}{de}\mathscr{I}_r(\eta + e\bar{\eta})\Big|^{e=0} = 2b_{hk}\bar{u}_h u_k + 2\Big[\xi_i(x)\bar{\eta}_i(x)\Big]_{x=a^1}^{x=a^2},$$

where ξ is the canonical associate of the solution η of the JDE. Since $\bar{\eta}_i(a^s) = c^s_{ih}\bar{u}_h$, (11.4)′ takes the form

$$(11.4)'' \qquad \frac{d}{de}\mathscr{I}_r(\eta + e\bar{\eta})\Big|^{e=0} = 2(b_{hk}u_k + c^2_{ih}\xi^2_i - c^1_{ih}\xi^1_i)\bar{u}_h.$$

We have in (11.4)″ a formula for the *first variation* of $\mathscr{I}_r$ *based* on a C_r''-admissible secondary extremal η and induced by a C_r''-admissible *variation* $\bar{\eta}$ from η.

Formula (11.4)″ implies the following.

THEOREM 11.1. *Let η be a C_r''-admissible secondary extremal which satisfies conditions C_r'' with an r-tuple u and let ξ be the canonical associate of η. The first variation of $\mathscr{I}_r$ based on η and induced by a C_r''-admissible variation $\bar{\eta}$ then vanishes for each such variation $\bar{\eta}$ if and only if*

$$(11.5) \qquad c^1_{ih}\xi^1_i - c^2_{ih}\xi^2_i = b_{hk}u_k \qquad (h = 1, \ldots, r)$$

Theorem 11.1 leads us to the following definition.

Definition 11.1. Conditions (11.5) are on any secondary extremal η which satisfies conditions C_r'' with an r-tuple u and which has a canonical associate ξ. They are termed *C_r''-transversality conditions* on η relative to $\mathscr{I}_r$.

Theorem 11.1 has the following corollary.

COROLLARY 11.1. *When r > 0, the C_r'' transversality conditions* (11.5) *are satisfied by any secondary extremal η for which* (11.3)′ *holds with an r-tuple u and which is such that $\mathscr{I}_r(\eta)$ affords a minimum to $\mathscr{I}_r$ relative to C_r''-admissible mappings $\bar{\eta}$.*

A Formal Analogy. The C_r''-transversality conditions (11.5) are termed *secondary*. They are formally analogous to the C_r-transversality conditions (9.11) on g in the special case in which the end conditions C_r leave the

x-coordinates of end points fixed. To make this clear set $Q(u) = b_{hk}u_h u_k$ and write (10.23) in the form

$$\mathscr{I}_r(\eta) = Q(u) + \int_{a^1}^{a^2} 2\Omega(x,\eta(x),\eta'(x))\,dx, \tag{11.6$'$}$$

subject to the endpoint conditions C_r'' : (11.3)$'$. The values of the primary functional (J,C_r), as given by (8.19), make it evident that $Q(u)$ in (11.6)$'$ is analogous to $\Theta(\alpha)$ in (8.19) and that the secondary preintegrand $2\Omega(x,\eta,\zeta)$ is analogous to the primary preintegrand $f(x,y,p)$. Moreover, the C_r-conditions $y_i^s = Y_i^s(\alpha)$ are analogous to the C_r''-conditions (11.3)$'$, so that $Y_{i\alpha_h}^s$ is analogous to $\eta_{iu_h}^s = c_{ih}^s$. With this understood one sees that the special primary transversality conditions (9.11) are formally analogous to the secondary transversality conditions (11.5), if the latter are written in the form

$$0 = 2b_{hk}u_k + 2\Big[c_{ih}^s\xi_i^s\Big]_{s=1}^{s=2} = Q_{u_h} + 2\Big[\eta_{iu_h}^s\Omega_{\zeta_i}\Big]_{s=1}^{s=2} \tag{11.6$''$}$$

for $h = 1, 2, \ldots, r$.

A special example is offered.

Example 11.1. If the end conditions C_r : (8.11) are those of Case I of Section 8 with $r = m$, then under the corresponding secondary conditions C_m'', the m-tuple $\eta(a^2)$ vanishes and the end values $(\eta(a^1),\xi(a^1))$ are biuniquely determined in (11.3)$'$ and (11.5) by the m-tuple u. In this case $\|c_{ih}^2\|$ is a null matrix and $\|c_{ih}^1\|$ an m-square matrix, as (10.7) shows. The rank of $\|c_{ih}^1\|$ is m by virtue of Lemma 10.1.

Case I will be studied more deeply as a part of the focal point theory of Chapter 5.

To prepare for the definition of the algebraic index of a critical extremal g of (J,C_r) we develop the analogue of the characteristic roots of a quadratic form. In general, there are no critical extremals of the functional $\mathscr{I}_r$. The functionals $\mathscr{I}_r^\lambda$, which we now introduce, do have critical extremals for certain values of λ, and these values are the analogues of characteristic roots of a quadratic form.

Accessory functionals $\mathscr{I}_r^\lambda$. A symmetric quadratic form $Q(z) = a_{ij}z_iz_j$ is to be regarded as defining a function Q. With Q and each real number λ there is associated a characteristic form, in classical form,

$$Q^\lambda(z) = a_{ij}z_iz_j - \lambda z_iz_i$$

to be regarded as defining a function Q^λ "accessory" to Q.

Similarly with each value $\mathscr{I}_r(\eta)$: (10.23) of the functional $\mathscr{I}_r$ of Definition 10.3 and with each real number λ we shall associate a value

$$\mathscr{I}_r^\lambda(\eta) = b_{hk}u_hu_k + \int_{a^1}^{a^2} (2\Omega(x, \eta(x), \eta'(x)) - \lambda\,\|\eta(x)\|^2)\,dx \tag{11.7}$$

of a new functional termed "accessory" to $\mathscr{I}_r$. The domain of $\mathscr{I}_r^\lambda$, like the domain of $\mathscr{I}_r$, is the set of all C_r''-admissible mappings η. Note that $\mathscr{I}_r^\lambda$ reduces to $\mathscr{I}_r$ when $\lambda = 0$.

Definition 11.2. When $r > 0$ a *critical secondary extremal* η of the above functional $\mathscr{I}_r^\lambda$ is the graph of a C^2-mapping $x \to \eta(x) : [a^1,a^2] \to E_m$ such that the conditions*

$$\begin{bmatrix} \dfrac{d}{dx}\Omega_{\zeta_i}(x,\eta,\eta') - \Omega_{\eta_i}(x,\eta,\eta') + \lambda\eta_i = 0 & (i = 1, \ldots, m) & (11.8)' \\ \eta_i^s = c_{ih}^s u_h \quad (s = 1, 2;\ i = 1, \ldots, m) & & (11.8)'' \\ c_{ih}^1\xi_i^1 - c_{ih}^2\xi_i^2 = b_{hk}u_k \quad (h = 1, \ldots, r) & & (11.8)''' \end{bmatrix}$$

are satisfied by η, the r-tuple u and canonical end coordinates ξ_i^1 and ξ_i^2 determined by η.

Conditions (11.8)′ are the Euler equations of the integral in (11.7).

Conditions (11.8)″ are the secondary end conditions C_r'' satisfied by each mapping in the domain of $\mathscr{I}_r^\lambda$ or $\mathscr{I}_r$.

Conditions (11.8)‴ are the C_r''-transversality conditions relative to $\mathscr{I}_r^\lambda$ on a C_r''-admissible mapping η (Definition 11.1). The conditions (11.8)‴ must be satisfied, according to Corollary 11.1, whenever $\mathscr{I}_r^\lambda(\eta)$ affords a minimum to $\mathscr{I}_r^\lambda$. As we shall see, a critical secondary extremal of $\mathscr{I}_r^\lambda$ does not in general afford a relative minimum of any sort of $\mathscr{I}_r^\lambda$.

Definition 11.3. *Characteristic solutions and roots of* (11.8). A nonnull C^2-mapping $x \to \eta(x) : [a^1,a^2] \to E_m$ which satisfies the conditions (11.8) with a value λ will be called a *characteristic solution* η (with *root* λ) of the conditions (11.8), or of $\mathscr{I}_r$. By the *multiplicity* of a characteristic root λ_0 is meant the number of linearly independent solutions η of (11.8) satisfying (11.8) with λ_0.

Hypotheses 11.1 *on* g. When $r = 0$ we suppose that g is conditioned as in the first paragraph of Section 8. When $r > 0$ we require, in addition, that g satisfy the C_r-transversality condition (9.1) and, with C_r'', the nontangency condition of Definition 10.1.

The following definition is a principal object of study of Section 11. It concerns a critical extremal g, conditioned as in Hypotheses 11.1 when $r > 0$. The brace { } indicates an alternative.

Definition 11.4. When $r > 0$ the *algebraic index* {*nullity*} of a critical extremal g of (J,C_r) is the number of negative {null} characteristic roots λ of

* The system (11.8) is termed *derived* in contrast with a *free* system (15.0). The system (11.8) is also termed *canonical* because the BC of (11.8) have their special form (11.8)″, (11.8)‴ in contrast with the general selfadjoint BC (29.4).

conditions (11.8), each root counted with its multiplicity. If the nullity of g vanishes, g is termed *nondegenerate*.

We shall see in Chapter 3 that there exist countably infinite many characteristic roots λ of conditions (11.8), but at most a finite number less than a prescribed constant. A corollary of the following theorem implies that characteristic roots λ of conditions (11.8) are bounded below.

Positive Definiteness of $\mathscr{I}_r^\lambda$. We shall say that $\mathscr{I}_r^\lambda$, $r > 0$, is *positive definite*, if $\mathscr{I}_r^\lambda(\eta) > 0$ for each nonnull η in the domain of $\mathscr{I}_r^\lambda$.

THEOREM 11.2. *Under Hypotheses* 11.1, *on the base extremal* g, *the functional* $\mathscr{I}_r^\lambda$, r > 0 *is positive definite for* $-\lambda$ *sufficiently large**.

Since the matrix $\|c_{ih}^s\|$ of Lemma 10.2 has the rank r one can write

$$\text{(11.9)} \quad \mathscr{I}_r^\lambda(\eta) = q(\eta) + \int_{a^1}^{a^2} (2\Omega(x, \eta(x), \eta'(x)) - \lambda \, \|\eta(x)\|^2) \, dx$$

(η C_r''-admissible)

where $q(\eta)$ is a quadratic form in a suitable subset of r of the $2m$ variables η_i^s. Any such form as $q(\eta)$ will satisfy a relation

$$\text{(11.10)} \qquad q(\eta) \geq -c[\|\eta(a^2)\|^2 + \|\eta(a^1)\|^2]$$

provided c is a sufficiently large positive constant. Let $x \to h(x) : [a^1, a^2] \to R$ be a mapping of class C^1 such that $h(a^1) = -1$ and $h(a^2) = 1$. Then (11.10) can be written in the form

$$\text{(11.11)} \qquad q(\eta) \geq -c \int_{a^1}^{a^2} \frac{d}{dx} (h(x) \, \|\eta(x)\|^2) \, dx$$

where η is any mapping in the domain of $\mathscr{I}_r$. From (11.9) and (11.11) we find that

$$\text{(11.12)} \quad \mathscr{I}_r^\lambda(\eta) \geq \int_{a^1}^{a^2} \left(2\Omega(x,\eta(x),\eta'(x)) - \lambda \, \|\eta(x)\|^2 - c \frac{d}{dx} (h(x) \, \|\eta(x)\|^2)\right) dx.$$

Under the integral sign in (11.12) we have a symmetric quadratic form Q^λ in the variables $\eta_1', \ldots, \eta_m', \eta_1, \ldots, \eta_m$, each evaluated at x. We shall complete the proof of Theorem 11.2 by proving the following.

(i). *If* $-\lambda$ *is sufficiently large, the quadratic form* Q^λ *is positive definite for each* x $\in$ [a^1,a^2].

We shall use the Kronecker rule for determining the index of the form Q^λ. See Dickson, p. 81. To that end set $A_0 = 1$ and let A_k, $k = 1, \ldots, 2m$ be the

* $\mathscr{I}_0^\lambda$ will be defined in Section 12 and Theorem 11.2 will be seen to be true when $r = 0$. A similar remark applies to Lemma 11.1 and Corollary 11.2.

principal minor of the matrix **a** of coefficients in the form Q^λ, the minor whose elements are those of the first k rows and columns of **a**. Kronecker's rule affirms that the index of the form Q^λ is the number of changes of sign in the sequence $A_0, A_1, \ldots, A_{2m}$ provided the matrix **a** is "regularly arranged." In particular, the matrix is regularly arranged if all of the A_k's are positive.

If one notes that the terms in the form Q^λ due to the introduction of new terms under the integral sign in (11.11) do not involve any terms quadratic in the variables $\eta_1', \ldots, \eta_m'$, one infers from the Legendre S-condition that the numbers $A_0, A_1, \ldots, A_m$ are all positive. Moreover, the remaining A_k's all become positive for $-\lambda$ sufficiently large. Hence for $-\lambda$ sufficiently large, the form Q^λ is positive definite and hence $\mathscr{I}_r^\lambda$ as well.

Theorem 11.2 is thus true.

The following lemma is needed in many places.

LEMMA 11.1. *If η is a characteristic solution of conditions* (11.8) *with a root λ and* r*-tuple* u, *then* $\mathscr{I}_{\mathrm{r}}(\eta) = 0$.

For $r > 0$ one can write $\mathscr{I}_r(\eta)$ in the form

$$\mathscr{I}_r^\lambda(\eta) = b_{hk}u_hu_k + \int_{a^1}^{a^2} (\eta_i(x)\Omega_{\eta_i}^x + \eta_i'(x)\Omega_{\zeta_i}^x - \lambda\eta_i(x)\eta_i(x))\,dx, \tag{11.13}$$

where the superscript x means evaluation with $(x,\eta,\zeta) = (x,\eta(x),\eta'(x))$. If the terms in the integrand involving η' are integrated by parts in the usual way, making use of the fact that η is a solution of the equations (11.8)′, one finds that

$$\mathscr{I}_r^\lambda(\eta) = b_{hk}u_hu_k + \xi_i^2\eta_i^2 - \xi_i^1\eta_i^1 \qquad \text{(subject to (11.8)}''\text{)}. \tag{11.14}$$

We now multiply the members of (11.8)‴ by u_h, sum as to h, and use the relations (11.8)″. We find thereby that the right member of (11.14) is 0, thereby establishing Lemma 11.1.

COROLLARY 11.2. *The characteristic roots $\bar\lambda$ of characteristic solutions $\bar\eta$ of conditions* (11.8) *are bounded below.*

According to Lemma 11.1 $\mathscr{I}_r^{\bar\lambda}(\bar\eta) = 0$. According to Theorem 11.2 there exists a number c so large that if $\lambda \leq -c$ then $\mathscr{I}_r^\lambda$ is positive definite. Since $\mathscr{I}_r^{\bar\lambda}$ is not positive definite we infer that $\bar\lambda > -c$.

Thus Corollary 11.2 is true.

When $r > 0$, $\mathscr{I}_r$ gives an evaluation (10.13) of a special second variation of (J,C_r). We shall show that the functionals $\mathscr{I}_r^\lambda$, introduced in (11.7) as accessory to $\mathscr{I}_r$, will arise in a similar way from a properly defined functional (J^λ,C_r) accessory to (J,C_r) and *based* on g.

The replacement f^λ *of* f *based on* g. For each value of λ and for (x,y,p) in the domain of the preintegrand f, set

$$f^\lambda(x,y,p) = f(x,y,p) - \frac{\lambda}{2}\|y - g(x)\|^2. \tag{11.15}$$

Given a C_r-admissible pair (χ,α), with $r > 0$ as in (8.19), $J(\chi,\alpha)$ of (8.19) can be replaced by

$$J^\lambda(\chi,\alpha) = \Theta(\alpha) + \int_{X^1(\alpha)}^{X^2(\alpha)} f^\lambda(x,\chi(x),\chi'(x))\,dx \tag{11.16}$$

defining thereby a functional (J^λ,C_r), *accessory* to (J,C_r), with the same domain as (J,C_r).

We continue with a lemma.

LEMMA 11.2. *The critical extremal,* g *of* (J,C_r), *is a critical extremal of* (J^λ,C_r) *for each value of* λ.

The Euler equations of the integral with preintegrand f^λ have the form

$$\frac{d}{dx} f_{p_i}(x,y,y') - f_{y_i}(x,y,y') + \lambda(y_i - g_i(x)) = 0 \qquad (i = 1, \ldots, m). \tag{11.17}$$

The curve g is given as an extremal when $\lambda = 0$, and is thereby an extremal for each value of λ. When $r > 0$ the satisfaction of the C_r-transversality conditions (9.1) by g implies the satisfaction by g of the conditions (9.1) with f replaced by f^λ.

Thus Lemma 11.2 is true when $r > 0$. When $r = 0$, the lemma is still true as we shall see.

A "Second Variation" of (J^λ,C_r). Given the* family $((\gamma,A))$ of C_r-admissible pairs $(\gamma^e,A(e))$, a review of the proof of (10.13), taking account of Lemma 11.2, shows that when $r > 0$

$$\frac{d^2}{de^2} J^\lambda(\gamma^e,A(e))\Big|^{e=0} = \mathscr{I}_r^\lambda(\eta), \tag{11.18}$$

where $\eta = \gamma_e\big|^{e=0}$. Thus the quadratic functional $\mathscr{I}_r^\lambda$ defined in (11.7), serves as in (11.18), to evaluate the *second variation* of (J^λ,C_r) when a family $((\gamma,A))$ is given.

The following extension of Definition 11.4 is here given. It is justified in Chapter 3 when we prove that the algebraic indices defined in this section equal the "*geometric indices*" there defined.

* See Definition 10.0.

Definition 11.5. When $r > 0$ the *algebraic index* {*nullity*} of g as a critical extremal of (J^λ, C_r) is the number of characteristic roots λ' of conditions (11.8) such that $\lambda' < \lambda$ $\{\lambda' = \lambda\}$, each root counted with its multiplicity.

Conditions (11.8) **when** g **is a periodic extremal.** We suppose that the pre-integrand f and extremal g have the period τ in the sense indicated in (8.9). For simplicity we suppose that $[a^1, a^2] = [0, \tau]$ and that $A^1 = (0, \mathbf{0})$ and $A^2 = (\tau, \mathbf{0})$, respectively, in the $(m + 1)$-plane of points (x, y). The end conditions satisfied by g have the form

$$\begin{aligned} &(11.19)' \quad \begin{cases} (x^1, y^1) = (X^1(\alpha), Y^1(\alpha)) = (0, \alpha) \\ (x^2, y^2) = (X^2(\alpha), Y^2(\alpha)) = (\tau, \alpha) \end{cases} \quad \text{(see Case III, Section 8)} \\ &(11.19)'' \end{aligned}$$

where α is an m-tuple in an open neighborhood $\mathbf{A}_m$ of the origin in R^m. The end conditions are of the general type termed conditions C_m in Section 8 but which here are, with less ambiguity, termed end conditions $C_m(P)$ (read *periodic* conditions C_m).

We shall take the end function $\alpha \to \Theta(\alpha)$, associated in (8.18) with the end conditions $C_m(P)$, as the function Θ with null values. When $\Theta(\alpha) \equiv 0$, g satisfies the transversality conditions (9.1) based on $C_m(P)$. Because of the periodicity of f and g, indicated in (8.9), these conditions reduce to $g'(0) = g'(\tau)$. Thus g is a critical extremal of $(J, C_m(P))$, or more generally of $(J^\lambda, C_m(P))$ for each λ. The elements b^s_{hk}, defined in (10.12)″, vanish.

The conditions (11.8), based on a periodic critical extremal g of $(J^\lambda, C_m(P))$, take the form

$$\begin{bmatrix} (11.20)' & \dfrac{d}{dx}\Omega_{\zeta_i}(x,\eta,\eta') - \Omega_{\eta_i}(x,\eta,\eta') + \lambda\eta_i = 0 & (i = 1, \ldots, m) \\ (11.20)'' & \eta_i(0) = \eta_i(\tau) = u_i & (i = 1, \ldots, m) \\ (11.20)''' & \xi_i(0) - \xi_i(\tau) = 0 & (i = 1, \ldots, m) \end{bmatrix}$$

These are the conditions (11.8) in which the m-square matrix $\|b_{hk}\|$ has null elements and in which the matrices $\|c^1_{ih}\|$ and $\|c^2_{ih}\|$ are the m-square matrix $\|\delta^h_i\|$, in accord with (10.7). The form Ω has the period τ in x.

Vector Spaces. Special vector spaces of solutions of the JDE have been used without calling these spaces vector spaces or invoking any of the classical theorems on such spaces. However, from this point on we shall invoke the general concepts of such spaces. The reader will find an excellent introduction to such spaces in Halmos [1].

Vector spaces of solutions of the JDE. Vector spaces over R of solutions of the JDE which satisfy prescribed boundary conditions such as (11.8)″ or

(11.8)‴, or both (11.8)″ and (11.8)‴, are a major object of consideration. A basic characteristic of such a space is its dimension ρ.

It follows from Corollary 2.2 of Appendix I that each solution η of the JDE (11.8)′, if restricted to the interval $[a^1,a^2]$, has component mappings $x \to \eta_i(x)$ of the form

$$\eta_i(x) = A_j p_{ij}(x) + B_j q_{ij}(x) : [a^1,a^2] \to R \qquad (i = 1, \ldots, m) \tag{11.21}$$

where A and B are m-tuples in R^m, uniquely determined by the solution η. A set of r solutions determined (as in (11.21)) by r $2m$-tuples

$$(A_1, \ldots, A_m, B_1, \ldots, B_m) = (A,B), \tag{11.22}$$

is linearly independent if and only if this set of r, $2m$-tuples (A,B) is linearly independent. More generally if H is a vector space of solutions of the JDE and V is the vector space of $2m$-tuples (A,B) which determine solutions $\eta \in H$, then

$$\dim H = \dim V. \tag{11.23}$$

In particular the vector space of all solutions of the JDE has the dimension $2m$.

We state a basic elementary lemma.

LEMMA 11.3. *If*

$$\mathrm{K}_{\mathrm{i}}(\eta^1_1, \ldots, \eta^1_{\mathrm{m}} : \eta^2_1, \ldots, \eta^2_{\mathrm{m}}) = 0 \qquad (\mathrm{i} = 1, \ldots, \mu) \tag{11.24}$$

is a set of $\mu \leq 2\mathrm{m}$ *linear homogeneous conditions on the end values at* a^1 *and* a^2 *of a solution of the* JDE, *then the vector space of the solutions of the* JDE *which satisfy the conditions* (11.24) *has a dimension* $\rho \geq 2\mathrm{m} - \mu$.

If a solution η satisfies the conditions (11.24), the $2m$-tuples (A,B) which determine η by way of (11.21), satisfy μ linear homogeneous conditions

$$L_i(A_1, \ldots, A_m; B_1, \ldots, B_m) = 0 \qquad (i = 1, 2, \ldots, \mu) \tag{11.25}$$

obtained by substituting the right members of (11.21) in the conditions (11.24). The solutions (A,B) of conditions (11.25) form a vector space V of dimension $\rho \geq 2m - \mu$. The corresponding vector space of solutions of the JDE has the same dimension ρ. Hence Lemma 11.3 is true.

For each solution η of the JDE we have defined a "canonical associate" ξ as in (11.2). We refer to the pair (η,ξ) as a solution of the JDE if η is a solution and ξ is its canonical associate. One seeks solutions (η,ξ) of the JDE which satisfy conditions (11.8)″ and (11.8)‴. In solving this problem the following lemma will be useful.

LEMMA 11.4. *Let* H *be a vector space of solutions* η *of the* JDE *of dimension* ρ. *If*

$$H_j(\eta^1,\eta^2,\xi^1,\xi^2) = 0 \qquad (j = 1, \ldots, r \leq \rho) \tag{11.26}$$

is a set of homogeneous linear conditions on the end values at a^1 *and* a^2 *of a solution* (η,ξ) *of the* JDE, *then the vector subspace* $\hat{H}$ *of* H *of solutions which satisfy the conditions* (11.26) *is such that*

$$\dim \hat{H} \geq \rho - r. \tag{11.27}$$

The proof of this lemma is similar to the proof of Lemma 11.3 and is left to the reader. Lemmas 11.3 and 11.4 are essential in the proof of Lemma 5.1 of Appendix I. The following exercises are relevant.

Exercises

Refer to conditions (11.8)″ and (11.8)‴ and verify the two following affirmations.

11.1. There exist $2m - r$ linearly independent linear forms $K_i(\eta^1,\eta^2)$ in the $2m$ variables $\eta_1^1, \ldots, \eta_m^1; \eta_1^2, \ldots, \eta_m^2$ such that the $2m - r$ conditions

$$K_i(\eta^1,\eta^2) = 0 \qquad (i = 1, \ldots, 2m - r) \tag{11.28}$$

are equivalent to the conditions C_r'' of (11.8)″ on the m-tuples η^1 and η^2.

11.2. There exist r linearly independent linear forms $H_j(\eta^1,\eta^2,\xi^1,\xi^2)$ in the $4m$ components of the m-tuples η^1, η^2, ξ^1, ξ^2 such that the forms K_i and H_j are linearly independent and the $(2m - r)$ conditions (11.28), supplemented by the r-conditions

$$H_j(\eta^1,\eta^2,\xi^1,\xi^2) = 0 \qquad (j = 1, \ldots, r) \tag{11.29}$$

are equivalent to the conditions (11.8)″ and (11.8)‴ with the parameters $u_1, \ldots, u_r$ eliminated.

11.3. Show that for η conditioned as in Corollary 11.1 $\mathscr{I}_r(\eta) = 0$.

12. Critical Secondary Extremals: Endpoints Fixed

In this important special case the x-parameterized curves of class D^1 which are compared with the extremal g, join the endpoints A^1 and A^2 of g. The endpoint conditions $(x^s,y^s) = A^s$, $s = 1, 2$, are denoted by C_0. The functional whose values are the integrals $\mathbf{J}(\gamma)$ of Chapter 1 are here denoted by (J,C_0). (J,C_0) is to be contrasted with (J,C_r) with $r > 0$ and with values given in (8.19). When $r = 0$ there is no endpoint function Θ. There is no C_0-transversality. The relations (4.2) hold and justify calling the extremal g with endpoints A^1 and A^2, a *critical extremal* of (J,C_0).

The *second variation* $\mathscr{I}_0$ of (J,C_0) is *based* on the extremal g. By definition $\mathscr{I}_0$ is a functional with values $\mathscr{I}_0(\eta)$ given by (4.1)′. $\mathscr{I}_0$ is to be contrasted with the quadratic functional of (J,C_r), $r > 0$, with values $\mathscr{I}_r(\eta)$ given by (10.23). The mappings η in the domain of $\mathscr{I}_0$ satisfy the secondary endpoint conditions

$$C_0'' : \eta(a^1) = \eta(a^2) = \mathbf{0}. \tag{12.0}$$

With each value $\mathscr{I}_0(\eta)$ of the functional $\mathscr{I}_0$ and each real number λ there is associated a value

$$\mathscr{I}_0^\lambda(\eta) = \int_{a^1}^{a^2} (2\Omega(x, \eta(x), \eta'(x)) - \lambda \, \|\eta(x)\|^2) \, dx \tag{12.1}$$

of a new functional $\mathscr{I}_0^\lambda$, to be termed *accessory* to $\mathscr{I}_0$. Definition 11.2 is paralleled by the following definition.

Definition 12.1. A *critical secondary extremal* η of $\mathscr{I}_0^\lambda$ is the graph of a C^2-mapping $x \to \eta(x) : [a^1,a^2] \to E_m$ satisfying the conditions,*

$$\frac{d}{dx} \Omega_{\zeta_i}(x,\eta,\eta') - \Omega_{\eta_i}(x,\eta,\eta') + \lambda\eta_i = 0 \qquad (i = 1, \ldots, m) \tag{12.2)'}$$

$$\eta(a^1) = \eta(a^2) = \mathbf{0} \tag{12.2)''}$$

Characteristic solutions η and roots λ of (12.2) are introduced as in Definition 11.3, by replacing conditions (11.8) by conditions (12.2) and $\mathscr{I}_r$, $r > 0$, by $\mathscr{I}_0$. We suppose the critical extremal g conditioned as in the first paragraph of Section 8. The following definition replaces Definition 11.4 when C_0 replaces C_r, $r > 0$.

Definition 12.2. The *algebraic index* {*nullity*} of the critical extremal g of (J,C_0) is the number of negative {null} characteristic roots λ of conditions (12.2), each root λ counted with its multiplicity. If the nullity of g vanishes, g is termed *nondegenerate*.

$\mathscr{I}_0^\lambda$ is termed positive definite if $\mathscr{I}_0^\lambda(\eta) > 0$ for each nonnull η in the domain of $\mathscr{I}_0^\lambda$.

THEOREM 12.1. *If the extremal* g *is conditioned as in the first paragraph of Section* 8, $\mathscr{I}_0^\lambda$ *is positive definite for* $-\lambda$ *sufficiently large.*

The proof of this theorem is a simplified version of the proof of Theorem 11.2.

LEMMA 12.1. *If* η *is a characteristic solution of conditions* (12.2) *with root* λ, *then* $\mathscr{I}_0^\lambda(\eta) = 0$.

The lemma follows from Lemma 4.2 on replacing $2\Omega(x,\eta,\zeta)$ by $2\Omega(x,\eta,\zeta) - \lambda\|\eta\|^2$.

Theorem 12.1 and Lemma 12.1 have the corollary.

* Termed derived and canonical.

COROLLARY 12.1. *The characteristic roots λ of characteristic solutions η of Conditions* (12.2) *are bounded below.*

When $r = 0$ we define f^λ as in (11.15), and replace the integral $\mathbf{J}(\gamma)$ of (1.4) by the integral

$$\mathbf{J}^\lambda(\gamma) = \int_{a^1}^{a^2} f^\lambda(x,\gamma(x),\gamma'(x))\,dx, \tag{12.3}$$

introducing thereby a functional (J^λ,C_0) *accessory* to the functional (J,C_0). (J,C_0) and (J^λ,C_0) have the same domains. If $J^\lambda(e)$ is the value of the integral $\mathbf{J}^\lambda(\gamma^e)$ along the curve of the family (1.6) with parameter e, one finds, as in (4.1)″ that

$$\frac{d}{de} J^\lambda(e)\Big|^{e=0} = 0, \qquad \frac{d^2}{de^2} J^\lambda(e)\Big|^{e=0} = \mathscr{I}_0^\lambda(\eta), \tag{12.4}$$

and accordingly calls $\mathscr{I}_0^\lambda$ the *second variation* of (J^λ,C_0).

λ-Extremals, secondary λ-extremals, λ-conjugate points. The graph of a C^2-solution of the Euler equations of the integral $\mathbf{J}^\lambda$ will be called a *λ-extremal.* The graph of a C^2-solution of the equations (12.2)′ will be called a *secondary λ-extremal*. The equations (12.2)′ are the JDE based on g as a λ-extremal. If one sets

$$\Omega^\lambda(x,\eta,\zeta) = \Omega(x,\eta,\zeta) - \frac{\lambda}{2}\,\|\eta\|^2 \tag{12.5}$$

one sees that Ω^λ is derivable from f^λ, as Ω of (4.0) was derived from f. That is, Ω^λ is "based" on g as a λ-extremal, while Ω is "based" on g as an ordinary extremal.

Conjugate points *relative* to g, as a λ-extremal, will be termed *λ-conjugate points*, as will their x-coordinates. λ-conjugate points are determined as in Section 5 on replacing solutions of the JDE (4.2) by solutions of the JDE (12.2)′. When one uses the Kneser determination of conjugate points, the families of ordinary extremals (5.15) are to be replaced by the corresponding family of λ-extremals.

In Chapters 3 and 4 we shall make essential use of broken λ-extremals E^λ defined as follows.

Definition 12.3. *Broken λ-extremals.* A *broken λ-extremal* E^λ is a continuous x-parameterized curve in X which is a sequence of a finite number of compact λ-extremals no one of which bears a point λ-conjugate to its initial point. A broken secondary λ-extremal is similarly defined. The condition that the component extremals of a broken extremal H be compact will on occasion be relaxed, in that the, initial {final} component extremal of H may be extended as an extremal in the sense of decreasing {increasing} x on a relatively compact interval which has no left {right} end point. If the extension is made at both end points of H the resultant extension H^* of H will be called an *open broken extremal.* The intermediate components of H^* (if any exist) remain compact.

Our base extremal g can be regarded as a broken λ-extremal for a prescribed λ. Given λ, g will qualify as a broken λ-extremal if subdivided into a finite sequence of subextremals no one of which bears a point λ-conjugate to its initial point.

We turn to Definition 11.5 and close Section 12 with an analogous definition relevant to the case $r = 0$.

Definition 12.4. When $r = 0$ the *algebraic index* {*nullity*} of g as a critical extremal of (J^λ, C_0) is the number of characteristic roots λ' of conditions (12.2) such that $\lambda' < \lambda\{\lambda' = \lambda\}$, counting each root with its multiplicity.

Part II

QUADRATIC INDEX FORMS, DERIVED AND FREE

CHAPTER 3

Derived Index Forms

Objective. In Sections 11 and 12 we have defined the "algebraic index and nullity" of a critical extremal g of the functional (J,C_r). The first objective of Chapter 3 is to disclose the geometric meaning of this index and nullity with the aid of "*index forms*" for g. The definition of an index form for g requires the construction of a framed family Γ^λ of broken λ-extremals near g (Definition 12.3). Broken extremals have been used by Hahn [1] and Signorini [1] and others in the study of minima. They have led to generalizations for dimensions $m > 2$ of the Sturm separation and comparison theorems (Morse [1] and [2]), as well as to topological interpretations of the index of a critical extremal.

Hypotheses 13.1. In Chapter 3 we continue with the preintegrands f and f^λ of Chapter 2, and with the *critical extremal* g, as conditioned under Hypotheses 11.1.

The quadratic index forms which we define in this chapter are termed *derived*. In Section 15 index forms are defined which do not presuppose the existence of a preintegrand and critical extremal g. These more general index forms are termed *free*.

13. The Framed Family Γ^λ of Broken λ-Extremals

A lemma is required. Given a point $x = c$ such that $a^1 \leq c < a^2$, let $c(\lambda)$ denote the first value $x > c$ on the interval $[a^1,a^2]$ which is λ-conjugate* to $x = c$, if any such point $c(\lambda)$ exists.

LEMMA 13.1. *If* $\mathrm{c}(\lambda_0)$ *exists, then for* $\lambda < \lambda_0$, $\mathrm{c}(\lambda)$ *either fails to exist, or* $\mathrm{c}(\lambda) \geq \mathrm{c}(\lambda_0)$.

* That is conjugate to $x = c$ relative to the DE (11.8)$'$.

Suppose the lemma false in that for some $\lambda < \lambda_0$, $c(\lambda) < c(\lambda_0)$. Set $c(\lambda) = a$. There then exists a nonnull solution $x \to v(x)$ of the differential equations (11.8)$'$, such that $v(c) = v(a) = \mathbf{0}$. Now

$$\int_c^a (2\Omega(x, v(x), v'(x)) - \lambda_0 \|v(x)\|^2)\, dx > 0 \tag{13.0}$$

by Theorem 7.4, since $a < c(\lambda_0)$. The integral remains positive if λ_0 is replaced by $\lambda < \lambda_0$; however the integral, with λ_0 replaced by λ, vanishes by Lemma 4.2.

From this contradiction we infer the truth of Lemma 13.1.

λ-Frames for g are essential in our construction of a family Γ^λ of broken λ-extremals near g and containing g.

Definition 13.1. *A λ-frame* Λ_p^g. A λ-frame Λ_p^g for an extremal g, given with endpoints A^1 and A^2, is defined by a set of m-planes,

$$\Pi_1, \ldots, \Pi_p \tag{13.1}$$

in E_{m+1} orthogonal to the x-axis at points with x-coordinates $c_1, \ldots, c_p$, respectively, such that

$$a^1 < c_1 < c_2 < \cdots < c_p < a^2. \tag{13.2}$$

We suppose that the m-planes (13.1) meet g in points

$$Q_1, \ldots, Q_p,$$

respectively. Setting $Q_0 = A^1$ and $Q_{p+1} = A^2$, we require that for q on the range $0, 1, \ldots, p$, the extremal subarc $E(Q_q, Q_{q+1})$ of g bear no point λ-conjugate to its initial point Q_q.

From Lemma 13.1 we infer the following.

LEMMA 13.2. *If $\lambda_* > \lambda$, a λ_*-frame for* g *is a λ-frame for* g.

Notation for Theorem 13.1. We shall refer to conditions C_r of Section 8, defined by mappings X^s, Y^s whose domain is the set of r-tuples α in an open neighborhood $\mathbf{A}_r$ of the origin $\mathbf{0}$ in R^r. We suppose that $(X^s(\mathbf{0}), Y^s(\mathbf{0}))$ gives the coordinates (x,y) of A^1 when $s = 1$ and of A^2 when $s = 2$.

The interval $\aleph$, conditioned in Theorem 2.1, is open and contains $[a^1,a^2]$. We shall refer to a subinterval $[a',a'']$ of $\aleph$ such that

$$a' < a^1 < a^2 < a''. \tag{13.3}$$

The broken λ-extremal which we shall introduce in Theorem 13.1 shall have C_r-admissible endpoints.

$$(X^1(\alpha), Y^1(\alpha)), \qquad (X^2(\alpha), Y^2(\alpha)), \tag{13.3$'$}$$

defined by r-tuples α with norms $\|\alpha\|$ so small that

$$a' < X^1(\alpha) < X^2(\alpha) < a''. \tag{13.3)''}$$

Theorem 13.1 is stated for the case $r > 0$, but is valid when $r = 0$, if the changes indicated in two footnotes are made. Theorem 13.1 characterizes a *framed family* Γ^λ *of broken* λ*-extremals.*

THEOREM 13.1. *Let* g *be a critical extremal of* (J,C_r) *with endpoints* A^1 and A^2. *Let* $\Pi_1, \ldots, \Pi_p$ *be the* m*-planes of a* λ*-frame* Λ_p^g *for* g *and let*

$$\mathbf{P} = (P_0, P_1, \ldots, P_p, P_{p+1}) \tag{13.4}$$

be a sequence of points in X *such that*

$$P_1 \in \Pi_1, \ldots, P_p \in \Pi_p$$

*while (*for* $r > 0$) *terminal points* $P_0 : [X^1(\alpha), Y^1(\alpha)]$ *and* $P_{p+1} : [X^2(\alpha), Y^2(\alpha)]$ *satisfy conditions* C_r : (8.11) *with some* r*-tuple* α, *so small in norm that* (13.3)″ *holds.*

Corresponding to a sufficiently small open neighborhood N *of* g *in* X *there exists a constant* $e > 0$ *so small that if the distances,*

$$P_0Q_0, P_1Q_1, \ldots, P_pQ_p, P_{p+1}Q_{p+1} \tag{13.5)†}$$

in E_{m+1} *are less than* e, *there exist in* N *unique* λ*-extremals* $E(P_q, P_{q+1})$, $q = 0, 1, \ldots, p$, *joining the successive points of the sequence* $\mathbf{P}$: (13.4) *and defining thereby a broken* λ*-extremal* $E(\mathbf{P})$, "*based*", *as we shall say, on the* λ*-frame* Λ_p^g *and end conditions* C_r.

Proof. The existence and uniqueness of the λ-extremals $E(P_q, P_{q+1})$ under the conditions of Theorem 13.1 follow from Theorem 5.1, on replacing g of Theorem 5.1 by the extremal subarc $E(Q_q, Q_{q+1})$ of g and the preintegrand f by f^λ. That there is no point on $E(P_q, P_{q+1})$, λ-conjugate to P_q, provided the distances (13.5) are sufficiently small, follows similarly from our supplement of Theorem 5.1.

Thus Theorem 13.1 is true.

Definition 13.2. The points $P_1, \ldots, P_p$ of $E(\mathbf{P})$, with their x-coordinates $c_1 < c_2 < \cdots < c_p$, will be called *division points* of $E(\mathbf{P})$. The set of points $P_0, P_1, \ldots, P_p, P_{p+1}$ will be called the *skeleton* of the broken λ-extremal $E(\mathbf{P})$.

We shall give x-parameterizations of the λ-extremals $E(P_q, P_{q+1})$ when these arcs are sufficiently near the corresponding subarcs $E(Q_q, Q_{q+1})$ of g. It is essential to show how these x-parameterizations depend upon coordinates

* When $r = 0$, $P_0 = A^1$, $P_{p+1} = A^2$ and α fails to exist.

† When $r = 0$ the first and last distances in (13.5) vanish.

of P_q and P_{q+1}. To that end we assign coordinate axes parallel to the y_i-axis in E_{m+1} to each m-plane Π_q and take the point Q_q on g as the *origin* in Π_q. Points

$$v^q = (v_1^q, \ldots, v_m^q) \qquad (q = 1, \ldots, p) \tag{13.6}$$

with rectangular coordinates v_i^q are thereby introduced in Π_q.

We distinguish between the *terminal* subarcs $E(P_0,P_1)$ and $E(P_p,P_{p+1})$ of $E(\mathbf{P})$ and the *intermediate* subarcs $E(P_q,P_{q+1})$ of $E(\mathbf{P})$. On the latter $q = 1, \ldots, p-1$. The x-intervals of the terminal subarcs of $E(\mathbf{P})$ are $[X^1(\alpha),c_1]$ and $[c_p,X^2(\alpha)]$, respectively.

We shall refer to an open origin-centered m-ball D_ρ^m of radius ρ in R^m.

Case I. $q = 1, \ldots, p-1, r \geq 0$. Let $E(P_q,P_{q+1})$ be a λ-extremal affirmed to exist in Theorem 13.1. Let v^q and v^{q+1} represent P_q and P_{q+1} in Π_q and Π_{q+1}, respectively. Corresponding to a point (x,y) on $E(P_q,P_{q+1})$ set $y = \gamma^q(x,v^q,v^{q+1})$. It follows from Theorem* 5.1 that if ρ is sufficiently small, the mapping

$$(x,v^q,v^{q+1}) \to \gamma^q(x,v^q,v^{q+1}) : [c_q,c_{q+1}] \times D_\rho^m \times D_\rho^m \to R^m \tag{13.7}$$

is of class C^2, as is the partial derivative γ_x^q.

Case II. *An extension of* $\mathrm{E(P_0,P_1)}$ *when* $\mathrm{r} > 0$. Here P_0 is a point $[X^1(\alpha), Y^1(\alpha)]$ satisfying Conditions C_r with an r-tuple α, restricted to an open origin-centered r-ball D_ρ^r of so small a radius ρ that the condition $P_0Q_0 < e$ of Theorem 13.1 is satisfied as well as the condition $a' < X^1(\alpha)$ of (13.3)″. Let the m-tuple v^1 representing $P_1 \in \Pi_1$ be restricted to an origin-centered m-ball D_ρ^m in Π_1, so that P_0 and P_1 are both conditioned by the choice of ρ. Suppose then that ρ is so small that the λ-extremal $E(P_0,P_1)$ is uniquely determined in N of Theorem 13.1 and is extendable as a λ-extremal $\hat{E}$ over the x-domain $(a',c_1]$.

If (x,y) is a point in $\hat{E}$ we set $y = \gamma^0(x,\alpha,v^1)$. It follows from Theorems 5.1 and 2.1 that if $\rho > 0$ is sufficiently small, the mapping

$$(x,\alpha,v^1) \to \gamma^0(x,\alpha,v^1) : (a',c_1] \times D_\rho^r \times D_\rho^m \to R^m \tag{13.8}$$

is of class C^2, as is the partial derivative γ_x^0.

Case III. *An extension of* $\mathrm{E(P_p,P_{p+1})}$, $\mathrm{r} > 0$. Case III is similar to Case II. The λ-extremal $E(P_p,P_{p+1})$ replaces the λ-extremal $E(P_0,P_1)$ while the m-tuple v^p replaces v^1. The conclusion of III is similar to that of Case II. It is that if $\rho > 0$ is sufficiently small, there exists a C^2-mapping

$$(x,v^p,\alpha) \to \gamma^p(x,v^p,\alpha) : [c_p,a'') \times D_\rho^m \times D_\rho^r \to R^m \tag{13.9}$$

* In applying Theorem 5.1, g of Theorem 5.1 is to be replaced by the extremal arc $E(Q_q,Q_{q+1})$ and f by f^λ.

which for a fixed m-tuple $v^p \in D_\rho^m$ and r-tuple $\alpha \in D_\rho^r$ has for graph a unique λ-extremal in N whose initial point P_p is the point $v^p \in \Pi_p$, which meets the terminal point $(X^2(\alpha), Y^2(\alpha)) = P_{p+1}$ of $E(\mathbf{P})$ and which is extended as a λ-extremal with x-domain $[c_p,a'')$.

When r $= 0$ a representation of "intermediate" λ-extremals of $E(\mathbf{P})$ is given under Case I. Terminal λ-extremal arcs of $E(\mathbf{P})$ when $r = 0$, are treated under Cases IV and V.

Case IV. $q = 0$, $r = 0$. Because $r = 0$, $P_0 = A^1$. Let $E(A^1,P_1)$ be a λ-extremal affirmed to exist in Theorem 13.1. Let v^1 represent P_1 in Π_1. Corresponding to a point (x,y) on $E(A^1,P_1)$ set $y = \gamma^0(x,v^1)$. It follows from Theorem 5.1 that if ρ is sufficiently small, the mapping

$$(x,v^1) \to \gamma^0(x,v^1) : [a^1,c_1] \times D_\rho^m \to R^m \tag{13.10}$$

is of class C^2, as is γ_x^0.

Case V. $q = p$, $r = 0$. This case is treated as is Case IV, interchanging P_0 and P_{p+1}, P_1, and P_p, v^1 and v^p, a^1, and a^2. Corresponding to a point (x,y) on $E(P_p,A^2)$, set $y = \gamma^p(x,v^p)$. Then for ρ sufficiently small the mapping

$$(x,v^p) \to \gamma^p(x,v^p) : [c_p,a^2] \times D_\rho^m \to R^m \tag{13.11}$$

is of class C^2, as is γ_x^p.

A representation of arcs $E(\mathbf{P})$ as a whole. When $r > 0$ an r-tuple α and points $v^1, \ldots, v^p$ in the respective m-planes $\Pi_1, \ldots, \Pi_p$ of a prescribed λ-frame for g, define a set

$$v = (v_1, \ldots, v_\mu) = (\alpha_1, \ldots, \alpha_r : v_1^1, \ldots, v_m^1; \ldots ; v_1^p, \ldots, v_m^p) \tag{13.12}$$

of $\mu = r + mp$ *coordinates of* $\mathbf{P}$. If v is restricted in R^μ to an origin-centered μ-ball D_ρ^μ of sufficiently small radius, the sequence

$$\mathbf{P}(v) = (P_0(\alpha),P_1(v^1), \ldots, P_p(v^p),P_{p+1}(\alpha)) \tag{13.13}$$

of points (13.4) thereby represented, defines a broken λ-extremal $E(\mathbf{P}(v))$ which is unique in N in accord with Theorem 13.1. Let $\hat{E}(\mathbf{P}(v))$ be the *proper extremal* extension* of $E(\mathbf{P}(v))$ defined by extending $E(P_0,P_1)$ and $E(P_p,P_{p+1})$ as under Cases II and III. The x-domain of $\hat{E}(\mathbf{P}(v))$ is (a',a'').

We can define a representation

$$(x,v) \to \hat{\Gamma}^\lambda(x,v) : (a',a'') \times D_\rho^\mu \to R^m \tag{13.14}$$

* By a *proper extremal extension* A of a broken extremal B is meant an "open" broken extremal which is identical with the broken extremal B except that the terminal extremals of A extend the terminal extremals of B. See Definition 12.3.

of $\hat{E}(\mathbf{P}(v))$ by setting

$$(13.15)\quad \begin{cases} \hat{\Gamma}^\lambda(x,v) = \gamma^0(x,\alpha,v^1), \ (a' < x \le c_1) \\ \hat{\Gamma}^\lambda(x,v) = \gamma^q(x,v^q,v^{q+1}), \ (c_q \le x \le c_{q+1}) \qquad (q = 1, \ldots, p-1) \\ \hat{\Gamma}^\lambda(x,v) = \gamma^p(x,v^p,\alpha), \ (c_p \le x < a'') \end{cases}$$

understanding that v is given as in (13.12). For reference in the following theorem set

$$(13.16)\qquad \Delta_j = I_j \times D^\mu_\rho \qquad (j = 0, 1, \ldots, p)$$

where I_j is the interval for x of the respective mappings γ^j in (13.15).

We summarize the results of Section 13 in Theorem 13.2, thereby continuing Theorem 13.1. Theorem 13.2 is stated for the case $r > 0$ but, as reformulated below, is valid when $r = 0$.

THEOREM 13.2. *The framed family* Γ^λ, $\mathrm{r} > 0$. *Given a* $\mathrm{C_r}$*-admissible* λ*-extremal* g *and* λ*-frame* $\Lambda^{\mathrm{g}}_{\mathrm{p}}$, *as in Theorem* 13.1, *then for a sufficiently small positive constant* ρ *the following is true.*

Each μ*-tuple* v *in* D^μ_ρ *of form* (13.12) *determines the skeleton*

$$(13.17)\quad \mathbf{P}(\mathrm{v}) = (\mathrm{P}_0(\alpha), \mathrm{P}_1(\mathrm{v}^1), \ldots, \mathrm{P}_{\mathrm{p}}(\mathrm{v}^{\mathrm{p}}), \mathrm{P}_{\mathrm{p}+1}(\alpha)) \qquad (\text{cf. } (13.13))$$

of a unique broken λ*-extremal* $\mathrm{E}(\mathbf{P}(\mathrm{v}))$ *which reduces to* g *when* $\mathrm{v} = \mathbf{0}$, *which is based on the* λ*-frame* $\Lambda^{\mathrm{g}}_{\mathrm{p}}$ *and is the graph* $\Gamma^{\lambda,\mathrm{v}}$, *in* N *of Theorem* 13.1, *of a* $\mathrm{C_r}$*-admissible mapping*

$$(13.18)\qquad \mathrm{x} \to \Gamma^\lambda(\mathrm{x},\mathrm{v}) : [\mathrm{X}^1(\alpha),\mathrm{X}^2(\alpha)] \to \mathrm{R}^{\mathrm{m}}$$

with the following properties.

There exists a mapping

$$(13.19)\qquad (\mathrm{x},\mathrm{v}) \to \hat{\Gamma}^\lambda(\mathrm{x},\mathrm{v}) : (\mathrm{a}',\mathrm{a}'') \times \mathrm{D}^\mu_\rho \to \mathrm{R}^{\mathrm{m}}$$

such that the restrictions

$$(13.20)\qquad \hat{\Gamma}^\lambda \,|\Delta_{\mathrm{j}};\ \hat{\Gamma}^\lambda_{\mathrm{x}}|\,\Delta_{\mathrm{j}} \qquad (\mathrm{j} = 0, \ldots, \mathrm{p})$$

are of class C^2 *and such that for each* $\mathrm{v} \in \mathrm{D}^\mu_\rho$ *the graph* $\hat{\Gamma}^{\lambda,\mathrm{v}}$ *of the partial mapping*

$$(13.21)\qquad \mathrm{x} \to \hat{\Gamma}^\lambda(\mathrm{x},\mathrm{v}) : (\mathrm{a}',\mathrm{a}'') \to \mathrm{R}^{\mathrm{m}}$$

is a proper extremal extension in X *of the graph* $\Gamma^{\lambda,\mathrm{v}}$ *of the mapping* (13.18).

Theorem 13.2 *in case* $\mathrm{r} = 0$. In case $r = 0$, $P_0 = A^1$, $P_{p+1} = A^2$, $\mu = mp$ and (13.12) should be replaced by the sequence

$$(13.22)\qquad (v_1, \ldots, v_\mu) = (v^1_1, \ldots, v^1_m; \ldots; v^p_1, \ldots, v^p_m)$$

and (13.6) by the skeleton

$$\mathbf{P}(v) = (A^1, P_1(v^1), \ldots, P_p(v^p), A^2). \tag{13.23}$$

The interval $[X^1(\alpha), X^2(\alpha)]$ in (13.18) should be replaced by $[a^1, a^2]$. The domains (13.16) should be defined by taking $I_0, \ldots, I_p$ as the respective intervals

$$[a^1, c_1],\ [c_1, c_2], \ldots, [c_{p-1}, c_p],\ [c_p, a^2]. \tag{13.24}$$

Reformulated with these changes, Theorem 13.2 holds when $r = 0$.

14. The "Geometric" Index of a Critical Extremal g

We shall begin by defining the *index* of an ordinary critical point.

Let $v = (v_1, \ldots, v_\mu)$ be a point in E_μ and let D_ρ^μ be an origin-centered open μ-ball of E_μ of radius ρ. Let

$$v \to G(v) : D_\rho^\mu \to R \tag{14.1}$$

be a C^2-mapping such that $\mathbf{0}$ is a critical point of G. Let $z = (z_1, \ldots, z_\mu)$ be a μ-tuple in D_ρ^μ and let e be a scalar such that $|e| < 1$. For fixed z the mapping $e \to G(ez) = \varphi(e)$, is of class C^2 for $|e| < 1$, with $\varphi'(0) = 0$. With i, j on the range $1, \ldots, \mu$, the quadratic form

$$\frac{d^2}{de^2} G(ez)\Big|^{e=0} = a_{ij} z_i z_j = Q(z) \tag{14.2}$$

is well-defined and termed the *index form* of the critical point $\mathbf{0}$ of G. The form Q is defined for all μ-tuples z. It is said to be *derived* from G.

Definition 14.0. By the *index* of a symmetric quadratic form Q in μ variables is meant the maximum of the integers k such that Q is negative definite on some k-plane meeting the origin. By the *nullity* of Q is meant the number of linearly independent critical points of Q.

The form Q will be termed *singular* if its nullity is positive or, equivalently, if the determinant of its coefficients vanishes.

Definition 14.1. *An Index Function G^λ.* G^λ is "based" on C_r and a λ-frame Λ_p^g. Set $\mu = r + mp$. Theorem 13.2, as explicitly stated when $r > 0$, and reformulated when $r = 0$, affirms the existence of a C_r-admissible, framed, broken extremal $\Gamma^{\lambda,v}$ which is the graph of a C_r-admissible mapping of the form

$$x \to \Gamma^\lambda(x,v) : [X^1(\alpha), X^2(\alpha)] \to R^m \qquad (\text{when } r > 0) \tag{14.3}$$

and of the form

$$x \to \Gamma^\lambda(x,v) : [a^1, a^2] \to R^m \qquad (\text{when } r = 0), \tag{14.4}$$

where α is the initial r-tuple of z when $r > 0$ and v is in the μ-ball D_ρ^μ of Theorem 13.2. For each $v \in D_\rho^\mu$ we define $G^\lambda(v)$ by setting

$$(14.5)' \quad G^\lambda(v) = \Theta(\alpha) + \int_{X^1(\alpha)}^{X^2(\alpha)} f^\lambda(x,\Gamma^\lambda(x,v),\Gamma_x^\lambda(x,v))\, dx = J^\lambda(\Gamma^{\lambda,v},\alpha) \qquad (\text{when } r > 0)$$

$$(14.5)'' \quad G^\lambda(v) = \int_{a^1}^{a^2} f^\lambda(x,\Gamma^\lambda(x,v),\Gamma_x^\lambda(x,v))\, dx = \mathbf{J}^\lambda(\Gamma^{\lambda,v}) \qquad (\text{when } r = 0).$$

The mapping

$$(14.6) \qquad v \to G^\lambda(v) \,:\, D_\rho^\mu \to R$$

will be called the *index function* G^λ based on g, C_r, and the γ-frame Λ_p^g.

When ambiguity as to its origin might arise, the above index function G^λ will be denoted by $G^\lambda\#(C_r,\Lambda_p^g)$, to be read G^λ, "based on C_r and Λ_p^g." We shall show that an index function G^λ is of class C^2 and has the origin in R^μ as a critical point. Index functions G^λ have been introduced for the purpose of giving the following definition.

Definition 14.2. *An index* form* Q^λ *for* g. For $r \geq 0$ the quadratic form Q^λ with values

$$(14.7) \qquad \frac{d^2}{de^2} G^\lambda(ez)\bigg|^{e=0} = Q^\lambda(z) \qquad (z \in R^\mu)$$

will be called the *index form* for g based on the end conditions C_r and λ-frame Λ_p^g and will on occasion be denoted by $Q^\lambda\#(C_r,\Lambda_p^g)$.

Theorem 14.1 below shows that the index and nullity of an index form $Q^\lambda\#(C_r,\Lambda_p^g)$ are independent of the choice of Λ_p^g among admissible λ-frames for g. Because of its origin we shall call $Q^\lambda\#(C_r,\Lambda_p^g)$ a *derived* index form. Derived index forms will be distinguished from *free* index forms to be more generally defined without use of an index function.

Definition 14.3. The index {nullity} common to the derived index forms $Q^\lambda\#(C_r,\Lambda_p^g)$ of a critical extremal g of (J^λ,C_r), $r \geq 0$, will be called the *geometric index {nullity}* of g.

A basic theorem follows.

THEOREM 14.1.† *The algebraic index {nullity} of* g (Definitions 11.5 and 12.4) *as a critical extremal of* $(\mathbf{J}^\lambda,\mathbf{C}_r)$ *equals the geometric index {nullity} of* g (Definition 14.3) *as a critical extremal of* $(\mathbf{J}^\lambda,\mathbf{C}_r)$.

* This index form is termed *derived* in contrast with the *free* index form introduced in Section 15.

† Theorem 14.1 has an analogue, Theorem 15.2, formulated for "free" index forms.

The proof of Theorem 14.1 will be completed in Section 15. Theorem 14.1 is true for $r \geq 0$. The proof will be explicit in the case $r > 0$ and will be indicated in case $r = 0$ by appropriate comments and footnotes. Before beginning the proof of Theorem 14.1 we state a useful corollary, a consequence of the fact that the geometric index, and hence the algebraic index of g, as a critical extremal of (J^λ, C_r), is finite.

COROLLARY 14.0. *The characteristic roots λ of conditions* (11.8) *or* (12.2) *less than a prescribed constant are finite in number and hence isolated.*

LEMMA 14.1 (i). *An index function $G^\lambda \#(C_r, \Lambda_p^g)$ of g is of class C^2.* (ii) *The origin in the domain R^μ of G^λ is a critical point of G^λ.*

Notation. We refer to the x-coordinates $c_1 < c_2 < \cdots < c_p$ of the respective m-planes $\Pi_1, \ldots, \Pi_p$ of the frame Λ_p^g. For v in the μ-ball D_ρ^μ of Theorem 13.2 and α the initial r-tuple of the μ-tuple v set

$$(14.8)' \quad L_0^\lambda(v) = \int_{X^1(\alpha)}^{c_1} f^\lambda(x, \Gamma^\lambda(x,v), \Gamma_x^\lambda(x,v))\, dx \qquad (r > 0)$$

$$(14.8)'' \quad L_q^\lambda(v) = \int_{c_q}^{c_{q+1}} f^\lambda(x, \Gamma^\lambda(x,v), \Gamma_x^\lambda(x,v))\, dx \qquad (q = 1, \ldots, p-1)$$

$$(14.8)''' \quad L_p^\lambda(v) = \int_{c_p}^{X^2(\alpha)} f^\lambda(x, \Gamma^\lambda(x,v), \Gamma_x^\lambda(x,v))\, dx \qquad (r > 0)$$

replacing the limit $X^1(\alpha)$ by a^1 and the limit $X^2(\alpha)$ by a^2 when $r = 0$. By virtue of the definition of G^λ in (14.5)

$$(14.9) \qquad G^\lambda(v) = \Theta(\alpha) + L_0^\lambda(v) + \cdots + L_p^\lambda(v) \qquad (v \in D_\rho^\mu)$$

where α is the initial r-tuple of v and $\Theta(\alpha)$ is to be deleted when $r = 0$.

Proof of (i). We begin with a proof of (A).

(A). *For* k *on the range* 0, 1, ..., p, *the integral L_k^λ is of class C^2 for* $v \in D_\rho^\mu$.

Proof of (A). The integrand of the integral L_k^λ depends on the μ-tuple v. When $r > 0$ the first r coordinates of v, namely $\alpha_1, \ldots, \alpha_r$, determine the limits $X^1(\alpha)$ and $X^2(\alpha)$ of the first and last of these integrals. The integrand of L_k^λ, $k = 0, 1, \ldots, p$ has the domain

$$\Delta_k = I_k \times D_\rho^\mu \qquad (k = 0, 1, \ldots, p)$$

as defined in (13.16) when $r > 0$. The concluding sentence of Theorem 13.2 affirms that the integrand of L_k^λ is of class C^2 on its domain Δ_k. Moreover X^1 and X^2 are of class C^2. Statement (A) follows.

We infer that (i) of Lemma 14.2 is true when $r > 0$. The proof when $r = 0$ is similar.

Proof of (ii). It follows from (13.12) that the coordinates of v, other than its first r coordinates $\alpha_1, \ldots, \alpha_r$, when $r > 0$, are coordinates v_i^q, where q has the range $1, \ldots, p$, and i the range $1, \ldots, m$. Letting a superscript $^\circ$ denote evaluation for $v = \mathbf{0}$, we find that for $q = 1, \ldots, p$

$$(14.10) \qquad -\frac{\partial^\circ L_q^\lambda}{\partial v_i^q} = f_{p_i}^\lambda(c_q, g(c_q), g'(c_q)) = \frac{\partial^\circ L_{q-1}^\lambda}{\partial v_i^q},$$

and if k is on the range $0, \ldots, p$ but is neither q nor $q - 1$, the partial derivative of L_k^λ with respect to v_i^q vanishes. It follows that all partial derivatives of G^λ with respect to the variables v_i^q vanish when $v = \mathbf{0}$. Lemma 14.1 (ii) follows when $r = 0$.

Let $d^\circ G^\lambda$ denote the differential of G^λ when $r > 0$, evaluated when $v = \mathbf{0}$. It follows from the results of the preceding paragraph that $d^\circ G^\lambda$ is a linear form in $d\alpha_1, \ldots, d\alpha_r$. By hypothesis g is a critical extremal of (J^λ, C_r) and so satisfies the C_r-transversalty condition (9.1). To establish Lemma 14.1 (ii) it is accordingly sufficient to establish the identity

$$(14.11) \quad d^\circ G^\lambda = d\Theta + [(f - p_i f_{p_i})\, dX^s + f_{p_i}\, dY_i^s]_{s=1}^{s=2} \qquad (\alpha = \mathbf{0})$$

in the differentials $d\alpha_1, \ldots, d\alpha_r$. The bracket on the right of (14.11), is evaluated as in (9.1), along g, with $\alpha = \mathbf{0}$. In accord with the definition in (11.15) of f^λ, it is unchanged if f and f_{p_i} are replaced by f^λ and $f_{p_i}^\lambda$, evaluated along g.

Verification of (14.11). Let $u = (u_1, \ldots, u_r)$ be an arbitrary r-tuple with $\|u\| < 1$. Let e be a scalar and let the μ-tuple v of (13.12) have the value

$$(14.12) \qquad v = (eu_1, \ldots, eu_r, 0, \ldots, 0) = V(e) \qquad (r > 0)$$

introducing $V(e)$ and restricting e by the condition $|e| < d$, with d so small that $V(e)$ is in the domain D_ρ^μ of v in Theorem 13.2. To obtain a formula for $G(V(e))$ we set

$$(14.13) \qquad \gamma(x,e) = \hat{\Gamma}^\lambda(x, V(e)) \qquad (a' < x < a'')$$

and $x^s(e) = X^s(eu)$ for $s = 1, 2$. By virtue of the definition of G^λ in (14.5), we find that for $|e| < d$

$$(14.14) \quad G(V(e)) \equiv \Theta(eu) + \int_{x^1(e)}^{x^2(e)} f^\lambda(x, \gamma(x,e), \gamma_x(x,e))\, dx \qquad (r > 0).$$

The graph of the partial mapping

$$x \to \gamma(x,e) : [x^1(e), x^2(e)] \to R^m \qquad \text{(cf. (10.1)'')}$$

is by definition of Γ^λ, a C_r-admissible broken λ-extremal, so that the identities

$$Y_i^s(eu) \equiv \gamma_i(X^s(eu),e) \qquad (i = 1, \ldots, m; s = 1, 2) \tag{14.15}$$

hold and imply the relations of form (9.8) when $e = 0$. The right member of (14.14) is now differentiated as to e, as in the proof of (9.1), with final evaluation when $e = 0$, and with f replaced throughout by f^λ. One thereby verifies (14.11) as an identity in the differentials

$$(d\alpha_1, \ldots, d\alpha_r) = (u_1, \ldots, u_r) \qquad (\|u\| < 1).$$

Thus Lemma 14.1 *is true.*

The following theorem on an "index form" of a critical extremal g of (J^λ, C_r) and the equivalent Corollary 14.1 below, are the principal instruments in proving and interpreting Theorem 14.1. For use in Theorem 14.2 we set

$$2\Omega^\lambda(x,\eta,\zeta) = 2\Omega(x,\eta,\zeta) - \lambda\eta_i\eta_i, \tag{14.16}$$

for $x \in \aleph$ and arbitrary m-tuples η and ζ.

THEOREM 14.2. *If* g *is a critical extremal of* (J^λ, C_r) *with** $r > 0$, *if* $Q^\lambda\#(C_r, \Lambda_p^g)$ *is the index form* (14.7) *of* g, *and if* $\mu = r + mp$, *then at a prescribed* μ*-tuple* $(z_1, \ldots, z_\mu)$ *with an initial* r*-tuple* $(u_1, \ldots, u_r)$,

$$Q^\lambda(z) = b_{hk}u_h u_k + \int_{a^1}^{a^2} 2\Omega^\lambda(x, h^\lambda(x,z), h_x^\lambda(x,z))\, dx \tag{14.17}$$

where for fixed z *and* x, *the* m*-tuple*

$$h^\lambda(x,z) = \frac{d}{de}\Gamma^\lambda(x,ez)\Big|^{e=0}, \qquad (a^1 \leq x \leq a^2) \tag{14.18}$$

and $\|b_{hk}\|$ *is the comparison matrix associated in Section* 10 *with* g *as a critical extremal of* (J, C_r).

In anticipation of the proof of Theorem 14.2, a mapping

$$(x,e) \to \gamma(x,e) : (a',a'') \times (-d,d) \to R^m \tag{14.19$'$}$$

was given in (10.24). Values $c_1, \ldots, c_p$ of x in (a',a'') were given, dividing (a',a'') into subintervals

$$(a',c_1], [c_1,c_2], \ldots, [c_{p-1},c_p], [c_p,a''). \tag{14.19$''$}$$

According to Definition 10.4 the mapping γ is said to be *piecewise of class* C_x^2 relative to the points $c_1, \ldots, c_p$ of division of (a',a''), if for each interval I_j in the set (14.19)$''$, the restriction

$$\gamma \,|\, (I_j \times (-d,d)) \tag{14.20}$$

* The theorem holds for $r = 0$ on setting $\mu = mp$ and deleting all references to u.

is of class C^2. Such a mapping is *here* defined by setting

$$\gamma(x,e) = \Gamma^\lambda(x,ez) \qquad (a' < x < a'') \tag{14.21}$$

for each $e \in (-d,d)$. If d is sufficiently small and if the values $c_1, \ldots, c_p$ of Definition 10.4 are the x-coordinates of the m-planes $\Pi_1, \ldots, \Pi_p$ of the λ-frame Λ_p^g, the mapping γ defined in (14.21) is piecewise of class C_x^2.

Moreover for d sufficiently small and for each $e \in (-d,d)$ the restricted partial mapping,

$$\gamma^e : x \to \gamma(x,e) : [x^1(e),x^2(e)] \to R^m \tag{14.22$'$}$$

of the mapping γ : (14.21), is a C_r-admissible broken λ-extremal of the framed family Γ^λ of Section 13. According to Theorem 10.2, if d is sufficiently small, the variation $\eta = \gamma_e|^{e=0}$ is such that for each $e \in (-d,d)$

$$\frac{d^2}{de^2} J^\lambda(\gamma^e,eu)\Bigg|^{e=0} = b_{hk}u_h u_k + \int_{a^1}^{a^2} 2\Omega^\lambda\Big(x,\eta(x),\eta'(x)\Big)\, dx. \tag{14.22$''$}$$

We conclude the proof of Theorem 14.2 by proving (i).

(i) *If* G^λ *is the index function* (Definition 14.1) *based on the λ-extremal* g *and λ-frame* $\Lambda_{\mathrm{p}}^{\mathrm{g}}$, *if* $\mathrm{r} > 0$ *and* $\mu = \mathrm{r} + \mathrm{mp}$, *and if* z *is a prescribed μ-tuple with initial* r*-tuple* u, *then for* d *sufficiently small*

$$\mathrm{J}^\lambda(\gamma^e,\mathrm{eu}) \equiv G^\lambda(\mathrm{ez}) \qquad (|\mathrm{e}| < \mathrm{d}) \tag{14.23$'$}$$

and

$$\gamma_e(\mathrm{x,e})\Bigg|^{\mathrm{e}=0} = \frac{\mathrm{d}}{\mathrm{de}}\,\Gamma^\lambda(\mathrm{x,ez})\Bigg|^{\mathrm{e}=0} \qquad (\mathrm{a}^1 \le \mathrm{x} \le \mathrm{a}^2). \tag{14.23$''$}$$

Proof of (14.23)$'$. Suppose first that $r > 0$. Let $u_1, \ldots, u_r$ be the initial r-tuple of the prescribed μ-tuple z. If $d > 0$ is sufficiently small and $|e| < d$, the values

$$x^1(e) = X^1(eu), \qquad x^2(e) = X^2(eu)$$

are well-defined, and the following identities in e are valid:

$$G^\lambda(ez) \equiv \Theta(eu) + \int_{x^1(e)}^{x^2(e)} f^\lambda\Big(x,\Gamma^\lambda(x,ez),\Gamma_x^\lambda(x,ez)\Big)\, dx \tag{14.24$'$}$$

$$\equiv \Theta(eu) + \int_{x^1(e)}^{x^2(e)} f^\lambda\Big(x,\gamma(x,e),\gamma_x(x,e)\Big)\, dx \tag{14.24$''$}$$

$$\equiv J^\lambda(\gamma^e,eu). \tag{14.24$'''$}$$

Of these identities, (14.24)$'$ follows from Definition 14.1 of G^λ. The identity (14.24)$''$ then follows from the definition of γ in (14.21), since $a' < x^1(e) < x^2(e) < a''$, if d is sufficiently small. The final identity is a consequence of the definition of $J(\chi,\alpha)$ in (8.19).

Thus the identity (14.23)′ is valid when $r > 0$, if d is sufficiently small. When $r = 0$ one refers to $\mathbf{J}^\lambda$ of (12.3) and readily establishes the identity $\mathbf{J}^\lambda(\gamma^e) \equiv G^\lambda(ez)$.

Proof of (14.23)″. This relation follows from the definition of γ in (14.21).

Completion of Proof of Theorem 14.2. From the identity (14.23)′ and the definition of the index form in (14.7), we infer that

$$\frac{d^2}{de^2} J^\lambda(\gamma^e,eu)\Big|^{e=0} \equiv \frac{d^2}{de^2} G^\lambda(ez)\Big|^{e=0} = Q^\lambda(z).$$

Relation (14.17) now follows with the aid of (14.22)″, taking (14.23)″ into account.

The proof of Theorem 14.2 is complete.

An interpretation of Theorem 14.2. The framed family Γ^λ of broken λ-extremals of Theorem 13.2 induces a framed family h^λ of broken secondary λ-extremals in terms of which Theorem 14.2 can be simply restated. This restatement is in terms of the quadratic functional $\mathscr{I}_r^\lambda$: (11.7) based on the critical extremal g of (J^λ, C_r). It has values

$$\mathscr{I}_r^\lambda(\eta) = b_{hk}u_h u_k + \int_{a^1}^{a^2} 2\Omega^\lambda\Big(x,\eta(x),\eta'(x)\Big)\, dx, \tag{14.25}$$

and domain the set of C_r''-admissible mappings η of $[a^1,a^2]$ into R^m satisfying conditions C_r'' : (10.10). Cf. Definition 10.3 of $\mathscr{I}_r$. When $r = 0$ reference to u is to be deleted.

Definition 14.4. By a *BSλ-extremal** E^λ is meant a continuous x-parameterized curve E^λ, defined for $x \in [a^1,a^2]$, and composed of a finite sequence of compact secondary λ-extremals on no one of which is there a point λ-conjugate (see Section 12) to the initial point of that secondary λ-extremal.

Definition 14.5. The set of endpoints of the $p + 1$ secondary λ-extremals of E^λ, omitting the initial and final point of E^λ, will be called the *division points* of E^λ. Cf. Definition 13.2.

When $r > 0$ let the μ-tuple $z_1, \ldots, z_\mu$ prescribed in Theorem 14.2, be written in the form

$$(z_1, \ldots, z_\mu) = (u_1, \ldots, u_r : z_1^1, \ldots, z_m^1; \ldots ; z_1^p, \ldots, z_m^p). \tag{14.26$'$}$$

With this understood we state the following lemma.

LEMMA 14.2. *If* $h^\lambda(x,z)$ *is defined as in Theorem* 14.2, *then for each μ-tuple* z *of form* (14.26)′ *the mapping into* R^m

$$x \to h^\lambda(x,z) = \frac{d}{de}\Gamma^\lambda(x,ez)\Big|^{e=0} \qquad (a^1 \leq x \leq a^2) \tag{14.26$''$}$$

* "BSλ-extremal" abbreviates the phrase a "broken secondary λ-extremal."

has a graph in E_{m+1} *which is a* BSλ*-extremal denoted by* $h^{\lambda,z}$ *and characterized as follows.*

(i) *When** $r > 0$, *the end points of* $h^{\lambda,z}$, *together with the initial* r*-tuple* u *of* z, *satisfy the end conditions* C_r'' : (11.8)$''$.

(ii) *The "division points" of* $h^{\lambda,z}$ *are the points with coordinates* $(x, y_1, \ldots, y_m)$ *in* E_{m+1} *of the respective forms,*

$$(c_1, z_1^1, \ldots, z_m^1), \ldots, (c_p, z_1^p, \ldots, z_m^p) \tag{14.27}$$

where $c_1, \ldots, c_m$ *are the* x*-coordinates of the* m*-planes* $\Pi_1, \ldots, \Pi_p$.

By virtue of Lemma 14.2 Theorem 14.2 can be equivalently stated as follows.

COROLLARY 14.1. *The domain of a derived index form* $Q^\lambda\#(C_r,\Lambda_p^g)$ *is* R^μ *where* $\mu = r + mp$. *For* $z \in R^\mu$ *and* $r > 0$

$$Q^\lambda(z) = \mathscr{I}_r^\lambda(\eta) \qquad (cf.\ (11.7)) \tag{14.28}$$

where $\mathscr{I}_r^\lambda$ *is the quadratic functional of* (J^λ,C_r) *based on* g *and graph* η *is the* BSλ*-extremal* $h^{\lambda,z}$ *determined by the* μ*-tuple* z *as in Lemma* 14.2.

The case $r = 0$. When $r = 0$ Corollary 14.1 is valid if $h^{\lambda,z}$ is determined by $z = (z^1, \ldots, z^p)$ as in Lemma 14.2, with $u_1, \ldots, u_r$ deleted and $h^\lambda(x,z) = \mathbf{0}$ when $x = a^1$ and a^2.

Theorem 14.1 of this section remains to be proved. It is primarily a theorem in the analysis of quadratic functionals.

In Section 15 we initiate the theory of such functionals, freed from consideration of the framed family Γ^λ of broken extremals of Theorem 13.2 and the index function G^λ, defined in (14.6). It is only by developing the quadratic theory by quadratic or linear processes that the requisite generality can be obtained.

The theorems of Section 13 and Section 14 serve a special purpose. They locally characterize a critical extremal in a way essential for global variational analysis. End products are the quadratic functionals and index forms which we now study in a more general setting.

* When $r = 0$, $h^\lambda(a^s,z) = \mathbf{0}$ for $s = 1, 2$.

CHAPTER 4

Free Index Forms

15. Free Linear Differential and Boundary Conditions

The three conditions (11.8) are derived (as opposed to free). That is, the differential equations (11.8)′ presuppose the derivation of the quadratic form Ω of (4.0), based on the extremal g. The secondary end conditions C_r'' of (11.8) are derived in Section 10 from the primary end conditions C_r. The r-square matrix $\|b_{hk}\|$ appearing in (11.8)‴, was derived in Section 10. The conditions (11.8) are thus accessory to the theory of critical extremals. They can be best studied if their essential characteristics are sharply enumerated free of any antecedent variational theory and in the context of the theory of selfadjoint differential equations and boundary conditions. This we now do.

Replacement of Ω *by* ω *of Appendix* I.—A second major departure from classical variational theory will be to replace the preintegrand $2\Omega(x,\eta,\zeta)$: (4.0) of the second variation $\mathscr{I}_0$: (4.1)′ by the quadratic form $2\omega(x,\eta,\zeta)$ in the m-tuples η and ζ of Appendix I. This replacement of Ω makes the resultant extension of the Sturm theorems true generalizations of the Sturm theorems in the plane.

The reader should read Section 1-4 of Appendix I.

Free Conditions Replacing Conditions (11.8). Given a value of λ and an integer $r > 0$, the conditions (11.8)′, (11.8)″, (11.8)‴ will be replaced by conditions (15.0)′, (15.0)″, (15.0)‴ respectively, denoted collectively by $W_r(\lambda)$ and termed *free*. We begin by replacing Ω^λ of (11.8)′ by the more general quadratic form ω^λ of (1.1) : Appendix I, and, when $r > 0$, present conditions $W_r(\lambda)$ as follows. As indicated at the end of Section 1 of Appendix I one could prefer (1.7) of Appendix I as a definition of ω^λ.

$W_r(\lambda)$: *An ensemble of 'free' conditions of positive dimension r*:

$$\begin{bmatrix} \dfrac{d}{dx}\,\omega^{\lambda}_{\zeta_i}(x,\eta,\eta') = \omega^{\lambda}_{\eta_i}(x,\eta,\eta') & (i = 1, \ldots, m) & (15.0)' \\ \eta^s_i = c^s_{ih} u_h \quad (s = 1, 2;\ i = 1, \ldots, m) & & (15.0)'' \\ c^1_{ih}\xi^1_i - c^2_{ih}\xi^2_i = b_{hk}u_k \quad (h = 1, \ldots, r) & & (15.0)''' \end{bmatrix}$$

defined for each $\lambda \in R$, where h, k have the range $1, \ldots, r$ and ξ is the "canonical associate" of η, defined as in (11.2), with ω replacing Ω, or in (1.3) of Appendix I.

In (15.0) the matrix $\|c^s_{ih}\|$, is prescribed, and of rank r, subject to the condition that it have $2m$ rows indexed by pairs (s,i) and r columns indexed by h. The constants b_{hk} are prescribed, subject to the condition that $b_{hk} = b_{kh}$. We term $\|b_{hk}\|$ the *comparison matrix* of (15.0).

Chapter 7 is concerned with all selfadjoint differential and boundary conditions of the type arising in ordinary variational theory. In this context the free conditions $W_r(\lambda)$: (15.0) are termed *canonical*,* as are the following conditions when $r = 0$.

$W_0(\lambda)$. *An ensemble of free conditions of dimension* $r = 0$:

$$\begin{bmatrix} \dfrac{d}{dx}\,\omega^{\lambda}_{\zeta_i}(x,\eta,\eta') = \omega^{\lambda}_{\eta_i}(x,\eta,\eta') & (i = 1, \ldots, m) & (15.1)' \\ \eta^s_i = 0 \quad (s = 1, 2;\ i = 1, \ldots, m) & & (15.1)'' \end{bmatrix}$$

defined for each $\lambda \in R$.

Definition 15.0. *Canonical systems** W_r, $r \geq 0$. The union for $\lambda \in R$ of the canonical conditions, $W_r(\lambda)$: (15.0) when $r > 0$, or of the canonical conditions, $W_0(\lambda)$: (15.1) when $r = 0$, will be denoted by W_r and termed a *canonical system* W_r of free conditions $W_r(\lambda)$.

A nonnull solution η of conditions $W_r(\sigma)$ is called a *characteristic solution* of the system W_r with *characteristic root* σ. It is of class C^1.

Selfadjoint BC.† The BC of the "derived" conditions (11.8) when $r > 0$, and (12.2) when $r = 0$, are selfadjoint in the sense of Section 29, as are the above "free" BC of (15.0) or (15.1). However, as we shall see, the above free BC include a set of canonical conditions "equivalent" to *arbitrary* selfadjoint BC.

The quadratic functional $\mathscr{I}^{\lambda}_r$, introduced in (11.7) when $r > 0$ and in (12.1) when $r = 0$, is called a *derived* functional because of its origin in

* The selfadjoint boundary conditions of form (29.4) are not in general *canonical* in the sense of Definition 15.0.

† BC abbreviates "boundary conditions."

Definition 10.3. *Free* functionals I_r^λ will now be defined. Their domains $\{W_r\}$ are independent of λ.

Definition 15.1. *A free functional $I_r^\lambda \# W_r$ with domain $\{W_r\}$*. The symbol # is read *based on*. Given a system W_r of free conditions $W_r(\lambda)$, the domain of a quadratic functional I_r^λ, based on W_r, shall be the vector space over R, denoted by $\{W_r\}$ of all D^1-mappings of $[a^1,a^2]$ into R^m which satisfy the end conditions (15.0)″ when $r > 0$ or (15.1)″ when $r = 0$. For $\eta \in \{W_r\}$ and an arbitrary number σ we assign the functional $I_r^\sigma \# W_r$ a value

$$(15.2)' \qquad I_r^\sigma(\eta) = \int_{a^1}^{a^2} 2\omega^\sigma(x,\eta(x),\eta'(x))\, dx \qquad (\text{when } r = 0)$$

and a value

$$(15.2)'' \qquad I_r^\sigma(\eta) = b_{hk}u_h u_k + \int_{a^1}^{a^2} 2\omega^\sigma(x,\eta(x),\eta'(x))\, dx \qquad (\text{when } r > 0)$$

where the comparison matrix $\|b_{hk}\|$ is prescribed in (15.0)‴ and the r-tuple u satisfies (15.0)″ with η, when $r > 0$.

We can now define a "*free*" *index form* Q^λ. When explicitness is required such a form will be denoted by $Q^\lambda \#(W_r,\Lambda_p)$. Definitions 15.2(a), 15.2(b), and 15.2(c) are required, supplemented by Lemma 15.1 and Corollary 4.1 of Appendix I.

Definition 15.2(a). *λ-Frames, Λ_p over (a^1,a^2).* Let $\Pi_1, \ldots, \Pi_p$ be a set of m-planes* in E_{m+1} orthogonal to the x-axis at points in (a^1,a^2) with x-coordinates $c_1 < c_2 < \cdots < c_p$, respectively, such that no one of the $p + 1$ successive intervals into which $[a^1,a^2]$ is divided by the values $c_1, \ldots, c_p$, contains a point λ-conjugate† to its initial point. We term such a set of m-planes a *λ-frame Λ_p over (a^1,a^2)*.

Definition 15.2(b). *Framed* BS*σ-extremals $h^{\sigma,z}$*. Given a σ-frame Λ_p over (a^1,a^2), set $\mu = r + mp$. For each μ-tuple z in R^μ, a BSσ-extremal‡ $h^{\sigma,z}$, more explicitly denoted by $h^{\sigma,z}\#(W_r,\Lambda_p)$, is uniquely determined by the following conditions. Use Corollary 4.1 of Appendix I.

(α) The mapping η of $[a^1,a^2]$ into R^m whose graph is $h^{\sigma,z}$, shall satisfy conditions (15.0)″ (with an r-tuple u when $r > 0$).

(β). The "division points" of $h^{\sigma,z}$, $r \geq 0$, shall be points $z^1, \ldots, z^p$ on the respective m-planes $\Pi_1, \ldots, \Pi_p$ of Λ_p and will be determined by the μ-tuple

* A point $z^i \in \Pi_i$ will be assigned coordinates

$$(x, y_1, \ldots, y_m) = (c_i, z_1^i, \ldots, z_m^i).$$

† Points in ℵ are λ-conjugate, if conjugate relative to the differential equations (15.0)′. Secondary λ-extremals are here graphs of C^1-solutions of (15.0)′.

‡ BS abbreviates broken secondary.

z on giving z a representation $z = (z^1, \ldots, z^p)$, when $r = 0$, and, when $r > 0$, a representation $z = (u : z^1, \ldots, z^p)$ with u an r-tuple.

Definition 15.2(c). *Values of* $Q^\sigma\#(W_r,\Lambda_p)$. Set $\mu = r + mp$. Given a μ-tuple z, let η be the mapping whose graph is the BSσ-extremal $h^{\sigma,z}\#(W_r,\Lambda_p)$. The value $Q^\sigma(z)$ of $Q^\sigma\#(W_r,\Lambda_p)$ is defined for each $z \in R^\mu$ by setting

$$Q^\sigma(z) = I_r^\sigma(\eta) \qquad (\text{graph } \eta = h^{\sigma,z}\#(W_r,\Lambda_p)). \tag{15.3}$$

The formal identity of (15.3) with (14.28) is to be noted.

Lemma 13.2 has a simple extension.

LEMMA 15.0. *A λ-frame is a σ-frame for $\sigma \leq \lambda$.*

The proof is similar to that of Lemma 13.2 replacing Ω by ω. It is understood that λ-conjugate points are conjugate points relative to the DE (15.0)$'$.

The nature of $Q^\sigma\#(W_r,\Lambda_p)$ as a quadratic form is made clear by the following lemma. In this lemma we fix λ and take $\sigma \leq \lambda$.

LEMMA 15.1. *Given* $\mathrm{W_r}$ *and a λ-frame* Λ_p, *set* $\mu = \mathrm{r} + \mathrm{mp}$. *If* $\sigma \leq \lambda$, Λ_p *is admissible as a σ-frame and* $\mathrm{Q}^\sigma\#(\mathrm{W_r},\Lambda_\mathrm{p})$ *is representable as a symmetric quadratic form*

$$\mathrm{Q}^\sigma(\mathrm{z}) = \mathrm{a_{ij}}(\sigma)\mathrm{z_i z_j} \qquad (\mathrm{z} \in \mathrm{R}^\mu) \tag{15.4}$$

for each $\sigma \leq \lambda$.

Proof of Lemma 15.1. That a λ-frame is also a σ-frame for $\sigma \leq \lambda$, follows from Lemma 15.0. For a given $\sigma \leq \lambda$ and μ-tuple z, the framed BSσ-extremal $h^{\sigma,z}\#(W_r,\Lambda_p)$ is the graph of a mapping

$$x \to H^\sigma(x,z) : [a^1,a^2] \to R^m \qquad (\text{in } \{W_r\}) \tag{15.5$'$}$$

in which for q on the range $1, \ldots, \mu$ and $\sigma \leq \lambda$,

$$H^\sigma(x,z) = Z^q(x,\sigma)z_q, \tag{15.5$''$}$$

where for the given σ, the graph of the mapping $x \to Z^q(x,\sigma)$ of $[a^1,a^2]$ into R^m is a C_r''-admissible BSσ-extremal with frame Λ_p, each of whose $p + 2$ vertices* except at most one is on the x-axis. (Cf. Appendix I, Lemma 3.3.)

The reader will note that when $r > 0$, the mapping (15.5)$'$ satisfies (15.0)$''$ with the initial r-tuple u of z. In any case, $Q^\sigma\#(W_r,\Lambda_p)$ can be evaluated with the aid of (15.3), setting $\eta(x) = Z^q(x,\sigma)z_q$ in $I_r^\sigma(\eta)$. The representation (15.4) follows.

* By the *vertices* of a C_r''-admissible BSσ-extremal with frame Λ_p are meant its endpoints and intersections with the m-planes $\Pi_1, \ldots, \Pi_p$ of Λ_p.

This completes the definition of $Q^\sigma\#(W_r,\Lambda_p)$ *of a free index form* Q^σ *based on* W_r *and* Λ_p. Q^σ *is uniquely determined by* σ, W_r, *and* Λ_p.

We shall need formulas for the partial derivatives of a free index form Q^λ with respect to coordinates of μ-tuples in its domain. Lemma 15.2 provides such formulas.

Introduction to Lemma 15.2. The domain of an index form $Q^\sigma\#(W_r,\Lambda_p)$ is R^μ, where $\mu = r + mp$. The partial derivatives of $Q^\sigma\#(W_r,\Lambda_p)$ will be evaluated at a point $\bar{z}$ in R^μ. This evaluation will be facilitated if we represent $\bar{z}$ in the form

$$(15.6)' \qquad \bar{z} = (\bar{z}_1^1, \ldots, \bar{z}_m^1; \ldots; \bar{z}_1^p, \ldots, \bar{z}_m^p) \qquad (\text{when } r = 0)$$

and in the form

$$(15.6)'' \qquad \bar{z} = (\bar{u}_1, \ldots, \bar{u}_r : \bar{z}_1^1, \ldots, \bar{z}_m^1; \ldots; \bar{z}_1^p, \ldots, \bar{z}_m^p) \qquad (\text{when } r > 0)$$

where $\bar{z}^1, \ldots, \bar{z}^p$ is a sequence of points in the respective m-planes $\Pi_1, \ldots, \Pi_p$ of Λ_p, and where $\bar{u}$ is the initial r-tuple of $\bar{z}$.

Let $\bar{\eta}$ be the mapping of $[a^1,a^2]$ into R^m whose graph is the framed BSσ-extremal $h^{\sigma,\bar{z}}\#(W_r,\Lambda_p)$. To evaluate the partial derivatives of $Q^\sigma\#(W_r,\Lambda_p)$ at the point $\bar{z}$, we shall need the associated canonical mappings

$$(15.7) \qquad x \to \bar{\xi}_i(x) = \omega^\sigma_{\zeta_i}(x,\bar{\eta}(x),\bar{\eta}'(x)) \qquad (i = 1, \ldots, m).$$

The m-tuples $\bar{\xi}(x)$ are well-defined for $x \in [a^1,a^2]$ except at most for the values $c_1, \ldots, c_p$ of x on the respective m-planes $\Pi_1, \ldots, \Pi_p$ of Λ_p. If $x = c_q$ is the q-th of these values, finite limiting μ-tuples $\bar{\xi}(c_q-)$ and $\bar{\xi}(c_q+)$ exist.

Notation. In Lemma 15.2 and its proof, $\bar{u}$ is the initial r-tuple of $\bar{z}$. The variables z_i^q are defined for $q = 1, \ldots, p$ and $i = 1, \ldots, m$. The integers h, k are on the range $1, \ldots, r$.

LEMMA 15.2. *For the index form* $Q^\sigma\#(W_r,\Lambda_p)$

$$(15.8)' \qquad \frac{\partial}{\partial z_i^q} Q^\sigma \bigg|^{z=\bar{z}} = 2(\bar{\xi}_i(c_q-) - \bar{\xi}_i(c_q+)) \qquad (r \geq 0)$$

$$(15.8)'' \qquad \frac{\partial}{\partial u_h} Q^\sigma \bigg|^{z=\bar{z}} = 2(b_{hk}\bar{u}_k + \bar{\xi}_i(a^2)c_{ih}^2 - \bar{\xi}_i(a^1)c_{ih}^1) \qquad (r > 0).$$

Proof of (15.8)$'$. Given $z \in R^\mu$, a BSσ-extremal $h^{\sigma,z}\#(W_r,\Lambda_p)$ is the graph of a mapping of form (15.5) such that

$$(15.9) \qquad H_i^\sigma(c_q,z) \equiv z_i^q, \qquad \frac{\partial}{\partial z_j^q} H_i^\sigma(c_q,z) \equiv \delta_i^j,$$

where the second set of relations in (15.9) follows from the first. The formulas (15.8)′ follow from the definition of $Q^\sigma(z)$ in (15.3) on differentiating under the integral sign in (15.2), integrating by parts in the usual way and making use of the second set of relations in (15.9).

Proof of (15.8)″. The formulas (15.8)″ are meaningless when $r = 0$. When $r > 0$ a BSσ-extremal $h^{\sigma,z}\#(W_r,\Lambda_p)$ is the graph of a mapping $x \to H^\sigma(x,z)$ of form (15.5). For fixed σ and z, end conditions (15.0)″ are satisfied by this mapping, so that

$$(15.10) \qquad H_i^\sigma(a^s,z) \equiv c_{ih}^s u_h, \qquad \frac{\partial}{\partial u_h} H_i^\sigma(a^s,z) \equiv c_{ih}^s \qquad (s = 1, 2)$$

where the second set of relations in (15.10) follows from the first. In (15.8)″ the terms $2b_{hk}\bar{u}_k$ are the partial derivatives of $b_{hk}u_h u_k$ in (15.2)″, evaluated when $u = \bar{u}$. The remaining terms on the right of (15.8)″ are obtained on differentiating with respect to u_h under the integral sign in (15.2)″, integrating by parts and using the second set of relations in (15.10).

This completes the proof of Lemma 15.2.

We continue with a basic theorem on a free index form Q^λ. Here $r \geq 0$.

THEOREM 15.1(i). *A necessary and sufficient condition that a free index form* $Q^\sigma\#(W_r,\Lambda_p)$ *be singular is that* σ *be a characteristic root of* W_r.

(ii). *The nullity* (Def. 14.0) *of a free index form* $Q^\sigma\#(W_r,\Lambda_p)$ *equals the multiplicity of* σ *as a characteristic root of* W_r.

Proof of (i). We prove (i) by proving (α) and (β).

(α). *The condition of* (i) *is necessary.* Set $\mu = r + mp$. By hypothesis there exists a nonnull μ-tuple $\bar{z}$ which is a critical point of Q^σ. The BSσ-extremal $h^{\sigma,\bar{z}}\#(W_r,\Lambda_p)$ is the graph of a mapping $\bar{\eta}$ of $[a^1,a^2]$ into R^m. We shall show that $\bar{\eta}$ is a characteristic solution of W_r with root σ.

Note first that $\bar{\eta}$ is nonnull with $\bar{z}$. The BSσ-extremal $h^{\sigma,\bar{z}}$ has no corners, since the left and hence the right members of (15.8)′ vanish when $\lambda = \sigma$. Hence $\bar{\eta}$ is a solution of the DE (15.0)′ when $\lambda = \sigma$. The proof of (α) is complete when $r = 0$.

When $r > 0$, conditions (15.0)″ are satisfied by $\bar{\eta}$ with the initial r-tuple $\bar{u}$ of $\bar{z}$, in accord with the definition of $h^{\sigma,\bar{z}}$ when $r > 0$. Conditions (15.0)‴ are satisfied by $\bar{\eta}$, $\bar{u}$ and the canonical mapping $\bar{\xi}$ defined by $\bar{\eta}$, by virtue of (15.8)″, since $\bar{z}$ is by hypothesis a critical point of Q^σ.

Thus (α) is true.

(β). *The condition of* (i) *is sufficient.* By hypothesis there exists a nonnull solution $\bar{\eta}$ of conditions $W_r(\sigma)$. Let $\bar{z}$ be a μ-tuple such that $h^{\sigma,\bar{z}}\#(W_r,\Lambda_p)$ is the graph of $\bar{\eta}$. Denote $\bar{z}$ by $\Phi(\bar{\eta})$. Then $\Phi(\bar{\eta})$ is nonnull, since $\bar{\eta}$ is nonnull. Moreover the right and hence the left members of (15.8)′ vanish for the given Q^σ. When $r = 0$ we conclude that $\bar{z}$ is a critical point of $Q^\sigma\#(W_0,\Lambda_p)$. When $r > 0$, the partial derivatives of $Q^\sigma\#(W_r,\Lambda_p)$ with respect to the initial coordinates $u_1, \ldots, u_r$ of z vanish when $z = \bar{z}$ by (15.8)″, since $\bar{\eta}$ satisfies conditions (15.0)‴ with $\bar{u}$ by hypothesis.

Then (β) is true.

Proof of (ii). The proof of (i) shows that (i) can be given an equivalent formulation: a point z in the domain of $Q^\sigma\#(W_r,\Lambda_p)$ is a critical point of Q^σ if and only if $h^{\sigma,z}\#(W_r,\Lambda_p)$ is the graph of a solution $\eta(z)$ of W_r with root σ. Moreover, the mapping thereby defined of critical points z of Q^σ into solutions $\eta(z)$ of W_r with root σ, is linear, biunique, and onto. Theorem 15.1(ii) follows.

Preparation for Index Theorem 15.2. The above Theorem 15.1 and the following Theorem 15.2, lead to a proof of Theorem 14.1. Lemma 15.3 below is needed in proving the Index Theorem 15.2.

In Lemma 15.3 we consider free index forms $Q^\sigma\#(W_r,\Lambda_p)$* for $\sigma \le \lambda$, a prescribed constant. Set $\mu = r + mp$. According to Lemma 15.1, for $\sigma \le \lambda$, Q^σ is a symmetric quadratic form with values

$$Q^\sigma(z) = a_{ij}(\sigma)z_i z_j \qquad (z \in R^\mu). \tag{15.11}$$

Concerning the coefficients in this form we state a lemma.

LEMMA 15.3. *The coefficients* $a_{ij}(\sigma)$ *in the representation* (15.11) *of* $Q^\sigma(z)$ *vary continuously with* σ *for* $\sigma \le \lambda$.

For $\mu = r + mp$, for $\sigma \le \lambda$ and $z \in R^\mu$, let $h^{\sigma,z}\#(W_r,\Lambda_p)$ be the BSσ-extremal introduced in Definition 15.2(b). If

$$x \to H^\sigma(x,z) : [a^1,a^2] \to R^m \tag{15.12}$$

is the D^1-mapping whose graph is $h^{\sigma,z}$, then for q on the range $1, \ldots, \mu$, for x, σ, and z as above,

$$H^\sigma(x,z) = Z^q(x,\sigma)z_q \qquad \text{(as in (15.5)″)} \tag{15.13}$$

where each of the mappings

$$(x,\sigma) \to Z^q(x,\sigma) : [a^1,a^2] \times (-\infty,\lambda] \to R^m \qquad (q = 1, \ldots, \mu) \tag{15.14$'$}$$

$$(x,\sigma) \to \frac{\partial}{\partial x} Z^q(x,\sigma) : [a^1,a^2] \times (-\infty,\lambda] \to R^m \qquad (q = 1, \ldots, \mu) \tag{15.14$''$}$$

is continuous and for fixed σ of class D^1. One verifies this affirmation by considering the restrictions of the mappings (15.14) to the closed intervals

* As in Lemma 15.1, Λ_p is a λ-frame independent of σ.

into which the interval $[a^1,a^2]$ is divided by the m-planes $\Pi_1, \ldots, \Pi_p$ of Λ_p and applying Corollary 4.1 of Appendix I.

With $h^{\sigma,z}$ so presented, Lemma 15.3 follows from the definition of $Q^\sigma(z)$ in (15.3).

INDEX THEOREM 15.2(i). *The nullity of a free index form* $Q^\lambda\#(W_r,\Lambda_p)$ *equals the multiplicity of* λ *as a characteristic root of the system* W_r.

(ii). *The index of* $Q^\lambda\#(W_r,\Lambda_p)$ *is the "count"* κ_λ *of characteristic roots* $\sigma < \lambda$ *of* W_r.

Proof of (i). Theorem 15.2(i) is implied by Theorem 15.1(ii).

Proof of (ii). The frame Λ_p conditioning Q^λ in Theorem 15.2 will serve as a σ-frame for each $\sigma \leq \lambda$. Theorem 15.2(ii) will follow from Theorem 3.1 of Appendix III, once the following lemma is verified.

LEMMA 15.4. *For* Λ_p *as in Theorem* 15.2, *the family* F *of quadratic forms* $Q^\sigma\#(W_r,\Lambda_p)$, *defined for each* $\sigma \leq \lambda$, *is a "model" family, as defined in Section* 1 *of Appendix* III.

One can identify the value λ given in Theorem 15.2 with the value λ in Lemma 15.3.

Condition I of Appendix III is then satisfied by F since the representation (15.11) of the quadratic forms Q^σ is one in which the coefficients $a_{ij}(\sigma)$ vary continuously with σ for $\sigma \leq \lambda$.

Condition II, that Q^σ be positive definite for σ negative and sufficiently large in absolute value, may be inferred from the proof of Theorem 11.2 replacing $\mathscr{I}_r^\lambda$ by I_r^λ in accord with (15.3).

Condition III of Appendix III will now be verified. Let σ and $\bar{\sigma}$ be such that $\sigma < \bar{\sigma} \leq \lambda$. Let a nonnull μ-tuple z be given. Condition III of Appendix III is satisfied if

$$Q^{\bar{\sigma}}(z) < Q^\sigma(z) \qquad (\bar{\sigma} > \sigma) \tag{15.15}$$

for each such σ, $\bar{\sigma}$, and nonnull z.

Proof of (15.15). Let η and $\bar{\eta}$ be the D^1-mappings of $[a^1,a^2]$ into R^m whose graphs are respectively the BSσ-extremals

$$h^{\sigma,z}\#(W_r,\Lambda_p), \qquad h^{\bar{\sigma},z}\#(W_r,\Lambda_p) \tag{15.15$'$}$$

of Definition 15.2(b). In terms of the functional $I_r^\sigma\#W_r$ of (15.2)

$$Q^{\bar{\sigma}}(z) = I_r^{\bar{\sigma}}(\bar{\eta}), \qquad Q^\sigma(z) = I_r^\sigma(\eta) \qquad \text{(by (15.3))} \tag{15.15$''$}$$

so that

$$(15.16)' \qquad Q^{\bar{\sigma}}(z) - Q^{\sigma}(z) = (I_r^{\bar{\sigma}}(\bar{\eta}) - I_r^{\bar{\sigma}}(\eta)) + (I_r^{\bar{\sigma}}(\eta) - I_r^{\sigma}(\eta)).$$

$I_r^{\bar{\sigma}}(\bar{\eta}) - I_r^{\bar{\sigma}}(\eta)$ is nonpositive by virtue of Corollary 7.1, applied to each of the $p + 1$ subintervals $[a,b]$ into which $[a^1,a^2]$ is subdivided by the λ-frame Λ_p, since by Corollary 7.1

$$\int_a^b \omega^{\bar{\sigma}}(x,\bar{\eta}(x),\bar{\eta}'(x))\, dx \leq \int_a^b \omega^{\bar{\sigma}}(x,\eta(x),\eta'(x))\, dx.$$

Moreover

$$(15.16)'' \qquad I_r^{\bar{\sigma}}(\eta) - I_r^{\sigma}(\eta) = (\sigma - \bar{\sigma}) \int_{a^1}^{a^2} C_{ij}(x)\eta_i(x)\eta_j(x)\, dx$$

if $\omega^{\sigma}(x,\eta,\zeta)$ is defined by (1.7) of Appendix I. One sets $C_{ij}(x) = \delta_i^j$, if $\omega^{\sigma}(x,\eta,\zeta)$ is defined by (1.1) of Appendix I. In both cases the difference (15.16)″ is negative, since $\sigma < \bar{\sigma}$ by hypothesis and η is nonnull.

By (15.16)′, $Q^{\bar{\sigma}}(z) - Q^{\sigma}(z) < 0$, so that (15.15) holds.

Lemma 15.4 follows.

Completion of Proof of Index Theorem 15.2. By virtue of Theorem 3.1 of Appendix III, index Q^{λ} equals the count of characteristic roots less than λ of the "model" family F. In the case at hand the model family is the family of quadratic forms $Q^{\sigma}\#(W_r,\Lambda_p)$ for $\sigma \leq \lambda$. By Theorem 15.1 the characteristic roots $\sigma < \lambda$ of F are identical with the characteristic roots $\sigma < \lambda$ of W_r and have the same multiplicities.

Theorem 15.2(ii) follows from Theorem 3.1 of Appendix III. Theorem 15.2(i) follows from Theorem 15.1.

Definition 15.3. To a "derived" index form $Q^{\sigma}\#(C_r,\Lambda_p^g)$ (Def. 14.2) of a critical extremal g of (J^{σ},C_r) we make formally correspond the "free" index form $Q^{\sigma}\#(W_r,\Lambda_p)$ for which the frame $\Lambda_p = \Lambda_p^g$, and for which the conditions $W_r(\lambda)$ are formally identical with the Conditions (11.8) when $r > 0$, and with the Conditions (12.2) when $r = 0$. This is a *mapping* of the set of derived index forms into the set of free index forms. It is not surjective.

Under Definition 15.3 the derived index form $Q^{\sigma}\#(C_r,\Lambda_p^g)$ of Theorem 14.2 and Corollary 14.1 is made to correspond to a free index form $Q^{\sigma}\#(W_r,\Lambda_p)$. As a consequence one has the numerical identity

$$(15.17) \qquad Q^{\sigma}\#(C_r,\Lambda_p^g) = Q^{\sigma}\#(W_r,\Lambda_p)$$

of the corresponding quadratic forms.

Completion of Proof of Theorem 14.1. The geometric index {nullity} (Definition 14.3) of g as a critical extremal of (J^{λ},C_r) is by definition the index {nullity} of a "derived" index form $Q^\lambda\#(C_r,\Lambda_p^g)$ for g and so the index {nullity} of the "corresponding" free index form $Q^\sigma\#(W_r,\Lambda_p^g)$ of Definition 15.3. The latter index and nullity are given by Index Theorem 15.2 and, so given, confirm Theorem 14.1.

The proof of Theorem 14.1 *is complete.*

An important consequence of Index Theorem 15.2 is that the index and nullity of free index forms $Q^\lambda\#(W_r,\Lambda_p)$ are independent of the choice of λ-frames Λ_p among admissible λ-frames Λ_p.

A convention. Now that Theorem 14.1 is proved we shall drop the qualifying adjectives *algebraic* and *geometric* in referring to the index of a critical extremal g of (J^σ,C_r). When $\sigma = 0$ this index will be called *ordinary*.

Theorems on free functionals I_r^λ. Free functionals I_r^λ will play an increasing role in the rest of this book. In a definition which is analogous to Definition 14.0 of the index and nullity of a quadratic form we shall define the index and nullity of a free functional $I_r^\sigma\#W_r$ and prove some first theorems on such functionals.

Definition 15.4. *The index and nullity of $I_r^\sigma\#W_r$.* By the *index* of $I_r^\sigma\#W_r$ is meant the maximum* of the integers k such that $I_r^\sigma\#W_r$ is negative definite on a vector subspace over R of dimension k of $\{W_r\}$ (Definition 15.1). By the *nullity* of $I_r^\sigma\#W_r$ is meant the multiplicity (in general 0) of σ as a characteristic root of the canonical conditions W_r (Definition 15.0).

THEOREM 15.3(i). *The nullity of a free functional* $\mathrm{I}_\mathrm{r}^\sigma\#\mathrm{W}_\mathrm{r}$ *equals the nullity of the corresponding free index forms* $\mathrm{Q}^\sigma\#(\mathrm{W}_\mathrm{r},\Lambda_\mathrm{p})$ *for each admissible σ-frame* Λ_p.

(ii). *The index of* $\mathrm{I}_\mathrm{r}^\sigma\#\mathrm{W}_\mathrm{r}$ *exists and equals the index of each corresponding index form* $\mathrm{Q}^\sigma\#(\mathrm{W}_\mathrm{r},\Lambda_\mathrm{p})$.

Statement (i) is an immediate consequence of the definition of nullity $I_r^\sigma\#W_r$ and of Theorem 15.1.

Set $\mu = r + mp$. The proof of Theorem 15.3(ii) makes use of two linear mappings both of which depend upon the choice of the frame Λ_p. One maps $\{W_r\}$ into R^μ and the other maps R^μ into $\{W_r\}$ in a special way. Recall that $\{W_r\}$: (Definition 15.1) is the domain of $I_r^\sigma\#W_r$.

The first mapping is denoted by Z. It is a linear mapping

$$\eta \rightarrow Z(\eta) : \{W_r\} \rightarrow R^\mu \tag{15.18}$$

defined as follows. The m-planes $\Pi_1, \ldots, \Pi_p$ have x-coordinates $c_1 < c_2 < \cdots < c_p$ respectively. A mapping $\eta \in \{W_r\}$ goes into a μ-tuple $Z(\eta)$ such that

$$Z(\eta) = [\eta(c_1), \ldots, \eta(c_p)] \qquad \text{(when } r = 0) \tag{15.18$'$}$$

* That this maximum exists as a finite number follows from Lemma 15.5.

and

$$(15.18)'' \qquad Z(\eta) = [u : \eta(c_1), \ldots, \eta(c_p)] \qquad (\text{when } r > 0)$$

where the r-tuple u satisfies (15.0)″ with η.

Lemma 15.5 below is one of two lemmas which together imply Theorem 15.3(ii). Lemma 15.5 is an immediate consequence of the definition (15.3) of the values of $Q^\sigma\#(W_r,\Lambda_p)$.

LEMMA 15.5. *If* $I_r^\sigma\#W_r$ *is negative definite on a vector* k*-subspace* L_k *of* $\{W_r\}$, *then* $Z(L_k)$ *is a* k*-plane* π_k *through the origin in* R^μ *on which* $Q^\sigma\#(W_r,\Lambda_p)$ *is negative definite.*

It follows from Lemma 15.5 that index $I_r^\sigma\#W_r$ exists, and that

$$(15.19) \qquad \text{Index } I_r^\sigma\#W_r \leq \text{Index } Q^\sigma\#(W_r,\Lambda_p)$$

for each admissible σ-frame Λ_p. It remains to show that the sign $<$ can be excluded in (15.19).

For $z \in R^\mu$ let the mapping in $\{W_r\}$ whose graph is the framed BSσ-extremal $h^{\sigma,z}\#(W_r,\Lambda_p)$ be denoted by $H^\sigma(\cdot,z)$. By virtue of Definition 15.2(b) and the linear form of (15.5)″, the mapping

$$(15.20) \qquad z \rightarrow H^\sigma(\cdot,z) : R^\mu \rightarrow \{W_r\}$$

is linear. From this and (15.3) we infer the following.

LEMMA 15.6. *If* π_k *is a* k*-plane meeting the origin in* R^μ *on which the free index form* $Q^\sigma\#(W_r,\Lambda_p)$ *is negative definite, then* $H^\sigma(\cdot,\pi_k)$ *is a vector* k*-subspace of* $\{W_r\}$ *on which* $I_r^\sigma\#W_r$ *is negative definite.*

It follows from Lemma 15.6 that

$$(15.21) \qquad \text{Index } Q^\sigma\#(W_r,\Lambda_p) \leq \text{Index } I_r^\sigma\#W_r.$$

From (15.19) and (15.21) we infer the truth of Theorem 15.3(ii).

Theorem 15.3 has the following corollary.

COROLLARY 15.1. *If the index* k *of a free index form* $Q^\sigma\#(W_r,\Lambda_p)$ *vanishes, then* $I_r^\sigma\#W_r$ *has no negative values.*

If $I_r^\sigma(\eta) < 0$ for some $\eta \in \{W_r\}$ the index k of $Q^\sigma\#(W_r,\Lambda_p)$ would be at least 1, contrary to hypothesis.

Corollary 15.1 implies another corollary of Theorem 15.3.

COROLLARY 15.2. *If* g *is a critical extremal of* (J^σ, C_r) *of index* 0 *and if* $\mathscr{I}_r^\sigma$ *is the quadratic functional of* (J^σ, C_r) *based on* g, *then* $\mathscr{I}_r^\sigma(\eta) \geq 0$ *for each* C_r''*-admissible* η.

Corollary 15.2 follows from Corollary 15.1 if one regards the given nonnegative index form $Q^\sigma\#(C_r,\Lambda_p^g)$ and derived quadratic functional $\mathscr{I}_r^\sigma$ as special choices of a free index form $Q^\sigma\#(W_r,\Lambda_p)$ and free functional $I_r^\sigma\#W_r$. This is permissible. One takes $\omega^\sigma(x,\eta,\zeta)$ as the form $\Omega^\sigma(x,\eta,\zeta)$ based on g and takes W_r as the corresponding system (11.8) or (12.2).

Theorem 15.3 and the Index Theorem 15.2 have the following corollary.

COROLLARY 15.3. (i) *The nullity of a free functional* $I_r^\lambda\#W_r$ *equals the multiplicity (in general zero) of* λ *as a characteristic root of the system* W_r, *confirming Definition* 15.4.

(ii) *The index of* $I_r^\lambda\#W_r$ *is the count of characteristic roots* $\sigma < \lambda$ *of* W_r.

The following definition and lemma are needed in applying Theorem 15.3.

Definition 15.5. By a *proper* BSσ-extremal is meant a BSσ-extremal which has at least one corner.* A vector k-subspace L_k of $\{W_r\}$ is termed a *space of proper* BSσ-extremals if each nonnull mapping in L_k has a graph which is a proper BSσ-extremal.

LEMMA 15.7. *If there exists a vector* k*-subspace* L_k *of* $\{W_r\}$ *of proper* BSσ*-extremals on which* $I_r^\sigma\#W_r$ *is never positive, there then exists a vector* k*-subspace* L_k' *of* $\{W_r\}$ *on which* $I_r^\sigma\#W_r$ *is negative definite.*

Proof. Let $\gamma^1, \ldots, \gamma^k$ be a base for L_k. Let a σ-frame Λ_p over (a^1,a^2) be chosen so that no corner on the graphs of the mappings $\gamma^1, \ldots, \gamma^k$ is on one of the m-planes $\Pi_1, \ldots, \Pi_p$ of Λ_p. The graph of no nonnull mapping in L_k then has a corner on one of the m-planes $\Pi_1, \ldots, \Pi_p$. If Λ_p is so chosen, then under the mapping (15.18), $Z(L_k)$ is a k-plane π_k on which the form $Q^\sigma\#(W_r,\Lambda_p)$ is negative definite. The image $H^\sigma(\cdot,\pi_k)$ of π_k under the mapping (15.20) will be a vector k-subspace of $\{W_r\}$ on which $I_r^\sigma\#W_r$ is negative definite.

Thus Lemma 15.7 is true.

The "first variation" of $I_r^\lambda\#W_r$. Corresponding to a characteristic solution γ of the system W_r with root λ, and a D^1-mapping η in the domain $\{W_r\}$ of I_r^λ, satisfying the conditions (15.0)″ with an r-tuple u when $r > 0$, the derivative,

$$\frac{d}{de} I_r^\lambda(\gamma + e\eta)\bigg|^{e=0}, \tag{15.22}$$

* By a *corner* of the graph of a D^1-map γ is meant a point $\gamma(x)$ at which $\gamma'(x)$ fails to exist.

is called the "*first variation*" *of* I_r^λ *along* γ *induced by* η. When $r = 0$, reference to the r-tuple u is deleted.

As one would expect, the following lemma is true.

LEMMA 15.8. *Under the conditions on the characteristic solution* γ *and the variation* η, *the "first variation" of* I_r^λ *along* γ, *induced by* η, *vanishes.*

Suppose first that $r > 0$. If γ satisfies the conditions (15.0) with an r-tuple v replacing u, one finds that the derivative (15.22) is twice the value

$$B_r^\lambda(\gamma,\eta) = b_{hk}u_h v_k + \int_{a^1}^{a^2} [\eta_i \omega_{\eta_i}^\lambda(x,\gamma,\gamma') + \eta_i' \omega_{\zeta_i}^\lambda(x,\gamma,\gamma')]\, dx \tag{15.23}$$

of a functional B_r^λ, bilinear in γ, η for γ, η conditioned as above. It should be noted that the r-tuples u and v are uniquely determined by η and γ and are linear in η and γ, respectively. When $r = 0$, $B_r^\lambda(\gamma,\eta)$ is defined by (15.23) on deleting the form $b_{hk}u_h v_k$.

If one simplifies the integral in (15.23) by the usual integration by parts and makes use of the conditions (15.0)″ on η and u when $r > 0$, as well as the conditions (15.0)″ and (15.0)‴ on γ and v, one finds that $B_r^\lambda(\gamma,\eta) = 0$. When $r = 0$, the vanishing of $B_0^\lambda(\gamma,\eta)$ is similarly verified, on making use of the fact that γ and η then vanish when $x = a^1$ and a^2.

Thus Lemma 15.8 is true.

The vanishing of $B_r^\lambda(\gamma,\eta)$ implies that $I_r^\lambda(\gamma) = 0$ for a characteristic solution γ or W_r with root λ, since

$$I_r^\lambda(\gamma) = B_r^\lambda(\gamma,\gamma) = 0. \tag{15.24}$$

For two characteristic solutions γ and χ of W_r with roots λ and σ, respectively,

$$B_r^\lambda(\gamma,\chi) - B_r^\sigma(\gamma,\chi) = (\sigma - \lambda)\int_{a^1}^{a^2} C_{ij}(x)\gamma_i(x)\chi_j(x)\, dx = 0 \tag{15.25}$$

if the Definition (1.7) of $\omega^\lambda(x,\eta,\zeta)$ in Appendix I is preferred. Here $C_{ij}(x) = \delta_i^j$, if the Definition (1.1) of Appendix I is preferred. When $\sigma \neq \lambda$, γ and χ are thus orthogonal in a classical sense. This orthogonality implies that solutions in a finite set of characteristic solutions of W_r, with different roots λ, are linearly independent.

In Chapter 5 we shall replace the underlying quadratic form $\omega^\lambda(x,\eta,\zeta)$ in (η,ζ) by a quadratic form involving the parameter λ in a way which, in general, is non-linear.* New methods will be required.

* When $m = 1$ the classical Sturm-Liouville theory involves the parameter λ both in a linear and a nonlinear way.

Exercises

15.1. Let $Q^0(z)$ be a real valued quadratic form in μ-variables $z_1, \ldots, z_\mu$. Let y be a μ-tuple such that $Q^0(z) \geq Q^0(y)$ for each μ-tuple z. Show that y is a critical point of Q^0 and that $Q^0(y) = 0$.

15.2. Making use of an index form $Q^0\#(W_0,\Lambda_p)$ with suitably chosen frame Λ_p, of Corollary 7.1, of Exercise 15.1 and the relation (15.8)′, prove Theorem 7.5.

16. Relative Minima under General End Conditions C_r

An extremal g affording a relative minimum to (J,C_r) is a special kind of "critical" extremal of (J,C_r). Its endpoints A^1 and A^2 satisfy the end conditions C_r. When $r > 0$, g satisfies the C_r-transversality conditions (9.1) in accord with Theorem 9.1, satisfying (9.1) with the set of r parameters $\alpha = \mathbf{0}$. The index k of g as a critical extremal of (J,C_r) is particularly relevant to our study. This index is a specialization of the index k_λ of g as a critical extremal of (J^λ,C_r). When $\lambda = 0$ this index is called the *ordinary* index of g. A first general theorem follows. It concerns a weak minimum, as defined in Section 3 when $r = 0$, and in Section 9 when $r > 0$.

THEOREM 16.1. *If* g *affords a weak minimum to* (J,C$_r$), *then the "ordinary" index* k *of* g, *as defined above, is zero.*

Notation. We refer to the framed family Γ^λ of broken extremals characterized in Theorem 13.2. This family gives rise to an "index function" G^λ in Definition 14.1 from which an "index form" Q^λ, based on g, is derived in (14.7). G^λ and Q^λ have been more explicitly denoted by $G^\lambda\#(C_r,\Lambda_p^g)$ and $Q^\lambda\#(C_r,\Lambda_p^g)$. We here specialize Γ^λ, G^λ and Q^λ by setting $\lambda = 0$, thereby obtaining Γ^0, G^0, and Q^0. We shall refer to the framed broken extremal $\Gamma^{\lambda,v}$ of the family Γ^λ and to its specialization $\Gamma^{0,v}$ when $\lambda = 0$.

Proof of Theorem 16.1. Suppose that $k > 0$, contrary to Theorem 16.1. Since k is the index of a quadratic form $Q^0\#(C_r,\Lambda_p^g)$ there exists a μ-tuple z, with $\mu = r + mp$ such that $Q^0(z) < 0$. Let $G^0\#(C_r,\Lambda_p^g)$ be the index function from which the form Q^0 is derived. If e is a scalar, the relations

$$(16.1)\qquad \frac{d}{de}\, G^0(ez)\bigg|^{e=0} = 0, \qquad \frac{d^2}{de^2}\, G^0(ez)\bigg|^{e=0} = Q^0(z) < 0 \qquad \text{(by (14.7))}$$

then hold when $Q^0(z) < 0$, in accord with Lemma 14.1. Fix z with $Q^0(z) < 0$. If d is a sufficiently small positive constant, then

$$(16.2)\qquad G^0(ez) < G^0(\mathbf{0}) \qquad (\text{for } 0 < |e| < d)$$

as a consequence of the relations (16.1). By virtue of the definition of G^0 in (14.5)

$$G^0(ez) = J(\Gamma^{0,ez}, e\alpha), \qquad G^0(\mathbf{0}) = J(g,\mathbf{0}) \qquad (\text{when } r > 0)$$

where α is the initial r-tuple of z, and

$$G^0(ez) = \mathbf{J}(\Gamma^{0,ez}), \qquad G^0(\mathbf{0}) = \mathbf{J}(g) \qquad (\text{when } r = 0).$$

Hence the inequality (16.2) is contrary to the hypothesis that $\mathbf{J}(g)$ and $J(g,\mathbf{0})$ afford weak minima to (J,C_r).

We infer the truth of Theorem 16.1.

Theorem 16.1 has the following corollary.

COROLLARY 16.1. *Under the conditions of Theorem* 16.1 *the following is true.*

(i) *If* $\mathscr{I}_r$ *is the quadratic functional of* (J,C_r) *based on* g, *then* $\mathscr{I}_r(\eta) \geq 0$ *for each* C_r''*-admissible mapping* η.

(ii) *If* $\mathscr{I}_r(\bar{\eta}) = 0$ *for a nonnull* C_r''*-admissible mapping* $\bar{\eta}$, *then* $\bar{\eta}$ *is a solution, with characteristic root* $\lambda = 0$, *of conditions* (11.8) *when* $r > 0$, *and of conditions* (12.2) *when* $r = 0$.

Proof of (i). By virtue of Theorem 16.1, index $g = 0$ and (i) follows from Corollary 15.2.

Proof of (ii). By (i), $\mathscr{I}_r(\eta) \geq 0$ for all C_r''-admissible η. If then $\mathscr{I}_r(\bar{\eta}) = 0$, $\bar{\eta}$ affords a minimum, relative to C_r''-admissible η, and accordingly relative to C_0''-admissible η. Hence $\bar{\eta}$ must satisfy the Euler equations for the integral with preintegrand Ω. When $r = 0$, (12.2)$'$ and (12.2)$''$ thus hold for $\lambda = 0$. When $r > 0$, (11.8)$'$ holds with $\lambda = 0$, and as affirmed in Corollary 11.1 in the case $r > 0$, η must also satisfy the C_r''-transversality conditions (11.8)$'''$ with the r-tuple u with which $\bar{\eta}$ satisfies the conditions (11.8)$''$. Thus (ii) is true.

Definition 16.1. *Proper, strong, relative minima of* (J,C_r). In case $r > 0$, C_r-admissible pairs (χ,α) are understood in the sense of Definition 8.2. Values $J(\chi,\alpha)$ of (J,C_r) are defined in (8.19). The pair $(\chi,\alpha) = (g,\mathbf{0})$ is C_r-admissible. We say that $J(g,\mathbf{0})$ affords a *strong, relative minimum* to (J,C_r) when $r > 0$, if

$$J(\chi,\alpha) \geq J(g,\mathbf{0}) \tag{16.3}$$

for each C_r-admissible pair (χ,α) for which the graph of χ is in some open neighborhood N_g of g in X. The minimum is termed *proper* if the equality holds in (16.3) only when $(\chi,\alpha) = (g,\mathbf{0})$. When $r = 0$, the sense in which g affords a strong relative minimum to $\mathbf{J}$ is defined in Section 3; $\mathbf{J}$ is afforded a *proper* minimum by g if $\mathbf{J}(\gamma) > \mathbf{J}(g)$ for all C_0-admissible curves γ different from g.

A Special Neighborhood V_e. Let v be a μ-tuple of form (13.12). By an open neighborhood V_e in R^μ of the μ-tuple $v = \mathbf{0}$ is meant the set of all μ-tuples (13.12) such that

$$\|\alpha\| < e, \|v^1\| < e, \ldots, \|v^p\| < e. \tag{16.4}$$

A major theorem follows.

THEOREM 16.2r, r > 0. *Let* g *be a critical extremal of* (J,C$_r$) *with* x-*domain* [a^1,a^2] *and endpoints* A^1, A^2, *satisfying Conditions* C$_r$: (8.11) *with an* r-*tuple* $\alpha = \mathbf{0}$. *Sufficient conditions that* J(g,**0**) *afford a proper, strong minimum to* (J,C$_r$), *relative to* C$_r$-*admissible curves* χ *in some open neighborhood* N$_g$ *of* g *in* X, *are that the Weierstrass and Legendre* S-*conditions hold, relative to* f *and* g, *and that the "ordinary" index and nullity of* g *be zero.*

Since the ordinary index and nullity of g vanish, the index form $Q^0\#(C_r,\Lambda_p^g)$ of an index function $G^0\#(C_r,\Lambda_p^g)$ is positive definite. Hence, if e is a sufficiently small positive constant and if v is a nonnull μ-tuple such that $v \in V_e$ then $G^0(v) > G^0(\mathbf{0})$, or equivalently, by definition of G^0 in (14.5)

$$J(\Gamma^{0,v},\alpha) > J(g,\mathbf{0}) \qquad (v \neq \mathbf{0}, v \in V_e) \tag{16.5*}$$

where α is the initial r-tuple of v.

By hypothesis the Weierstrass and Legendre S-conditions hold for g and f. The extremal g is subdivided by the m-planes of the frame Λ_p^g into $p + 1$ successive extremals g^q, $q = 1, \ldots, p + 1$, on no one of which is the initial point conjugate to a point on the extremal. It follows from Theorems 5.1† and 7.3 that there exists an *open neighborhood* N of g in X such that, if $\varepsilon > 0$ is sufficiently small, and if $v \in V_\varepsilon$, then $\Gamma^{0,v}$ exists, is in N and such that each of the $p + 1$ extremal subarcs γ^q of $\Gamma^{0,v}$ affords a proper strong minimum to **J** relative to x-parameterized curves of class D^1 which join the endpoints of γ^q in N. We take $\varepsilon < e$, where e is the constant for which (16.5) holds, and let $\{\Gamma^{0,v}\}_\varepsilon$ denote the class of broken extremals $\Gamma^{0,v}$ for which $v \in V_\varepsilon$.

The choice of N_g. Let $\hat{\chi}$ be a C_r-admissible curve in X. By the Λ_p^g-*skeleton* of $\hat{\chi}$ is meant the sequence $P_0, P_1, \ldots, P_p, P_{p+1}$ of points on $\hat{\chi}$ of which P_0 and P_{p+1} are the endpoints of $\hat{\chi}$, while the points $P_1, \ldots, P_p$ are the intersections of $\hat{\chi}$ with the respective m-planes $\Pi_1, \ldots, \Pi_p$ of the frame Λ_p^g. Let $N_g \subset N$ be so small an open neighborhood of g that if $\hat{\chi}$ is a C_r-admissible curve in N_g, then the Λ_p^g-skeleton of $\hat{\chi}$ is the Λ_p^g-skeleton of some broken extremal in $\{\Gamma^{0,v}\}_\varepsilon$.

* We are denoting the mapping χ whose graph is $\Gamma^{0,v}$, by $\Gamma^{0,v}$.

† And the Supplement to Theorem 5.1.

Let χ be an arbitrary C_r-admissible curve in N_g. Given χ, there exists a unique broken extremal in $\{\Gamma^{0,v}\}_\varepsilon$ (to be denoted by $\Gamma_\chi^{0,v}$) such that χ and $\Gamma_\chi^{0,v}$ have a common Λ_p^g-skeleton. It follows from the minimizing properties in N of the $p + 1$ extremal components of $\Gamma_\chi^{0,v}$ that

$$\text{(16.6)*} \qquad \mathbf{J}(\chi) \geq \mathbf{J}(\Gamma_\chi^{0,v}).$$

The equality holds in (16.6) if and only if $\chi = \Gamma_\chi^{0,v}$. Since the endpoints of χ and $\Gamma_\chi^{0,v}$ are the same and satisfy conditions C_r with the same r-tuple α, it follows from (16.5) and (16.6) that

$$\text{(16.7)} \qquad J(\chi,\alpha) \geq J(\Gamma_\chi^{0,v},\alpha) \geq J(g,\mathbf{0}),$$

where the equality holds if and only if $(\chi,\alpha) = (\Gamma_\chi^{0,v},\alpha) = (g,\mathbf{0})$.

This establishes Theorem 16.2r when $r > 0$.

When $r = 0$ we have the analogous theorem:

THEOREM 16.2r, $r = 0$. *Let* g *be a critical extremal of* $(\mathrm{J},\mathrm{C}_0)$ *with endpoints* A^1 *and* A^2. *Sufficient conditions that* $\mathbf{J}(\mathrm{g})$ *afford a proper, strong minimum to* $\mathbf{J} = (\mathrm{J},\mathrm{C}_0)$ *relative to* C_0*-admissible curves in some neighborhood* N_g *of* g *in* X *are that the Weierstrass and Legendre* S*-condition hold relative to* g *and that the ordinary index and nullity of* g *be zero.*

An obvious modification of the proof when $r > 0$ establishes the theorem when $r = 0$.

Positive definite functionals $I_r^\lambda \# W_r$. Theorems on the positive definiteness of the free functionals $I_r^\lambda \# W_r$ will be formulated and proved.

THEOREM 16.3. *The functional* $\mathrm{I}_\mathrm{r}^\lambda \# \mathrm{W}_\mathrm{r}$ *is positive definite, provided* $-\lambda$ *is sufficiently large.*

The proof of Theorem 11.2 will serve to establish Theorem 16.3 when $r > 0$. When $r = 0$ one deletes $q(\eta)$ from $\mathscr{I}_r^\lambda$ and from the proof, and reasons as in the case $r > 0$. In both cases one replaces Ω and $\mathscr{I}_r^\lambda$, respectively, by ω and $I_r^\lambda \# W_r$.

A final theorem is a special consequence of Corollary 15.3. A brack indicates an alternative.

THEOREM 16.4. *A quadratic functional* $\mathrm{I}_\mathrm{r}^\lambda \# \mathrm{W}_\mathrm{r}$ *for fixed* λ *is positive definite* {*nonnegative*}, *if and only if each characteristic root* σ *of* W_r *is such that* $\sigma > \lambda\{\sigma \geq \lambda\}$.

* Use definition (1.4) of $\mathbf{J}$ with the limits a^1 and a^2 appropriately chosen.

The following lemma is needed in our study of focal points in Chapter 5.

LEMMA 16.1. *Let* $\|\beta_{hk}\|$ *be an arbitrary* m-*square symmetric matrix of real constants. If* $b \in (a^1,a^2]$ *and if* $b - a^1$ *is sufficiently small, then for* h, k *on the range* 1, . . . , m

$$\beta_{hk}\eta_h(a^1)\eta_k(a^1) + \int_{a^1}^{b} 2\omega(x,\eta(x),\eta'(x))\,dx > 0 \tag{16.8}$$

for each nonnull mapping $x \to \eta(x) : [a^1,b] \to R^m$ *of class* D^1 *such that* $\eta(b) = \mathbf{0}$.

To establish Lemma 16.1 we require an extension of notation entailed by a replacement of the interval $[a^1,a^2]$ by an interval $[a^1,b]$. The following definition suffices.

Definition 16.2. *Notation on replacing* a^2 *by* b. The definition in Section 15 of the entities

$$W_r, \Lambda_p, I_r^\sigma \# W_r, h^{\sigma,z}\#(W_r, \Lambda_p), Q^\sigma\#(W_r, \Lambda_p) \tag{16.9}$$

depended upon the choice of the domain $[a^1,a^2]$ for x. These entities will be supposed redefined in the same way with the interval $[a^1,a^2]$ replaced by an interval $[a^1,b]$ such that $a^1 < b$ and the modified entities then denoted, respectively, by

$$W_{r,b}, \Lambda_{p,b}, I_{r,b}^\sigma \# W_{r,b}, h_b^{\sigma,z}\#(W_{r,b}, \Lambda_{p,b}), Q_b^\sigma\#(W_{r,b}, \Lambda_{p,b}) \tag{16.10}$$

We note that Theorem 16.4 is true if W_r is replaced by $W_{r,b}$ and $I_r^\sigma \# W$ by $I_{r,b}^\sigma \# W_{r,b}$.

In our proof of Lemma 16.1 we shall make use of conditions $W_{m,b}$ specialized so as to have the following form:

$$\left[\begin{array}{ll} \dfrac{d}{dx}\,\omega_{\zeta_i}^\lambda(x,\eta,\eta') - \omega_{\eta i}^\lambda(x,\eta,\eta') = 0 & (i = 1, \ldots, m) \\ \eta_i(a^1) = u_i, \eta_i(b) = 0 \quad (i = 1, \ldots, m) & \\ \xi_i(a^1) = \beta_{ik}u_k \quad (i = 1, \ldots, m) & \end{array}\right] \tag{16.11}$$

where the constants β_{ik} are those of (16.8). These conditions bear on λ and a D^1-mapping η of $[a^1,b]$ into R^m. In the above conditions $W_{m,b}$, the $2m \times m$ matrix $\|c_{ih}^s\|$, which appears in the model (15.0) of conditions W_r, is represented by an m-square submatrix $\|c_{ih}^1\| = \|\delta_i^h\|$ of rank m and an m-square submatrix $\|c_{ih}^2\|$ of null elements. Under conditions $W_{m,b}$, $\eta(a^1)$ is arbitrary, since the m-tuple u is arbitrary. Theorem 16.4 will imply Lemma 16.1 once we have proved (A).

(A) *If* $b \in (a^1,a^2]$ *and* $b - a^1$ *is sufficiently small, there is no characteristic solution of the system* $W_{m,b}$ *with root* $\lambda \leq 0$.

Proof of (A). From Theorem 16.3 we infer the following. If σ is a *negative* number with a sufficiently large absolute value, $I_m^\sigma \# W_m$ is positive definite. An examination of this functional shows that for the same fixed σ, $I_{m,b}^\sigma \# W_{m,b}$ is again positive definite for *each* $b \in (a^1,a^2]$, since conditions $W_{m,b}$ include the condition $\eta(b) = \mathbf{0}$. It follows from Theorem 16.4 that there exist no characteristic solutions of $W_{m,b}$ with root $\lambda \leq \sigma$. This σ appears in (B) below.

To complete the proof of (A) it suffices to prove the following.

(B) *If* $b - a^1$ *is positive and sufficiently small, there is no characteristic solution of the system* $W_{m,b}$ *with a root* $\lambda \in [\sigma,0]$.

Proof of (B). Let $(16.11)^\lambda$ denote the conditions (16.11) for a fixed prescribed λ with the conditions $\eta(b)$ deleted. With each solution η of conditions $(16.11)^\lambda$ we adjoin the "canonical associate" ξ of η and, by abuse of language, refer to (η,ξ) as a solution of $(16.11)^\lambda$. In this sense each solution (η,ξ) of the conditions $(16.11)^\lambda$ is linearly dependent on the columns of a matrix

$$\begin{Vmatrix} \eta_{ij}(x,\lambda) \\ \xi_{ij}(x,\lambda) \end{Vmatrix} \qquad \text{(range } i, j = 1, \ldots, m) \tag{16.12}$$

of $2m$ rows and m columns in which the j-th column is a solution of conditions $(16.11)^\lambda$ satisfying the initial conditions,

$$\eta_{ij}(a^1,\lambda) = \delta_i^j, \qquad \xi_{ij}(a^1,\lambda) = \beta_{ij} \qquad (i = 1, \ldots, m). \tag{16.13}$$

The elements in the matrix (16.12) vary continuously with $(x,\lambda) \in [a^1,a^2] \times R$, as the analysis of Section 4 of Appendix I shows. Hence the m-square determinant $|\eta_{ij}(x,\lambda)|$ varies continuously with $(x,\lambda) \in [a^1,a^2] \times R$.

Moreover the determinant $|\eta_{ij}(a^1,\lambda)| \equiv 1$ for $\lambda \in R$. Continuity considerations show that, for each constant b such that $0 < b - a^1 < e$ with e sufficiently small, and for σ fixed as above,

$$\det |\eta_{ij}(x,\lambda)| > 0, \qquad ((x,\lambda) \in [a^1,b] \times [\sigma,0]). \tag{16.14}$$

When $0 < b - a^1 < e$, (16.14) holds. Hence (B) holds for b so conditioned.

By virtue of the choice of σ preceding (B), (A) holds when (B) holds. Thus (A) is true when $0 < b - a^1 < e$.

For $0 < b - a^1 < e$, Theorem 16.4* now implies that $I_{m,b}^0 \# W_{m,b}$ is positive definite. For each mapping η admitted in (16.8), the left member of (16.8) gives the value at η of $I_{m,b}^0 \# W_{m,b}$. Cf. (15.2)″.

Lemma 16.1 *follows.*

* With $[a^1,a^2]$ replaced by $[a^1,b]$ and $\lambda = 0$.

The norm $\|\eta\|$. When λ enters into the definition of $\omega^\lambda(x,\eta,\zeta)$, as in (1.1) of Appendix I, we shall assign a norm $\|\eta\|$ to each mapping $\eta \in \{W_r\}$, the domain of $I_r^\lambda \# W_r$, by setting

$$\|\eta\|^2 = \int_{a^1}^{a^2} \eta_i(x)\eta_i(x)\, dx. \tag{16.15}$$

A simple consequence of Theorem 16.4 follows. Corollary 16.2 is well known.

COROLLARY 16.2. *The value of the smallest characteristic root of a system* W_r *is the minimum of the quadratic functional* $I_r^0 \# W_r$ *when evaluated on mappings* $\eta \in \{W_r\}$ *with* norm 1.

Corollary 16.2 is subject, a priori, to the condition that a smallest characteristic root of W_r exists. That W_r has an unbounded sequence $\lambda_1 < \lambda_2 < \cdots$, of characteristic roots will be established in Chapter 10 under conditions much more general than those here imposed on the system W_r. Under these more general conditions the analogue of Theorem 16.4 still holds, but Corollary 16.2 in general is not true.

Proof of Corollary 16.2. It follows from Theorem 16.4 that if σ is the smallest characteristic root of W_r, then $I_r^\sigma(\eta) \geq 0$ for each D^1-mapping η of $[a^1,a^2]$ into R^m in the domain $\{W_r\}$ common to the quadratic functionals $I_r^\lambda \# W_r$ for $\lambda \in R$. By Definition 15.1 of $I_r^\lambda \# W_r$

$$I_r^\sigma(\eta) = I_r^0(\eta) - \sigma \int_{a^1}^{a^2} \eta_i(x)\eta_i(x)\, dx \qquad (\eta \in \{W_r\}) \tag{16.15$'$}$$

so that if σ is the smallest characteristic root of the system W_r and if $\eta \in \{W_r\}$ has the norm 1, then $I_r^o(\eta) \geq \sigma$. If however $\bar{\eta}$ is a characteristic solution with root σ, then $I_r^\sigma(\bar{\eta}) = 0$, as the proofs of Lemma 11.1 and 12.1 show. If $\|\bar{\eta}\| = 1$, we infer from (16.15)$'$ that $I_r^o(\bar{\eta}) = \sigma$, thereby completing the proof of Corollary 16.2.

Example 16.1. In this example we consider the functional (J,C_r) of (8.19) with a null external function Θ. We suppose that

$$f(x,y,p) = (1 + p_1^2 + \cdots + p_m^2)^{1/2}, \tag{16.16}$$

and that the critical extremal g is the segment of the x-axis on which $a^1 \leq x \leq a^2$. Relative to this extremal the end conditions are the general end conditions C_r of Definition 8.1 with $r > 0$. The end parameters $\alpha_1, \ldots, \alpha_r$ define an r-tuple α. In the end conditions the m-tuple $\alpha = \mathbf{0}$ represents the end points of g. In particular $X^s(\mathbf{0}) = a^s$ for $s = 1, 2$.

One sees that the preintegrand f is positive regular (Definition 7.1), so that the Weierstrass and Legendre S-conditions of Section 7 are satisfied (Lemma 7.1). From (10.12) we find that

$$b_{hk} = X^2_{hk}(\mathbf{0}) - X^1_{hk}(\mathbf{0}) \qquad (h, k = 1, \ldots, r). \tag{16.17}$$

We note also that corresponding to f and the above extremal g

$$\int_{a^1}^{a^2} 2\Omega(x,\eta,\eta')\, dx = \int_{a^1}^{a^2} \|\eta'(x)\|^2\, dx \tag{16.18}$$

so that the accessory functional of (11.7) has values

$$\mathscr{I}^\lambda_r(\eta) = b_{hk}u_h u_k + \int_{a^1}^{a^2} (\|\eta'(x)\|^2 - \lambda\, \|\eta(x)\|^2)\, dx. \tag{16.19}$$

We state a consequence of Theorem 16.2r.

(A) *If the quadratic form with* r*-square* $\|\mathrm{b}_{\mathrm{hk}}\|$ *is positive definite, there exists an open neighborhood* $\mathrm{N_g}$ *of* g *such that in* Ex. 16.1

$$\int_{X^1(\alpha)}^{X^2(\alpha)} \mathrm{f}(\mathrm{x},\gamma(\mathrm{x}),\gamma'(\mathrm{x}))\, \mathrm{dx} > \mathrm{a}^2 - \mathrm{a}^1 \tag{16.20}$$

for each $\mathrm{C_r}$*-admissible mapping* γ *with end point parameters* α, *provided graph* $\gamma \subset \mathrm{N_g}$ *and graph* $\gamma \neq$ *graph* g.

We verify the conditions of Theorem 16.2r in the case $r > 0$. In particular the Weierstrass and Legendre S-conditions are satisfied since f is positive regular. According to Definition 11.4 the ordinary index and nullity of g are zero if there are no characteristic roots $\lambda \leq 0$ of the derived system (11.8) based on g and the primary end conditions C_r. According to Lemma 11.1 a characteristic solution η of conditions (11.8) with root λ and r-tuple u is such that $\mathscr{I}^\lambda_r(\eta) = 0$. It follows from (16.19) that this equality is impossible if $\lambda \leq 0$ and if the form $b_{hk}u_h u_k$ is positive definite. Hence the ordinary index and nullity of g are zero.

Thus (A) is true in accord with Theorem 16.2r, $r > 0$.

Part III

FOCAL CONDITIONS AND STURM-LIKE THEOREMS

CHAPTER 5

The Focal Point Theorem

17. Focal Conditions, Derived and Free

This chapter begins with a study of a critical extremal g of (J,C_r) when the end conditions C_r : (8.17) leave the second endpoint (x^2,y^2) fixed at A^2 and require that the first endpoint (x^1,y^1) rest on a manifold M_r not tangent to g. The conditions that the critical extremal g satisfy the C_r-transversality conditions (9.1) will be made to reflect properties of f, g, and M_r exclusively, by assuming that the *point function* Θ *associated with* C_r *when* $r > 0$ *is identically zero*. Under these conditions an extremal g which is C_r-admissible and satisfies the C_r-transversality conditions (9.1) will be said to be *transverse* to M_r at A^1. Here $0 < r \leq m$.

OBJECTIVE I. *To define the focal points of* $\mathrm{M_r}$ *on* g *and to evaluate the "ordinary" index and nullity of* g *as a critical extremal of* $(\mathrm{J},\mathrm{C_r})$ *in terms of these focal points.*

To formulate this objective more precisely our notation needs to be supplemented.

Definition 17.1. *End Condition* $\mathrm{C_r(M_r)}$. To distinguish the general end conditions C_r : (8.17) from the special end conditions determined as above by the manifold M_r, we shall denote the endpoint conditions C_r in the latter case by $C_r(M_r)$. Consistently with Definition 8.1, conditions $C_r(M_r)$ require the following.

The manifold M_r is given by a regular C^2-mapping,

$$(17.1) \qquad \alpha \to (X^1(\alpha), Y^1(\alpha)) : \mathbf{A}_r \to X \qquad \text{(see Section 1 for } X)$$

into X of an open neighborhood $\mathbf{A}_r$ in R^r of a base point $\alpha = \mathbf{0}$. Under the conditions $C_r(M_r)$, each endpoint $(x^1,y^1) = (X^1(\alpha), Y^1(\alpha))$ for a unique

$\alpha \in \mathbf{A}_r$ and the endpoint $(x^2,y^2) = (X^2(\alpha), Y^2(\alpha)) \equiv A^2$ for $\alpha \in \mathbf{A}_r$. The values of $(J,C_r(M_r))$ are defined by (8.19) with $\Theta(\alpha) \equiv 0$.

The *secondary end conditions* $C''_r(M_r)$. The end conditions $C_r(M_r)$ induce secondary end conditions $C''_r(M_r)$: (11.8)″ of the form

$$\eta_i^1 = c_{ih}^1 u_h, \qquad \eta_i^2 = 0 \qquad (i = 1, \ldots, m) \tag{17.2}$$

where $\|c_{ih}^1\|$ is an $m \times r$ matrix of rank r defined by (10.7) when $s = 1$. The matrix $\|c_{ih}^2\|$ of (10.7) is here the null matrix.

The quadratic functional $\mathscr{I}_r$ based on a critical extremal g of $(J,C_r(M_r))$ will be denoted by $\mathscr{I}_r\#C_r(M_r)$ in accord with Definition 10.3. It has values,

$$\mathscr{I}_r(\eta) = b_{hk}u_h u_k + \int_{a^1}^{a^2} 2\Omega(x,\eta(x),\eta'(x))\,dx, \tag{17.3}$$

for each $C''_r(M_r)$-admissible η and r-tuple u determined by η^1 in (17.2). The r-square matrix $-\|b_{hk}\|$ is symmetric and given by (10.12)″ when $s = 1$. The three conditions $(11.8)|^{\lambda=0}$ that η be a "secondary critical extremal" of $\mathscr{I}_r\#C_r(M_r)$, are the JDE's based on g, the conditions (17.2) and the $C''_r(M_r)$-transversality conditions (11.5) on η, relative to $\mathscr{I}_r\#C_r(M_r)$. The conditions $(11.8)|^{\lambda=0}$ can be decomposed into the condition $\eta(a^2) = \mathbf{0}$ and conditions termed *derived focal conditions based on* g *and* M_r.

Definition 17.2. *Focal conditions based on* g *and* M_r require that η be a C^1-mapping of $[a^1,a^2] \to R^m$ such that

$$\left[\begin{array}{ll} \dfrac{d}{dx}\Omega_{\zeta_i}(x,\eta(x),\eta'(x)) = \Omega_{\eta_i}(x,\eta(x),\eta'(x)) & (i = 1, \ldots, m) \quad (17.4)' \\ \eta_i^1 = c_{ih}^1 u_h \qquad (i = 1, \ldots, m) & \quad (17.4)'' \\ c_{ih}^1 \xi_i^1 = b_{hk} u_k \qquad (h = 1, \ldots, r) & \quad (17.4)''' \end{array}\right]$$

where u is an arbitrary r-tuple and ξ_i^1, $i = 1, \ldots, m$, "canonical coordinates" of η when $x = a^1$. See (11.2). BC, (17.4)″ and (17.4)‴, are called "*derived focal conditions*" because of the derivation of the conditions (17.4)″ from the conditions $C_r(M_r)$, and the derivation of the matrix $\|b_{hk}\|$ in Section 10.

Solutions (η,ξ) *of* (17.4). It will be notationally advantageous to associate each solution η of conditions (17.4) with the mapping $x \to \xi(x)$ of $[a^1,a^2]$ into R^m with canonical values $\xi_i(x) = \Omega_{\zeta_i}(x,\eta(x),\eta'(x))$, $(i = 1, \ldots, m)$ introduced in (11.2) and then term (η,ξ) a *solution* of (17.4). Solutions (η,ξ) of (17.4) have the basic property that their values $(\eta(x),\xi(x))$ vanish identically for $x \in [a^1,a^2]$, if they vanish for any one value of $x \in [a^1,a^2]$.

The focal conditions (17.4) should be sharply distinguished from the conditions defined by the conditions (17.4) *and* the condition $\eta(a^2) = \mathbf{0}$. The conditions (17.4) so supplemented have nonnull solutions (η,ξ), only exceptionally, while the focal conditions (17.4) are satisfied by an m-parameter family of linearly independent solutions. This property of the focal conditions is affirmed in Lemma 17.1.

LEMMA 17.1. *The vector space* H *over* R *of solutions* η *of the* JDE (17.4)′ *which satisfy the focal conditions* (17.4)″ *and* (17.4)‴ *has the dimension* m.

Let V denote the vector space of $2m$-tuples

$$(\eta^1,\xi^1) = (\eta^1_1, \ldots, \eta^1_m; \xi^1_1, \ldots, \xi^1_m)$$

which satisfy conditions (17.4)″ and (17.4)‴ with some r-tuple u. Let $\hat{\eta}$ be the canonical associate of a solution $\eta \in H$. We say that a solution $\eta \in H$ and a $2m$-tuple $(\eta^1,\xi^1) \in V$ are *mated* if $(\eta(a^1),\hat{\eta}(a^1)) = (\eta^1,\xi^1)$. It is clear that there is a bijective linear correspondence between H and V in which $\eta \in H$ corresponds to its mate in V. Hence dim H = dim V. To prove Lemma 17.1 it suffices to define a base B for V which consists of m $2m$-tuples $(\eta^1,\xi^1) \in V$.

A Base for V. Let $\theta^{(1)}, \ldots, \theta^{(m-r)}$ be a base for m-tuples ξ^1 which satisfy conditions (17.4)‴ with a null r-tuple u. Let $\mathbf{0}$ be a null m-tuple. The $2m$-tuples (η^1,ξ^1) of form

$$(\mathbf{0},\theta^{(1)}), \ldots, (\mathbf{0},\theta^{(m-r)}) \tag{17.5'}$$

are linearly independent and satisfy conditions (17.4)″ and (17.4)‴ with a null r-tuple u. They will be taken as m-r pairs (η^1,ξ^1) in the base B of V to be defined.

The Set (17.5)″. Let $\gamma^{(1)}, \ldots, \gamma^{(r)}$ be m-tuples which serve as a base for the m-tuples η^1 which satisfy (17.4)″ with an r-tuple u. Let $u^{(1)}, \ldots, u^{(r)}$ be the unique nonnull r-tuples with which $\gamma^{(1)}, \ldots, \gamma^{(r)}$ respectively satisfy (17.4)″. For $i = 1, \ldots, r$ the conditions (17.4)‴, with $u = u^{(i)}$ therein, are satisfied by an m-tuple $\xi^1 = \delta^{(i)}$ (possibly null). The set of $2m$-tuples

$$(\gamma^{(1)},\delta^{(1)}), \ldots, (\gamma^{(r)},\delta^{(r)}) \tag{17.5''}$$

are independent, since the $\gamma^{(i)}$'s are independent. As pairs (η^1,ξ^1) they satisfy conditions (17.4)″ and (17.4)‴. The set of r pairs (17.5)″ and $m - r$ pairs (17.5)′ form a base for V, as we now prove.

If (η^1,ξ^1) is an arbitrary $2m$-tuple satisfying conditions (17.4)″ and (17.4)‴ with an r-tuple u, there exists a unique linear combination $(\bar{\eta}^1,\bar{\xi}^1)$ of the pairs (17.5)″ such that the first m-components of the $2m$-tuple $(\eta^1,\xi^1) - (\bar{\eta}^1,\bar{\xi}^1)$ are null. This difference is linearly dependent on the pairs (17.5)′.

Thus the pairs (17.5)′ and (17.5)″ together form a base for V. Hence Lemma 17.1 is true.

A *Convention*. We shall give a meaning to Definition 17.2 when $r = 0$ by setting $M_0 = A^1$ and understanding that *focal conditions based* on g and M_0 require that η be a solution of the JDE (17.4)′ satisfying conditions $\eta_1(a^1) = \cdots = \eta_m(a^1) = 0$ in place of conditions (17.4)″ and the conditions (17.4)‴. We shall term a conjugate point on g of A^1 at which $x \mathrel{\hat{=}} c$, a *focal point* of M_0, and assign it a *multiplicity* equal to the number of linearly independent solutions of the JDE which vanish when $x = a^1$ and $x = c$. The point A^1 is not a conjugate point of itself on g, but it is counted as a focal point of M_0 of multiplicity m.

Definition 17.3. *Focal points of* $\mathrm{M_r}$ *on* g. Let η be a nonnull solution of the focal conditions (17.4) such that $\eta(c) = \mathbf{0}$ for some value $c \in (a^1,a^2]$. We then term the point on g at which $x = c$ a *focal point* of M_r on g and assign this focal point a *multiplicity* equal to the number of linearly independent solutions of the focal conditions (17.4) which vanish when $x = c$.

In Chapter 5 we give an evaluation of the ordinary index and nullity of g when $C_r = C_r(M_r)$. The theorem follows. Here $r \geq 0$.

THEOREM 17.1. *On focal points of* $\mathrm{M_r}$. *If* g *is a critical extremal of* $(\mathrm{J,C_r(M_r)})$ *with domain* $[\mathrm{a^1,a^2}]$ *for* x, *and endpoints* $\mathrm{A^1}$ *and* $\mathrm{A^2}$, *then the following is true.*

(i) *The "ordinary"* nullity of* g *is the multiplicity of* $\mathrm{A^2}$ *as a focal point of* $\mathrm{M_r}$.

(ii) *The "ordinary"* index of* g *is the count of focal points of* $\mathrm{M_r}$ *on* g *strictly between the endpoints of* g, *counting each focal point with its multiplicity.*

The proof of Theorem 17.1 will be concluded in Section 19. The following by-product of the proof of Theorem 17.1 covers the case $C_r = C_0$.

THEOREM 17.2. *Conjugate points on* g. *If* g *is a critical extremal of* $(\mathrm{J,C_0})$ *with* x-*domain* $[\mathrm{a^1,a^2}]$ *and endpoints* $\mathrm{A^1}$ *and* $\mathrm{A^2}$ *the following is true.*

(i) *The ordinary nullity of* g *is the multiplicity of* $\mathrm{A^2}$ *as a conjugate point of* $\mathrm{A^1}$.

(ii) *The ordinary index of* g *is the "count" of conjugate points of* $\mathrm{A^1}$ *on* g *strictly between the endpoints of* g, *counting each conjugate point of* $\mathrm{A^1}$ *with its multiplicity.*

One of the aids in the proof of Theorem 17.1 is the concept of "conjugate families" of solutions (η,ξ) of the JDE (17.4)′. See Bolza [1], p. 626, and Hadamard [1], p. 278. Such families were introduced by von Escherich in his study of minima in the problem of Lagrange. See Section 18.

* For meaning of "ordinary" see first paragraph of Section 16.

Our study of focal conditions with the aid of von Escherich families will be both simpler and more relevant to the extended Sturm theory if we replace the "derived" focal conditions (17.4) by free focal conditions.

OBJECTIVE II. *To define free focal conditions and prove the general Focal Point Theorem* 17.3.

Definition 17.4. **Free focal conditions** V_r, $r \geq 0$. For $r > 0$ the derived focal conditions, (17.4)′, (17.4)″, (17.4)‴ will be replaced by conditions termed *free* and of the form:

$$\begin{bmatrix} (17.6)' & V', \dfrac{d}{dx}\omega_{\zeta_i}(x,\eta(x),\eta'(x)) = \omega_{\eta_i}(x,\eta(x),\eta'(x)) & (i = 1, \ldots, m) \\ (17.6)'' & V''_r, \eta^1_i = c^1_{ih}u_h & (i = 1, \ldots, m) \\ (17.6)'''_* & V'''_r, c^1_{ih}\xi^1_i = b_{hk}u_k & (h = 1, \ldots, r) \end{bmatrix}$$

where ω is defined in Appendix I, $\|c^1_{ih}\|$ is a prescribed matrix of rank $r \leq m$ of m rows and r columns, and $\|b_{hk}\|$ is a prescribed symmetric matrix of r-rows and columns. These conditions will be called a set V_r of *free focal conditions of dimension* r. The domain of x is the interval $\aleph$.

Definition 17.4 (continued). *Focal conditions* V_0 *of dimension* $r = 0$. The theory of conjugate points can be included in the theory of focal points. We understand that focal conditions V_0 are the conditions

$$\begin{cases} (17.7)' & \dfrac{d}{dx}\omega_{\zeta_i}(x,\eta,\eta') = \omega_{\eta_i}(x,\eta,\eta') \quad (i = 1, \ldots, m) \\ (17.7)'' & \eta_1(a^1) = \cdots = \eta_m(a^1) = 0 \end{cases}$$

The following lemma extends Lemma 17.1 and has a similar proof when $r > 0$.

LEMMA 17.2. *The vector space over* R *of solutions of any set of free focal conditions* V_r : (17.6) *or* V_0 : (17.7) *has the dimension* m.

When the free focal conditions V_r replace the derived focal conditions (17.4), Definition 17.3 will be reformulated as follows. Here $r \geq 0$.

Definition 17.5. *Focal points of free focal conditions* V_r. Let η be a nonnull solution of the free focal conditions V_r such that $\eta(c) = \mathbf{0}$ for some $c \in \aleph$.

* $\xi_i(x)$ and ξ^1_i are defined as in (11.2) with Ω replaced by the form ω.

We then term $x = c$ a *focal point** of V_r in $\aleph$ and assign this focal point a multiplicity equal to the number of linearly independent solutions of the focal conditions V_r which vanish when $x = c$.

It will be noted that the focal points of V_0 in $\aleph$, other than a^1, are the *conjugate* points of $x = a^1$ in $\aleph$ relative to the DE (17.6)′.

Theorems 17.1 and 17.2 are special consequences of the general Focal Point Theorem 17.3 below. A definition is required.

Definition 17.6. *Systems $\mathscr{W}_r$ based on focal conditions* V_r ($r \geq 0$). Systems W_r of free canonical conditions are defined in Section 15. Corresponding to a set of focal conditions V_r : (17.6) a system $\mathscr{W}_r$ of free canonical conditions *based* on V_r is defined as follows. The DE of $\mathscr{W}_r$ are *accessory* to those of V_r. That is, if the DE of V_r have the form (17.6)′ the DE of $\mathscr{W}_r$ shall have the form

$$(17.8) \qquad \frac{d}{dx}\,\omega_{\zeta_i}(x,\eta,\eta') - \omega_{\eta_i}(x,\eta,\eta') + \lambda\eta_i = 0 \qquad (i = 1, \ldots, m).$$

The BC of $\mathscr{W}_r$ shall have the form of the BC of V_r, *supplemented* by the conditions,

$$(17.8)' \qquad \eta_1(a^2) = \cdots = \eta_m(a^2) = 0.$$

Definition 17.7. *Systems $\mathscr{W}_{r,b}$ based on focal conditions* V_r. Let b be a value in $\aleph$ such that $a^1 < b \leq a^2$. A system $\mathscr{W}_{r,b}$ based on V_r is defined as is a system $\mathscr{W}_r$ based on V_r, except that the conditions (17.8)′ are replaced by conditions

$$(17.8)'' \qquad \eta_1(b) = \cdots = \eta_m(b) = 0.$$

The basic theorem of Chapter 5 follows. Here $r \geq 0$.

THEOREM 17.3. *Focal Point Theorem.*† *For a system $\mathscr{W}_r$ of canonical conditions based on a set of free focal conditions* V_r : (17.6) *or* (17.7), *the following is true.*

(i) *The multiplicity of $\lambda = 0$ as a characteristic root of the system $\mathscr{W}_r$ is the multiplicity of* $x = a^2$ *as a focal point of the focal conditions* V_r.

(ii) *The count‡ of negative characteristic roots λ of $\mathscr{W}_r$ equals the count‡ of focal points of* V_r *in* (a^1,a^2).

The proof of Theorem 17.3 will be completed in Section 19. Theorem 17.1 is a corollary of Theorem 17.3, on taking the free focal conditions V_r of

* Termed *ordinary*. Termed *algebraic* if enumerated as in Definition 21.1.

† First proved in Morse [1], pp. 55–61, and in part in earlier papers.

‡ Each characteristic root and each focal point is counted with its multiplicity.

Theorem 17.3 as the derived focal conditions (17.4) based on g and M_r, with $\Omega(x,\eta,\zeta)$ serving as $\omega(x,\eta,\zeta)$.

When $r = 0$ Theorem 17.3 can be restated as follows.

CONJUGATE POINT THEOREM 17.4(i). *The multiplicity of* $\lambda = 0$ *as a characteristic root of the system** W_0 *is the multiplicity of* $x = a^2$ *as a conjugate point of* $x = a^1$.

(ii). *The count of negative characteristic roots* λ *of the system* W_0 *is the count of conjugate points of* $x = a^1$ *in* (a^1,a^2).

Relative minima in the one variable end point problem. We are assuming that the manifold M_r is regular, meets the extremal g at its initial end point A^1 and is not tangent to g at A^1. We are also assuming that the Legendre S-condition (7.9) is satisfied along g. We state a classical theorem. It is a consequence of preceding theorems of a more general character. Here $r > 0$.

THEOREM 17.5. *If* g *affords a weak minimum to* $(J,C_r(M_r))$, *then the following is true:*

(i) *The manifold* M_r *is "transverse" to* g *at* A^1;

(ii) *There is no point prior to the end point* A^2 *of* g *which is a focal point of* M_r *on* g.

Condition (i) is a consequence of the transversality Theorem 9.1, taking account of the convention in the first paragraph of this section that the point function Θ, associated with C_r when $r > 0$, is taken identically zero.

Our proof of (ii) is not the classical proof. We infer (ii) from Theorems 16.1 and 17.1. Under the hypotheses of Theorem 17.5, Theorem 16.1 implies that the index k of g as a critical extremal of $(J,C_r(M_r))$ has the value $k = 0$. Theorem 17.1(ii) then implies that there are no focal points of M_r on g strictly between the end points of g.

Thus Theorem 17.5 is true.

We state a second classical theorem. It is an immediate consequence of the general Theorem 16.2r and Theorem 17.1 on the focal points of M_r, $r > 0$, on g.

THEOREM 17.6. *In order that the extremal* g *afford a proper strong, relative minimum to* $(J,C_r(M_r))$ *it is sufficient that* M_r *be transverse to* g, *that the Weierstrass and Legendre* S-*conditions hold relative to* f *and* g *and that there be no focal points of* M_r *on* g.

* Equivalently, a system $\mathscr{W}_0$ based on focal conditions V_0 with the same DE as W_0.

18. Conjugate Families (von Escherich)

Let (η, ξ) and $(\bar{\eta}, \bar{\xi})$ be C^1-solutions of the JDE (17.6)′ for $x \in \aleph$. It is a consequence of the JDE in the form (1.5) of Appendix I that

$$\eta_i(x)\bar{\xi}_i(x) - \bar{\eta}_i(x)\xi_i(x) \equiv K, \qquad \text{a constant.} \tag{18.1}$$

If $K = 0$ the solutions (η, ξ) and $(\bar{\eta}, \bar{\xi})$ are called *mutually conjugate* solutions of (17.6)′ by von Escherich.

We shall verify the following lemma.

LEMMA 18.1. *In a system* S *of* k *mutually conjugate linearly independent solutions of the* JDE (17.6)′, k *is at most* m.

Suppose that $k \geq m$. We shall prove that $k = m$. To that end let

$$\mathbf{H}(x) = \left\| \begin{matrix} \eta_{ij}(x) \\ \\ \xi_{ij}(x) \end{matrix} \right\| \tag{18.2}$$

be a matrix of $2m$ rows indexed by i, and m columns indexed by j, in which the j-th column gives the values at x of the j-th solution (η, ξ) of the system S. Then the rank of the matrix $\mathbf{H}(x)$ is m for each $x \in \aleph$; were there a linear relation between the columns of the matrix $\mathbf{H}(x)$ for $x = c$, then a linear relation with the same coefficients would hold between these columns for each $x \in \aleph$.

Let (η, ξ) be a prescribed solution in the set S. We shall show that the $2m$ tuple

$$(\eta_1(x), \ldots, \eta_m(x); \xi_1(x), \ldots, \xi_m(x)) \tag{18.3}$$

is linearly dependent on the columns of the matrix $\mathbf{H}(x)$.

Since the solution (η, ξ) is mutually conjugate to each solution of the system S,

$$\eta_{ij}(a^1)\xi_i(a^1) - \xi_{ij}(a^1)\eta_i(a^1) = 0 \qquad (j = 1, \ldots, m). \tag{18.4}$$

We have here m equations satisfied by $2m$ variables $\eta_1(a^1), \ldots, \eta_m(a^1)$; $\xi_1(a^1), \ldots, \xi_m(a^1)$. Since the rank of the matrix $\mathbf{H}(a^1)$ is m, these variables are linearly dependent on any m linearly independent solutions of the m equations (18.4), in particular on the solutions afforded by the m columns of the matrix $\mathbf{H}(x)$ when $x = a^1$. The linear dependence of the $2m$-tuple (18.3) on the columns of the matrix $\mathbf{H}(x)$ is thereby established when $x = a^1$, and follows when $x \in \aleph$ without altering the coefficients of linear dependence when $x = a^1$.

This establishes Lemma 18.1.

Definition 18.1. *Conjugate bases and conjugate families*. A system of m linearly independent mutually conjugate solutions (η,ξ) of the JDE (17.6)′ will be called a *conjugate base*. The set of all solutions of (17.6)′ linearly dependent on the m solutions of a conjugate base will be called a *von Escherich or conjugate family*.

A determinant $|\eta_{ij}(x)|$ *of a conjugate family*. If the m columns of a matrix $\mathbf{H}(x)$: (18.2) represent m linearly independent solutions (η,ξ) of a von Escherich family, the m-square determinant $|\eta_{ij}(x)|$ will be called a *determinant* of the family. It is seen that any two determinants of a von Escherich family are nonnull constant multiples of one another.

Definition 18.2. *Focal points of a von Escherich family*. If a determinant $|\eta_{ij}(x)|$ of a conjugate family vanishes when $x = c$ then c will be called a *focal point* of the family. By the *multiplicity* of a focal point $x = c$ of the family is meant the number of linearly independent solutions (η,ξ) of the family such that $\eta(c) = \mathbf{0}$.

A determinant $|\eta_{ij}(x)|$ of a conjugate family is said to vanish to the k-th *order* at $x = a$, if for $|x - a| < e$ and e sufficiently small, the determinant

$$|\eta_{ij}(x)| = (x - a)^k E(x) \tag{18.5}$$

where $E(a) \neq 0$ and the mapping $x \to E(x) \in R$ is continuous for $|x - a| < e$. It is conceivable that $|\eta_{ij}(x)|$ might vanish to any order at $x = a$ or might vanish identically. The essential facts are given in the following theorem of Morse. See Morse [1], p. 47.

THEOREM 18.1. *The order of vanishing at* $x = a$ *of a determinant* $|\eta_{ij}(x)|$ *of a von Escherich family equals the multiplicity* s *of* $x = a$ *as a focal point of the family.*

We suppose $s > 0$ and $a = 0$. By hypothesis there are s linearly independent solutions of the family which vanish when $x = 0$. Let us take a new conjugate base with determinant $|y_{ij}(x)|$ in which the first s columns determine these s solutions. If $s < m$, the last $m - s$ columns of $\|y_{ij}(0)\|$ will form a matrix of rank $m - s$, since s is the multiplicity of $x = 0$ as a focal point of the family. For $s \leq m$ one can factor out an x from each of the first s columns of $|y_{ij}(x)|$ and so write $|y_{ij}(x)| = x^s E(x)$ where $E(x)$ is continuous for x near 0. We shall prove that $E(0) \neq 0$.

Proof that $E(0) \neq 0$. We note that $E(0)$ is obtained by differentiating the elements in the first s columns of $|y_{ij}(x)|$ and then evaluating the resultant determinant $|e_{ij}(x)|$ when $x = 0$. One has $E(0) = |e_{ij}(0)|$.

The case $s = m$. In this case each element $y_{ij}(0) = 0$. Hence $|y'_{ij}(0)| \neq 0$. Otherwise, the columns of $|y_{ij}(x)|$ would be linearly dependent. We see that $|y'_{ij}(0)| = E(0)$ so that $E(0) \neq 0$ when $s = m$.

The case $0 < s < m$. We shall show that the supposition that $E(0) = 0$ is incompatible with the Legendre S-condition on g.

To that end let $c_1, \ldots, c_m$ be m constants, as yet undetermined, and define solutions (u,ξ) and $(\bar{u},\bar{\xi})$ of the von Escherich family be setting

$$u_j(x) = \sum_{k=1}^{s} c_k y_{jk}(x) \qquad (j = 1, \ldots, m) \tag{18.6}$$

$$\bar{u}_j(x) = \sum_{k=s+1}^{m} c_k y_{jk}(x) \qquad (j = 1, \ldots, m). \tag{18.7}$$

Since $E(0) = 0$ by hypotheses, there is a nonnull set of constants $c_1, \ldots, c_m$ such that $u'(0) = \bar{u}(0)$. We shall show that $\bar{u}(0) \neq \mathbf{0}$.

Recall that $u(0) = \mathbf{0}$. Were $\bar{u}(0) = \mathbf{0}$, then $u'(0) = \mathbf{0}$, hence $u(x) \equiv \mathbf{0}$ and finally $\bar{u}(x) \neq \mathbf{0}$, since the m-tuple $\mathbf{c}$ is nonnull. Hence $\bar{u}(0) \neq \mathbf{0}$, since the multiplicity of $x = 0$ as a focal point is at most s.

The contradiction. Since $u(0) = \mathbf{0}$ the condition that u and $\bar{u}$ be mutually conjugate implies that for i, j on the range $1, \ldots, m$

$$R_{ij}(0)\bar{u}_i(0)u'_j(0) = 0. \tag{18.8}$$

Since $u'(0) = \bar{u}(0)$, and $\bar{u}(0) \neq \mathbf{0}$, (18.8) is contrary to the Legendre S-condition. Hence $E(0) \neq 0$.

Thus Theorem 18.1 is true.

Let $\mathbf{H}(x)$ of (18.2) be the $2m$ by m *matrix* of values of m linearly independent solutions (η,ξ) of a von Escherich family. A simple family with x-domain $\aleph$ is one in which the matrix $|\eta_{ij}(a^1)|$ is composed of null elements. All solutions (η,ξ) of this family are such that $\eta(a^1) = \mathbf{0}$. We term this von Escherich family *the family with pole* at $x = a^1$.

The following affirms the identity of von Escherich families with families satisfying *free* focal conditions. In the free focal conditions of Lemma 18.2 it is understood that the preintegrand ω is prescribed as in Appendix I and fixed. The dimension $r \geq 0$ can be prescribed with $r \leq m$. A matrix $\|c^1_{ih}\|$ of rank $r > 0$ and the symmetric matrix $\|b_{hk}\|$ can be prescribed.

LEMMA 18.2(i). *The solutions* (η,ξ) *of a set of free focal conditions* V_r *form a von Escherich family with domain the interval* $\aleph$.

(ii). *Conversely each von Escherich family* $\mathscr{F}$ *with domain* $\aleph$ *is the family of solutions of suitably chosen free focal conditions* V_r *at* a^1.

Proof of (i). The solutions of a set V_0 of focal conditions form a von Escherich family with pole at $x = a^1$. Suppose then that a set V_r is given with $r > 0$.

When $r > 0$, solutions (η,ξ) and $(\bar{\eta},\bar{\xi})$ of focal conditions V_r satisfy conditions of form (17.6)″ and (17.6)‴ with r-tuples u and $\bar{u}$, respectively. From the conditions (17.6)″ and (17.6)‴ we find that

$$\bar{\eta}_i^1 \xi_i^1 = b_{hk} u_k \bar{u}_h, \qquad \eta_i^1 \bar{\xi}_i^1 = b_{hk} \bar{u}_k u_h. \tag{18.9}$$

Since $b_{hk} = b_{kh}$ for h, k on the range $1, \ldots, r$, we infer that $\bar{\eta}_i^1 \xi_i^1 = \eta_i^1 \bar{\xi}_i^1$. Thus the solutions of V_r constitute a family $\mathscr{F}$ of von Escherich in accord with (18.1), Lemma 17.2, and Definition 18.1.

Lemma 18.2 (i) follows.

Proof of (ii). Let a von Escherich family $\mathscr{F}$ be given with a $2m \times m$ matrix $\mathbf{H}(x)$: (18.2) representing linearly independent mutually conjugate solutions of (17.6)′. If the rank r of $\|\eta_{ij}(a^1)\|$ is zero, then $\mathscr{F}$ has a pole at $x = a^1$ and is the family of solutions of V_0.

When the rank of $\|\eta_{ij}(a^1)\|$ is $r > 0$ we can suppose that the matrix $\|\eta_{ij}(a^1)\|$ has been so chosen that its last $m - r$ columns are null m-tuples. For i on the range $1, \ldots, m$ and h, k on the range $1, \ldots, r$ set

$$c_{ih}^1 = \eta_{ih}(a^1), \; b_{hk} = \eta_{ih}(a^1)\xi_{ik}(a^1). \tag{18.10}$$

Then rank $\|c_{ih}^1\| = r$. That $b_{hk} = b_{kh}$ is a consequence of the mutual conjugacy of the solutions of the family $\mathscr{F}$ represented by the h and k-th columns of $\mathbf{H}(x)$. Recall that $\mathbf{H}(x)$ has m columns.

Let V_r be free focal conditions of form (17.6) in which the conditions (17.6)″ and (17.6)‴ are determined by the elements c_{ih}^1 and b_{hk} defined in (18.10). To prove that $\mathscr{F}$ is the ensemble of solutions of focal conditions it is sufficient to show that each column of the matrix $\mathbf{H}$ satisfies the conditions V_r. For $\mu = 1, \ldots, r$ the μ-th column of the matrix $\mathbf{H}(a^1)$ satisfies the conditions V_r'' and V_r''' with an r-tuple u given by the μ-th row of the r-square matrix $\|\delta_h^\mu\|$. For $j = r + 1, \ldots, m$, the j-th column of the matrix $\mathbf{H}(a^1)$ satisfies the conditions V_r'' and V_r''' with a null r-tuple u, since $\eta_{ih}(a^1)\xi_{ij}(a^1) = 0$ for $h = 1, \ldots, r$; $j = r + 1, \ldots, m$, by virtue of the mutual conjugacy of the h-th and j-th columns of the matrix $\mathbf{H}(x)$.

Thus Lemma 18.2 is true.

Lemma 18.2(i) is true if the set of free focal conditions V_r are replaced by a set of derived focal conditions (17.4), as the proof of (i) shows. However, Lemma 18.2(ii) is not always true if free focal conditions are replaced by derived focal conditions, as examples will show. In saying this it is understood that derived focal conditions presuppose a given preintegrand f, an external function Θ, and extremal g, all fixed, and permit an arbitrary choice of conditions $C_r(M_r)$ such that when $r > 0$, g is a critical extremal of $(J, C_r(M_r))$.

Lemma 18.2(i) permits us to state a corollary of Theorem 18.1 essential for the proof in Section 19 of the Focal Point Theorem 17.3.

COROLLARY 18.1. *The focal points in* $\aleph$ *of a set of free focal conditions* V_r *are isolated. In particular the conjugate points of* $x = a^1$ *in* $\aleph$ *are isolated and bounded from* $x = a^1$.

It is clear that Corollary 18.1 is also true if a^1 is replaced by an arbitrary point in $\aleph$ as base of V_r.

The following lemma will be used in proving our extension of the Sturm Separation Theorem. Lemma 18.3 concerns a special case of Lemma 18.2(ii).

LEMMA 18.3. *Let* $\mathscr{F}$ *be a von Escherich family with* x-*domain* $\aleph$ *and with* $x = a^1$ *not a focal point of* $\mathscr{F}$. *The mappings* $\eta \in \mathscr{F}$ *are then the solutions of focal conditions of the form* (17.6) *in which* $\|c^1_{ih}\|$ *is the* m-*square matrix* $\|\delta^j_i\|$. *Thus* $\eta^1_i = u_i$ *and* $\xi^1_i = b_{ij}u_j$, $(i, j = 1, \ldots, m)$ *for a suitable choice of an* m-*tuple* u *and elements* b_{ij} *determined by* $\mathscr{F}$.

$\mathscr{F}$ is defined by a conjugate base $\mathbf{H}(x)$ of form (18.2). Since a^1 is not a focal point of $\mathscr{F}$ we can suppose that $\mathbf{H}(x)$ has been chosen so that $\|\eta_{ij}(a^1)\|$ is the m-square matrix $\|\delta^j_i\|$. If then m-square matrices $\|c^1_{ih}\|$ and $\|b_{hk}\|$ are defined as in (18.10), the resultant focal conditions (17.4) define the mappings of $\mathscr{F}$ and are of the form affirmed to exist in Lemma 18.3. That $b_{ij} = b_{ji}$ is shown immediately following (18.10).

Let F be prescribed as a von Escherich family of solutions of the JDE with x domain $\aleph$. Establish the following.

Exercise 18.1. If $x = a \in \aleph$ is not a focal point of F, show that the vector space over R of the JDE generated by F and the von Escherich family F_a with pole a, is the vector space of all solutions of the JDE.

19. Proof of Focal Point Theorem 17.3

The main object of study from Section 15 to Section 19 is to relate various entities which we shall identify by code names in order to indicate briefly the nature of the principal theorems. The code names are as follows.

A.	Char* roots of systems W_r	Definition 15.0
	Functionals I^λ_r	Definition 15.1
	Index forms Q^σ	Definition 15.2
	Focal points of conditions V_r	Definition 17.5
	σ-Conjugate points	Definition 19.1

* Char abbreviates "characteristic."

Each of the theorems which we shall list is concerned with relations between a pair of these five entities. In the case of index forms Q^σ, the geometric index and nullity of the form Q^σ are involved. In the case of I_r^λ the index and nullity of I_r^σ, as introduced in Definition 15.4, are involved. The following list of theorems associates with each theorem the entities in the list A which it relates under the specific conditions of the theorem.

Theorem 15.2	Index forms Q^σ : char roots of W_r
Theorem 15.3	Functionals I_r^σ : index forms Q^σ
Corollary 15.3	Functionals I_r^σ : char roots of W_r
Theorem 17.3	Focal points of V_r : char roots of W_r
Theorem 17.4	Conjugate points : char roots of W_0
Theorem 19.1	Focal points of V_r : index forms Q^σ
Theorem 19.2	σ-Conjugate points : char roots of W_0
Theorem 19.3	σ-Conjugate points : index forms Q^σ
Theorem 19.4	Focal points : conjugate points

When two theorems listed involve a common entry X in the list A the presumption is that there exists a theorem in which entry X is eliminated. For example, Theorems 15.2 and 15.3 imply Corollary 15.3, while Theorems 15.2 and 19.2 imply Theorem 19.3. In obtaining the new theorem with X eliminated it may be necessary to choose properly the parameters such as r and σ.

In this book *index* forms often serve as mediators between other entities such as conjugate points and characteristic roots. However, in the sequel to this book on Variational Topology, the index forms have the greatest topological content. They give geometric meanings to such analytic entities as the count of conjugate points and characteristic roots.

We return to the proof of the Focal Point Theorem 17.3, beginning by proving a theorem which, by virtue of Index Theorem 15.2, is equivalent to Theorem 17.3(ii).

THEOREM 19.1. *If* Λ_p *is a* 0*-frame over* (a^1,a^2), *if* V_r, $r \geq 0$, *is a set of focal conditions* (17.6) *or* (17.7) *and if* $\mathscr{W}_r$ *is a system of canonical conditions based on* V_r (Definition 17.6), *then the index of** $Q^0\#(\mathscr{W}_r,\Lambda_p)$ *equals the count of focal points of* V_r *in* (a^1,a^2).

In proving Theorem 19.1 it will simplify our analysis if the x-coordinates $c_1 < \cdots < c_p$ of the m-planes $\Pi_1, \ldots, \Pi_p$ of Λ_p are chosen so that the

* For the definition of $Q^\sigma\#(W_r,\Lambda_p)$ see Section 15.

successive differences in the sequence $a^1 < c_1 < \cdots < c_p < a^2$ equal a constant d. This is possible if p is sufficiently large. We suppose, moreover, that d is so small a positive constant that no subinterval of $[a^1,a^2]$ of length d contains a point conjugate to its initial point relative to the DE,

$$\frac{d}{dx}\,\omega_{\zeta_i}(x,\eta,\eta') - \omega_{\eta_i}(x,\eta,\eta') = 0 \qquad (i = 1, \ldots, m). \tag{19.1}$$

We begin the proof of Theorem 19.1 by showing that this theorem is true if the interval $[a^1,a^2]$ is replaced by an interval $[a^1,b]$ such that $b - a^1$ is positive and sufficiently small and $Q^0\#(\mathscr{W}_r,\Lambda_p)$ replaced by a corresponding index form $Q^0_b\#(\mathscr{W}_{r,b},\Lambda_{p,b})$ of the type introduced in Definition 16.2. The frame $\Lambda_{p,b}$ is chosen in the following special way.

The Frame $\Lambda_{p,b}$. For b prescribed in (a^1,a^2), let

$$a^1 < c_1(b) < \cdots < c_p(b) < b$$

be a sequence of values which divides the interval $[a^1,b]$ into successive intervals of equal length (necessarily less than d). The frame $\Lambda_{p,b}$ which replaces Λ_p is supposed defined by m-planes,

$$\Pi^b_1, \ldots, \Pi^b_p \tag{19.2}$$

on which x has the successive values $c_1(b) < \cdots < c_p(b)$.

The Index Form Q^0_b. The system $\mathscr{W}_{r,b}$, based on V_r, has been introduced in Definition 17.7. It is defined, as is $\mathscr{W}_r$, on replacing the interval $[a^1,a^2]$ by $[a^1,b]$. Set $\mu = r + mp$. The index form

$$Q^0_b\#(\mathscr{W}_{r,b},\Lambda_{p,b}) \tag{19.3$'$}$$

is defined as is $Q^0\#(\mathscr{W}_r,\Lambda_p)$, on replacing the interval $[a^1,a^2]$ by $[a^1,b]$. Q^0_b has values (cf. Definitions 15.2 and 16.2), defined for $z \in R^\mu$ by setting

$$Q^0_b(z) = I^0_{r,b}(\eta), \qquad (\text{graph } \eta) = h^{0,z}_b\#(\mathscr{W}_{r,b},\Lambda_{p,b}). \tag{19.3$''$}$$

The following lemmas will be needed to prove Theorem 19.1.

LEMMA 19.0. *Let* Q^t *be a real-valued symmetric quadratic form in* μ *variables* $z_1, \ldots, z_\mu$ *with coefficients which vary continuously with* t *for* t *on an interval* $(\tau - e, \tau + e)$. *If the form* Q^τ *has the nullity* ν, *possibly* 0, *and if* $|t - \tau|$ *is sufficiently small, then for fixed* τ

$$(\textit{index } \mathrm{Q}^\tau) \leq \textit{index } \mathrm{Q}^t \leq (\textit{index } \mathrm{Q}^\tau) + \nu. \tag{19.4}$$

The proof will be left to the reader. Cf. Morse and Cairns [1], Section 3.

LEMMA 19.1. *For* $\mathrm{b} \in (\mathrm{a}^1,\mathrm{a}^2]$, Q^0_b *is* a symmetric quadratic form with values*

$$\mathrm{Q}^0_\mathrm{b}(\mathrm{z}) = \mathrm{c}_{\mathrm{ij}}(\mathrm{b})\mathrm{z}_\mathrm{i}\mathrm{z}_\mathrm{j} \qquad (\mathrm{z} \in \mathrm{R}^\mu) \tag{19.5}$$

and coefficients $\mathrm{c}_{\mathrm{ij}}(\mathrm{b})$ *which vary continuously with* $\mathrm{b} \in (\mathrm{a}^1,\mathrm{a}^2]$.

For fixed b, Q^0_b is a symmetric quadratic form in the μ-tuple z, as follows from Lemma 15.1, setting $a^2 = b$ and $\sigma = 0$.

The continuity of the coefficients $c_{ij}(b)$ in (19.5) may be verified as follows. From the mapping (3.21) of Lemma 3.3 of Appendix I we infer that the mapping η, determined by z in (19.3)″, has values $\eta(x)$ which are linear in the μ coordinates z_j with coefficients $A^j(x,b)$ with the following properties. The graph of the mapping

$$x \to A^j(x,b) : [a^1,b] \to R^m \tag{19.5$'$}$$

is a BSE† G_b whose vertices have x-coordinates $a^1, c_1(b), \ldots, c_p(b), b$. The terminal vertex of G_b is on the x-axis, as are all the other $p + 1$ vertices of G_b with at most one exception. Points and slopes on each one of the $p + 1$ secondary extremal arcs of G_b vary continuously with their x-coordinates and with b. With this understood the formula for Q^0_b in (19.3)″ implies the truth of Lemma 19.1.

We shall begin the proof of Theorem 19.1 by verifying the following:

I. *For* b *on any subinterval of* $(\mathrm{a}^1,\mathrm{a}^2]$ *which contains no focal point of* V_r, Q^0_b *is nonsingular with an invariable index.*

II. *If* $\mathrm{b} - \mathrm{a}^1$ *is positive and sufficiently small, index* $\mathrm{Q}^0_\mathrm{b} = 0$.

III. *If there are no focal points of* V_r *in* $(\mathrm{a}^1,\mathrm{a}^2)$, *then Theorem* 19.1 *is true.*

Proof of I. By virtue of Theorem 15.1(i), applied with $\sigma = 0$ and a^2 replaced by b, Q^0_b is singular for $b \in (a^1,a^2]$, if and only if $x = b$ is one of the focal points of V_r in $(a^1,a^2]$. The continuity of the coefficients in (19.5) then implies that index Q^0_b is constant on any subinterval of $(a^1,a^2]$ which contains no focal point of V_r.

Proof of II. The representation of $Q^0_b(z)$ in (19.3)″ implies a representation of $Q^0_b(z)$ by the left member of (16.8) for proper choice of the constants β_{hk} (possibly zero). It follows from Lemma 16.1 that if $b - a^1$ is positive and sufficiently small, then Q^0_b is positive definite.

Thus II is true.

* Strictly, is *representable* as a symmetric quadratic form.

† BSE abbreviates a "broken secondary extremal."

Proof of III. It follows from I and II that Index $Q_b^0 = 0$ for $b \in (a^1,a^2]$, if there are no focal points of V_r in $(a^1,a^2]$. If there are no focal points of V_r in (a^1,a^2) but $x = a^2$ is a focal point, it follows, again from I and II, that index $Q_b^0 = 0$ for $b \in (a^1,a^2)$. Moreover, index $Q_b^0 = 0$, even when $b = a^2$; one infers this from the left inequality in (19.4), on taking b as the parameter t, and τ as a^2.

Thus III is true.

In the proof of Theorem 19.1 there remains the case in which there are focal points of V_r in (a^1,a^2). In this case let $b_1 < b_2 < \cdots < b_n$ be the x-coordinates of these focal points and let $\nu_1, \nu_2, \ldots, \nu_n$ be the corresponding multiplicities. Set

$$k = \nu_1 + \nu_2 + \cdots + \nu_n. \tag{19.6}$$

We shall complete the proof of Theorem 19.1 by proving separately, by different methods, that under the conditions of this paragraph

$$\text{index } Q^0\#(\mathscr{W}_r,\Lambda_p) \leq k \tag{19.7}$$

and

$$\text{index } Q^0\#(\mathscr{W}_r,\Lambda_p) \geq k. \tag{19.8}$$

Methods of Proof. We shall prove (19.7) by inductive reasoning, taking b successively as $b_1, b_2, \ldots, b_n, a^2$. Relation (19.8) will be established by making use of the equality

$$\text{index } I_r^0\#\mathscr{W}_r = \text{index } Q^0\#(\mathscr{W}_r,\Lambda_p) \tag{19.9}$$

implied by Theorem 15.3, and defining a linear subspace L_k^* in the domain $\{\mathscr{W}_r\}$ of $I_r^0\#\mathscr{W}_r$ on which this functional is negative definite.

Proof of (19.7). To begin the inductive proof of (19.7), note that for $a^1 < b < b_1$ index $Q_b^0 = 0$, by virtue of III. Let s have the range $1, \ldots, n$. Proceeding inductively, set $\nu_0 = 0$ and assume that for $a^1 < b < b_s$ and $0 < s < n$,

$$\text{index } Q_b^0 \leq \nu_0 + \nu_1 + \cdots + \nu_{s-1}. \tag{19.10}$$

This relation is valid when $s = 1$. By Theorem 15.1(ii), nullity $Q_{b_s}^0 = \nu_s$. It follows from (19.10) and the left inequality in (19.4), with $t = b$ and $\tau = b_s$ therein, that (19.10) is valid when $b = b_s$. From (19.10), from I and the right inequality in (19.4), we infer that for $a^1 < b < b_{s+1}$

$$\text{index } Q_b^0 \leq \nu_0 + \nu_1 + \cdots + \nu_s, \tag{19.11}$$

and then from the left inequality in (19.4) that (19.11) holds when $b = b_{s+1}$.

By virtue of this induction, index $Q_b^0 \leq k - \nu_n$ when $a^1 < b \leq b_n$. It follows from Lemma 19.0 that index $Q_b^0 \leq k$ when $a^1 < b \leq a^2$.

Thus (19.7) is true.

Proof of (19.8). Lemma 19.2 below will imply (19.8). Given V_r and $\mathscr{W}_r$, Lemma 19.2 is concerned with the functional $I_r^0\#\mathscr{W}_r$ (Cf. Definition 15.1.) The formulation of Lemma 19.2 requires a definition of basic importance.

Definition 19.0. *The* V_r*-nuclear subspace* $\{V_r\}$ *of the domain* $\{\mathscr{W}_r\}$. The nuclear subspace $\{V_r\}$ is defined by giving a base. Its dimension will be the count k : (19.6) of the focal points of V_r in (a^1,a^2).

For each s on the range $1, \ldots, n$ there are, by hypothesis, ν_s linearly independent solutions of focal conditions V_r which vanish when x has the focal value b_s. These ν_s solutions of focal conditions V_r will be indexed by j on the range $1, \ldots, \nu_s$ and denoted by $\gamma^{s,j}$. Let $\eta^{s,j}$ be the mapping of $[a^1,a^2]$ into R^m with null values $\mathbf{0}$ for $b_s \leq x \leq a^2$ and with

$$\eta^{s,j}(x) = \gamma^{s,j}(x) \qquad (a^1 \leq x \leq b_s). \tag{19.12}$$

The mappings $\eta^{s,j}$ are defined for $s = 1, \ldots, n$ and $j = 1, \ldots, \nu_s$. They are k in number, linearly independent, and serve as a base for a vector subspace $\{V_r\}$ of the domain $\{\mathscr{W}_r\}$ of $I_r^0\#\mathscr{W}_r$ termed V_r*-nuclear.*

LEMMA 19.2(i). *The nonnull mappings in the* V_r*-nuclear subspace* $\{V_r\}$ *of the domain of* $I_r^0\#\mathscr{W}_r$ *are* BSE *which are "proper" in the sense of Definition* 15.5.

(ii). *The functional* $I_r^0\#\mathscr{W}_r$ *vanishes on* $\{V_r\}$.

The truth of (i) is immediate. We turn to (ii).

The vanishing of $I_r^0\#\mathscr{W}_r$ *on* $\{V_r\}$. We begin with the case $r > 0$. When $r > 0$, for each $\eta \in \{\mathscr{W}_r\}$

$$I_r^0(\eta) = b_{hk}u_hu_k + \int_{a^1}^{a^2} 2\omega(x,\eta(x),\eta'(x))\,dx \qquad \text{(cf. (15.2))} \tag{19.13}$$

where the constants b_{hk} are taken from V_r''' and the r-tuple u is determined in V_r'' by the m-tuple $\eta(a^1)$. If η is a nonnull mapping in $\{V_r\}$, $\eta'(x)$ fails to exist at most when x is one of the focal values $b_1 < b_2 < \cdots < b_n$. For $i = 1, \ldots, m$, canonical coordinates $\xi_i(x)$ are given by $\omega_{\zeta_i}(x,\eta(x),\eta'(x))$ except at most when x has one of the values $b_1, \ldots, b_n$. At these values of x the left and right limits of each ξ_i exists. If the integral in (19.13) is evaluated on each of the closed subintervals into which $[a^1,a^2]$ is subdivided by the values $b_1, \ldots, b_n$, then, by a suitable integration by parts of the integral in (19.13) one finds that for $\eta \in \{V_r\}$

$$(19.14) \qquad I_r^0(\eta) = \{b_{hk}u_hu_k - \eta_i(a^1)\xi_i(a^1)\} + \sum_{s=1}^{n} \Big[\eta_i(x)\xi_i(x)\Big]_{b_s^+}^{b_s^-} \; *$$

The term in the brace vanishes, because η satisfies conditions V_r'' and V_r'''' with a unique r-tuple u. For s on the range $1, \ldots, n$ let T_s denote the term in (19.14) indexed by s.

The Vanishing of T_s. Given s one can represent the given $\eta \in \{V_r\}$ as a sum $\eta = w + v$ of two mappings in $\{V_r\}$ so chosen that $w(x) \equiv 0$ for $x \geq b_s$ and the graph of v has no corner when $x = b_s$. Then

$$(19.15) \qquad T_s = [v_i(x)\omega_{\zeta_i}(x,w(x),w'(x))]^{x=b_s^-}.$$

Set $b_0 = a^1$. Set the interval $(b_{s-1},b_s) = \beta_s$. The restrictions $v \mid \beta_s$ and $w \mid \beta_s$ can be continued as solutions of V_r over the x-domain $[a^1,a^2]$ and, so continued, are "mutually conjugate" in accord with Lemma 18.2(i). Hence for $x \in (b_{s-1},b_s]$

$$(19.16) \qquad v_i(x)\omega_{\zeta_i}(x,w(x),w'(x)) = w_i(x)\omega_{\zeta_i}(x,v(x),v'(x)).$$

From (19.16) and the vanishing of $w(b_s)$ we infer that $T_s = 0$.

Thus $I_r^0\#\mathscr{W}_r$ vanishes for $\eta \in \{V_r\}$ so that Lemma 19.2 is true when $r > 0$. When $r = 0$ one deletes the terms $b_{hk}u_hu_k$ from (19.13) and the brace from (19.14) and proves (for $\{V_r\}$ defined as above with $r = 0$) that Lemma 19.2 still holds.

This completes the proof of Lemma 19.2.

Proof of (19.8) *Concluded.* From Lemma 19.2 and Lemma 15.7 we infer that there exists a linear k-subspace L_k' of $\{\mathscr{W}_r\}$ on which $I_r^0\#\mathscr{W}_r$ is negative definite. Hence the index of this functional is at least k, and according to (19.9), the same is true of $Q^0\#(\mathscr{W}_r,\Lambda_p)$. Thus (19.8) is true when there are focal points of V_r in (a^1,a^2).

Proof of Theorem 19.1 *Concluded.* As we have seen in III, this theorem is true when there are no focal points of V_r on (a^1,a^2). Relations (19.7) and (19.8) have been established when there are focal points of V_r on (a^1,a^2), so that in this case

$$(19.17) \qquad \text{index } Q^0\#(\mathscr{W}_r,\Lambda_p) = k = \nu_1 + \nu_2 + \cdots + \nu_n.$$

Thus Theorem 19.1 is true in all cases.

* The $-$ and $+$ signify that limits, respectively from the left and right, are to be taken at b_s.

Completion of proof of Focal Point Theorem 17.3. The proof of Theorem 17.3(ii) involves the system $\mathscr{W}_r$ based on V_r (Definition 17.6) and three integers:

N_1. The count of focal points of V_r in (a^1,a^2).

N_2. The index of $Q^0\#(\mathscr{W}_r,\Lambda_p)$.

N_3. The count of negative characteristic roots λ of $\mathscr{W}_r$.

Of these numbers, $N_1 = N_2$, by Theorem 19.1, and $N_2 = N_3$, by Theorem 15.2. Hence $N_1 = N_3$, confirming Focal Point Theorem 17.3(ii).

Theorem 17.3(i) is true by virtue of the definition of the terms involved, in particular Definition 17.6 of the system $\mathscr{W}_r$ based on V_r, and Definition 17.5 of focal points of V_r and their multiplicities.

This completes the proof of Focal Point Theorem 17.3.

Implications of the Focal Point Theorem 17.3. Theorem 17.3 implies Theorem 17.1 and leads to three theorems on conjugate points.

λ-Conjugate points have been defined in Section 12 in terms of the derived DE(12.2)′. We shall replace λ by σ and the DE (12.2)′ by the DE

$$\frac{d}{dx}\,\omega^\sigma_{\zeta_i}(x,\eta,\eta') = \omega^\sigma_{\eta i}(x,\eta,\eta') \qquad (i = 1, \ldots, m) \tag{19.18}$$

of the system W_0 of conditions (15.1) and give a formal definition.

Definition 19.1. *σ-Conjugate points relative to a system* W_0. Points $x = c$ and $x = c'$ in $\aleph$ are termed *σ-conjugate relative* to W_0 if conjugate relative to the DE (19.18).

The above proof of Theorem 19.1, if restricted to the focal conditions V_0 of dimension 0, can be carried over to the case where the DE (19.1) are replaced by the DE (19.18) for some fixed σ. The following theorem is implied.

THEOREM 19.2(i). *For fixed σ the multiplicity of σ as a characteristic root of the system* W_0 *of conditions* (15.1) *is the multiplicity of* $x = a^2$ *as a σ-conjugate point of* $x = a^1$ *relative to* W_0.

(ii) *The count of characteristic roots of* W_0 *less than σ equals the count of σ-conjugate points of* $x = a^1$ *in* (a^1,a^2) *relative to* W_0.

When $\sigma = 0$, Theorem 19.2 implies Theorem 17.2.

Theorems 15.2 and 19.2 combine to give Theorem 19.3, eliminating reference to characteristic roots of W_0.

THEOREM 19.3(i). *The nullity of a free index form* $Q^\sigma\#(W_0,\Lambda_p)$ *equals the multiplicity of* $x = a^2$ *as a σ-conjugate point of* $x = a^1$ *relative to* W_0.

(ii) *The index of* $Q^\sigma\#(W_0,\Lambda_p)$ *equals the count of σ-conjugate points of f*$x = a^1$ *in* (a^1,a^2) *relative to* W_0.

The following theorem is essential in our study of comparison theorems of Sturmian type.

THEOREM 19.4. *The count of focal points of focal conditions* V_r : (17.6) *in* $(a^1,a^2]$ *with* $r > 0$, *is at least as great as the count of conjugate points of* $x = a^1$ *in* $(a^1,a^2]$ *relative to the* DE (17.6)′ *of* V_r.

The following definition is useful in the proof of this theorem.

Definition 19.2. *The superindex* k *of a quadratic form* Q. The maximum k of the integers t such that there exists a t-plane meeting the origin on which $Q \leq 0$, is called the *superindex* k of Q. Equivalently $k = \kappa + \nu$ where κ and ν are, respectively, the index and nullity of Q.

Proof of Theorem 19.4. Let V_0 be the focal conditions with the same DE as the conditions V_r : (17.6). We introduce the systems $\mathscr{W}_r$ and $\mathscr{W}_0$"based" respectively on the focal conditions V_r and V_0 (Definition 17.6).

For a suitably chosen 0-frame Λ_p over (a^1,a^2) the index forms

$$Q^0\#(\mathscr{W}_r,\Lambda_p), \qquad Q^0\#(\mathscr{W}_0,\Lambda_p) \tag{19.19}$$

are well-defined and will be denoted respectively by Q and q. Set $n = mp$ and $\mu = r + n$. The variables of Q are μ-tuples and those of q, n-tuples. Let $u = (u_1, \ldots, u_r)$ and $s = (s_1, \ldots, s_n)$ be respectively the initial r-tuple and terminal n-tuple in a μ-tuple

$$z = (u_1, \ldots, u_r : s_1, \ldots, s_n) \tag{19.20}$$

of Q. One can take the n-tuples s as the variables of q. Set $Q(z) = P(u,s)$.

Let $\mathbf{0}$ be a null r-tuple. The definition in Section 15 of the index forms (19.19) shows that $P(\mathbf{0},s) \equiv q(s)$. From this identity we infer that the superindex K of Q and the superindex k of q are such that $K \geq k$. This relation, now to be interpreted, implies Theorem 19.4.

From Theorems 15.2 and 17.3 we see that K is the count of focal points of V_r in $(a^1,a^2]$. Theorem 19.3, with $\sigma = 0$ therein, shows that k is the count of conjugate points of $x = a^1$ in $(a^1,a^2]$, relative to the DE (17.6)′.

Theorem 19.4 follows from the relation $K \geq k$.

A positive definite functional $I_r^0\#\mathscr{W}_r$. Corresponding to the focal conditions (17.6) we need to obtain conditions that the functional $I_r^0\#\mathscr{W}_r$, $r > 0$, be positive definite or nonnegative. This functional has a domain $\{\mathscr{W}_r\}$ which consists of the D^1-mappings η of $[a^1,a^2]$ into R^m which satisfy the conditions V_r'' : (17.6)″ with an r-tuple u and the conditions $\eta_1(a^2) = \cdots = \eta_m(a^2) = 0$.

For such an η

$$(19.21) \qquad I_r^0(\eta) = b_{hk}u_h u_k + 2\int_{a^1}^{a^2} \omega(x,\eta(x),\eta'(x))\, dx \qquad (r > 0).$$

We state a theorem.

THEOREM 19.5. *The functional* $\mathrm{I}_r^0 \# \mathscr{W}_r$ *with domain* $\{\mathscr{W}_r\}$ *and values* (19.21) *has the following properties when* $r > 0$.

(i) *It is nonnegative if and only if* V_r *has no focal point in* (a^1,a^2).

(ii) *It is positive definite if and only if* V_r *has no focal point in* $(a^1,a^2]$.

(iii) *It vanishes (in case* V_r *has no focal point in* (a^1,a^2)*) if and only if* η *satisfies the focal conditions* V_r *and* $\eta(a^2) = \mathbf{0}$.

Proof of (i) *and* (ii). Statements (i) and (ii) follow from Theorem 16.4, with $\lambda = 0$ therein, and from the Focal Point Theorem 17.3.

Proof of (iii). An index form $Q^0 \# (\mathscr{W}_r, \Lambda_p)$ exists. Set $\mu = r + mp$. For an arbitrary μ-tuple z let the D^1-mapping η of $[a^1,a^2]$ into R^m such that

$$(19.22) \qquad (\text{graph } \eta) = h^{0,z} \# (\mathscr{W}_r, \Lambda_p) \qquad \text{(see Definition 15.2)}$$

be denoted by $\eta(z)$. Recall that

$$(19.23) \qquad Q^0(z) = I_r^0(\eta(z)) \qquad \text{(by (15.3))}.$$

By Theorem 15.1, Q^0 is singular if and only if there exists a characteristic solution $\hat{\eta}$ of $\mathscr{W}_r$ when $\lambda = 0$, or equivalently, by Lemma 15.2, if and only if a μ-tuple y, such that $\hat{\eta} = \eta(y)$, is a critical point of Q^0. When V_r has no focal point in (a^1,a^2), or equivalently, when Q^0 and I_r^0 are nonnegative, a μ-tuple y is a critical point of Q^0, if and only if $Q^0(y) = 0$, or equivalently, if and only if $I_r^0(\eta(y)) = 0$. (Cf. Exericse 15.1.)

Statement (iii) follows.

The case $r = 0$. The functional $I_0^0 \# W_0$ has values

$$(19.24) \qquad I_0^0(\eta) = 2\int_{a^1}^{a^2} \omega(x,\eta(x),\eta'(x))\, dx \qquad \text{(cf. (15.2)')}$$

where η is a D^1-mapping of $[a^1,a^2]$ into R^m such that $\eta(a^1) = \eta(a^2) = \mathbf{0}$. The counterpart of Theorem 19.5 when $r = 0$ is as follows.

THEOREM 19.6. *The functional* $\mathrm{I}_0^0 \# \mathrm{W}_0$ *with values* (19.24) *has the following properties:*

(i) *It is nonnegative if and only if* $x = a^1$ *has no conjugate point in* (a^1,a^2).

(ii) *It is positive definite if and only if* $x = a^1$ *has no conjugate point in* $(a^1,a^2]$.

(iii) *It vanishes (in case* $x = a^1$ *has no conjugate point in* (a^1,a^2)) *if and only if* η *is a solution of the* DE (19.1) *and* $\eta(a^1) = \eta(a^2) = \mathbf{0}$.

The proof of Theorem 19.5 suffices to prove Theorem 19.6 if $r > 0$ is replaced by $r = 0$, (19.21) by (19.24), $\mathscr{W}_r$ by W_0 and focal conditions V_r by the condition that η be a C^1-solution of the DE (19.1) such that $\eta(a^1) = \mathbf{0}$, or equivalently conditions $\tilde{V}_0$.

Theorem 19.5 has a special corollary of interest. It concerns the focal points of the m-plane Π_{a^1} on which $x = a^1$.

Definition 19.3. *Focal points of* Π_{a^1}. The family of solutions η of the DE (17.6)′ such that the canonical associate ξ of η vanishes when $x = a^1$, is a von Escherich family, termed *transverse* to Π_{a^1}, and denoted by F^{a^1}. The focal points of F^{a^1} are said to be *focal* points of Π_{a^1}.

Focal conditions V_m of form (17.6), whose solutions are the solutions of F^{a^1}, can be given the form

$$(19.25)\qquad \begin{cases} \dfrac{d}{dx}\,\omega_{\zeta_i}(x,\eta(x),\eta'(x)) = \omega_{\eta_i}(x,\eta(x),\eta'(x)) \\ \eta_i(a^1) = \delta_i^j u_j \qquad \text{(range } i, j = 1, \ldots, m) \\ \xi_i(a^1) = 0 \qquad (i = 1, \ldots, m) \end{cases}$$

where u is an arbitrary m-tuple.

The corresponding functional (19.21) has values

$$(19.26)\qquad I_m^0(\eta) = \int_{a^1}^{a^2} 2\omega(x,\eta(x),\eta'(x))\,dx$$

and has for domain D the set of D^1-mappings η of $[a^1,a^2]$ into R^m such that $\eta(a^2) = \mathbf{0}$.

From Theorem 19.5 we infer the following.

COROLLARY 19.1. *The functional whose values are given by the integral* (19.26) *and whose domain* D *is the set of* D^1*-mappings* η *of* $[a^1,a^2]$ *into* R^m *which vanish when* $x = a^2$, *has the following properties:*

(i). *The integral* (19.26) *is nonnegative if and only if the* m*-plane* Π_{a^1} *has no focal point in* (a^1,a^2).

(ii). *The integral is positive definite if and only if* Π_{a^1} *has no focal point in* $(a^1,a^2]$.

(iii). *In case* Π_{a^1} *has no focal point in* (a^1,a^2), *the integral* (19.26) *vanishes if and only if* η *is in the family* F^{a^1} *and* $\eta(a^2) = \mathbf{0}$.

Corollary 19.1 is naturally supplemented by the following theorem.

THEOREM 19.7. *The functional with values*

$$\int_{a^1}^{a^2} 2\omega(x,\eta(x),\eta'(x))\,dx \tag{19.27}$$

whose domain D *is the set of* D^1*-mappings* η *of* $[a^1,a^2]$ *into* R^m *such that* $\eta(a^2) = \mathbf{0}$, *has an index, in the sense of Definition* 15.4, *which is the count of focal points of the* m-*plane* Π_{a^1} *in the interval* (a^1,a^2).

Theorem 19.7 follows from Theorem 17.3(ii) and Corollary 15.3 (ii), on setting $r = m$, $\lambda = 0$ and identifying $W_m(\lambda)$, when $\lambda = 0$, with the focal conditions (19.25), adjoined to the conditions $\eta(a^2) = \mathbf{0}$.

CHAPTER 6

Sturm-Like Theorems In E_{m+1}

20. The Separation Theorem in E_{m+1}

The classical Sturm Separation Theorem is concerned with a single second order DE in the (x,y)-plane E_2. The extended theorem is concerned with a selfadjoint system of m second order DE in a space E_{m+1} of coordinates $x, y_1, \ldots, y_m$. Such a system is given by (1.5) in Appendix I, or by (17.6)′, and has the form

$$(20.1) \quad \frac{d}{dx}(R_{ij}(x)\eta_j' + Q_{ij}(x)\eta_j) = (Q_{ji}(x)\eta_j' + P_{ij}(x)\eta_j) \qquad (i = 1, \ldots, m).$$

The coefficients are conditioned as in Appendix I. The domain of x is the open interval $\aleph$.

To prepare for our generalization we shall give the classical Sturm Theorem a new but equivalent form. Let u and $\hat{u}$ be two linearly independent solutions of the Sturmian DE, that is the equation (20.1) when $m = 1$.

A_1. *The Sturm Separation Theorem in* E_2. *If* τ *is a relatively compact subinterval of* $\aleph$, *the number of zeros of* u *in* τ *differs from the number of zeros of* $\hat{u}$ *in* τ *by at most* 1.

Theorem 20.1 below is a precise formulation of our Separation Theorem in E_{m+1}. As an introduction to this theorem we shall state a simplified version of Theorem 20.1 which implies the classical theorem when $m = 1$.

A_m. *A Separation Theorem in* E_{m+1}. *Let* τ *be a relatively compact subinterval of the open interval* $\aleph$ *on which the coefficients in the* DE (20.1) *are defined. If* F *and* $\hat{F}$ *are two von Escherich families of solutions of the* DE (20.1), *the count of focal points of* F *in* τ *differs from the count of focal points of* $\hat{\text{F}}$ *in* τ *by at most* m.

A sharper separation theorem, our principal separation theorem, follows. The integer ρ is such that $0 \leq \rho \leq m$.

THE EXTENDED SEPARATION THEOREM 20.1. *If two von Escherich families* F *and* $\hat{\text{F}}$ *of solutions of the* DE (20.1) *have exactly* ρ *linearly independent solutions in common, then the "count" of focal points of* F *in any relatively compact subinterval* τ *of* $\aleph$ *differs from the corresponding count for* $\hat{\text{F}}$ *by at most* m $-$ ρ. See Morse [2], Theorem 7.

The following lemma on quadratic forms is an aid in proving Theorem 20.1.

LEMMA 20.1. *Let* A(z) *and* B(z) *be two quadratic forms in the variables* z = (z$_1$, . . . , z$_\mu$). *Set*

$$\text{B(z)} - \text{A(z)} = \text{D(z)}. \tag{20.2}$$

If the indices of A(z), B(z), D(z) *and* $-$D(z) *are respectively* α, β, N *and* P, *then*

$$\alpha - \text{P} \leq \beta \leq \alpha + \text{N}. \tag{20.3}$$

The form $A(z)$ will be negative definite* on an α-plane π in R^μ meeting the origin. There will exist a similar $(\mu - P)$-plane π_1 such that $D(z) \leq 0$ for $z \in \pi_1$. Now π and π_1 intersect in a hyperplane π_2, possibly the origin, such that $\dim \pi_2 \geq \alpha - P$. From (20.2) we infer that $B(z)$ is negative definite on π_2, that is negative for $z \in \pi_2 - \mathbf{0}$. Hence $\beta \geq \alpha - P$. On writing $A(z) - B(z) = -D(z)$ one proves similarly that $\alpha \geq \beta - N$.

The relations (20.3) follow.

Proof of Theorem 20.1. It follows from Corollary 18.1 that the focal points of a von Escherich family in $\aleph$ are isolated. Since τ is relatively compact, the counts of focal points of F and $\hat{F}$, respectively, in τ are finite. Let the subinterval $[a^1,a^2]$ of $\aleph$ be chosen so that neither a^1 nor a^2 is a focal point of F or $\hat{F}$, and that (a^1,a^2) contains just those focal points of F and of $\hat{F}$ which are in τ.

Since $x = a^1$ is not a focal point of F, Lemma 18.3 implies that the mappings in F are the solutions of free focal conditions V_m of form (17.6) in which the BC have the form

$$\eta_i(a^1) = u_i; \qquad \xi_i(a^1) = b_{ij}u_j \qquad (i = 1, \ldots, m). \tag{20.4}$$

* A convention is needed. Every quadratic form will be said to be negative definite on a k-plane π for which $k = 0$.

Similarly the mappings in $\hat{F}$ are the solutions of free focal conditions $\hat{V}_m$ of the form (17.6) in which the BC have the form

$$\eta_i(a^1) = u_i; \qquad \xi_i(a^1) = \hat{b}_{ij}u_j \qquad (i = 1, \ldots, m). \tag{20.5}$$

The range of i and j is $1, \ldots, m$ and the matrices $\|b_{ij}\|$ and $\|\hat{b}_{ij}\|$ are determined, respectively, by F and $\hat{F}$. The DE (17.6)′ common to V_m and $\hat{V}_m$ are understood to be the DE (20.1).

A Difference Form (20.7). A necessary and sufficient condition that a solution η of the DE (20.1) be a member of both families F and $\hat{F}$, is that for an m-tuple $u = \eta(a^1)$

$$b_{ij}u_j = \hat{b}_{ij}u_j \qquad (i = 1, \ldots, m). \tag{20.6}$$

By hypothesis the number of linearly independent solutions η common to F and $\hat{F}$ is ρ. Hence the number of linearly independent m-tuples u for which (20.6) holds is ρ. Equivalently the *nullity* of the difference form

$$\Delta(u) = (\hat{b}_{ij} - b_{ij})u_iu_j \tag{20.7}$$

is ρ. Let N and P be, respectively, the indices of the quadratic forms $\Delta(u)$ and $-\Delta(u)$. Then

$$P + N = m - \rho. \tag{20.8}$$

Two Relevant Index Forms. Systems $\mathscr{W}_m$ and $\hat{\mathscr{W}}_m$ of free conditions (15.0) "based" on V_m and $\hat{V}_m$ are uniquely determined in the sense of Definition 17.6. The DE have the form (20.1), while the BC are those of V_m and $\hat{V}_m$, respectively, supplemented by the condition $\eta(a^2) = \mathbf{0}$. Let Λ_p be a 0-frame over (a^1,a^2) relative to the DE (20.1). Set $\mu = m + mp$. In Section 15 the index forms

$$Q^0\#(\mathscr{W}_m,\Lambda_p), \qquad Q^0\#(\hat{\mathscr{W}}_m,\Lambda_p) \tag{20.9}$$

are defined for each μ-tuple z and will be denoted by $A(z)$ and $B(z)$ respectively.

If the m-tuple u is the initial m-tuple of z, the defining conditions (15.3) and (15.2) show that

$$B(z) - A(z) = (\hat{b}_{ij} - b_{ij})u_iu_j = \Delta(u). \tag{20.10}$$

If α and β are the indices of $A(z)$ and $B(z)$ respectively, and P and N the indices of $\Delta(u)$ and $-\Delta(u)$ respectively, as defined above, then (20.3) holds. From (20.3) so interpreted and from (20.8), it follows that

$$|\beta - \alpha| \le P + N = m - \rho. \tag{20.11}$$

According to Theorem 19.1 the indices α and β of the forms in (20.9) are the counts of focal points in (a^1,a^2) of V_m and $\hat{V}_m$, respectively. By virtue of our choice of the interval $[a^1,a^2]$, α and β are, respectively, the counts of focal points of the respective von Escherich families F and $\hat{F}$ in τ.

Separation Theorem 20.1 is implied by (20.11).

Two Corollaries of the Separation Theorem. As was pointed out in Section 17, the conjugate points of $x = a^1$ relative to the DE (17.6)′ are the focal points, other than $x = a^1$, of the von Escherich family F_{a^1} of all solutions of the DE (17.6)′ which vanish when $x = a^1$. The focal points of F_{a^1} include the point $x = a^1$, counted as a focal point of F_{a^1} of multiplicity m. Our Separation Theorem has the following corollary.

COROLLARY 20.1. *Let* $[a,b]$ *be a subinterval of* $\aleph$. *Let* ν_a *and* ν_b *be, respectively, the multiplicities (possibly* 0) *of* b *and* a *as conjugate points of* a *and* b. *Let* κ_a *and* κ_b *be the "count," respectively, of conjugate points of* $x = a$ *and* $x = b$ *on* (a,b). *Then*

$$\nu_a = \nu_b, \qquad \kappa_a = \kappa_b. \tag{20.12}$$

That $\nu_a = \nu_b$ is immediate. Set $\nu = \nu_a = \nu_b$. To show that $\kappa_a = \kappa_b$, let F_a and F_b be, respectively, the von Escherich families of solutions of the DE (20.1) which vanish when $x = a$ and $x = b$. We shall apply the extended Separation Theorem to the families F_a and F_b.

On the interval $[a,b)$ the counts of focal points of F_a and F_b are respectively $m + \kappa_a$ and $\kappa_b + \nu$. By the Separation Theorem

$$|(m + \kappa_a) - (\kappa_b + \nu)| \leq m - \nu. \tag{20.13}$$

It follows that $\kappa_a \leq \kappa_b$. On replacing the interval $[a,b)$ by the interval $(a,b]$ we find similarly that $\kappa_a \geq \kappa_b$.

Thus $\kappa_a = \kappa_b$ and the proof of Corollary 20.1 is complete.

Theorem 19.3 with $\sigma = 0$ therein, and Corollary 20.1, combine to give a second Corollary of the Separation Theorem, interchanging the roles of a^2 and a^1 in Theorem 19.3.

COROLLARY 20.2(i). *The nullity of a free index form* $Q^0\#(W_0,\Lambda_p)$ *equals the multiplicity of* $x = a^1$ *as a conjugate point of* $x = a^2$.

(ii) *The index of* $Q^0\#(W_0,\Lambda_p)$ *equals the count of conjugate points of* $x = a^2$ *in* (a^1,a^2).

More generally one sees that Corollary 20.2 remains valid if Q^σ replaces Q^0 and σ-conjugate points, relative to W_0 (Definition 19.1), replace conjugate points relative to the DE (17.6)′.

Historical note. The extended Separation Theorem 20.1 was stated as Theorem 8.3 on page 104 of Morse [1]. Bliss and Schoenberg have stated a

Separation Theorem on page 788 of [1] which involves "adjacent" conjugate points. This theorem is readily inferred from Theorem 20.1 but the converse is not true since focal points of von Escherich families cannot, in general, be characterized in terms of conjugate points.

21. Comparison of Free Focal Conditions*

In this section we shall extend the first Sturmian comparison theorem, as presented in Ince [1], p. 228, or Bôcher [1], p. 59. In the general case the comparison is between two sets of focal conditions of form (17.6) or (17.7). The focal points are on the x-axis in E_{m+1}. In the classical case $m = 1$.

The Conditions to be Compared. We shall compare two sets of focal conditions V_r and $\hat{V}_\rho$ as defined in Section 17. As in the classical case, there are three cases to be considered:

(21.0)

$$\textit{Case I}: r = \rho > 0; \qquad \textit{Case II}: r = \rho = 0; \qquad \textit{Case III}: r = 0,\ \rho > 0.$$

In Section 22 we shall see how the data in the classical case lead naturally to this division of cases.

When $r > 0$, conditions V_r shall have the form (17.6) and $\hat{V}_\rho$ a similar form, (ρ replacing r)

$$\begin{bmatrix} \hat{V}'_\rho : \dfrac{d}{dx}\,\hat{\omega}_{\zeta_i}(x,\eta,\eta') = \hat{\omega}_{\eta_i}(x,\eta,\eta') \qquad (i = 1, \ldots, m) & (21.1)' \\ \hat{V}''_\rho : \eta^1_i = \hat{c}^1_{ih} u_h \qquad (i = 1, \ldots, m) & (21.1)'' \\ \hat{V}'''_\rho : \hat{c}^1_{ih}\xi^1_i = \hat{b}_{hk} u_k \qquad (h = 1, \ldots, \rho) & (21.1)''' \end{bmatrix}.$$

The representations of V_r and $\hat{V}_\rho$ are conditioned as in (17.6). Conditions $\hat{V}_0$ of dimension 0 consist of the DE of $\hat{V}_\rho$ and the BC

$$\hat{V}''_0 : \eta_1(a^1) = \cdots = \eta_m(a^1) = 0 \qquad \text{(cf. (17.7))}. \tag{21.2}$$

The quadratic form $\hat{\omega}(x,\eta,\zeta)$ upon which the DE of $\hat{V}_\rho$ are based, is conditioned as in Appendix I. The focal conditions V_0 are represented by conditions (17.7).

To give Definition 21.1 below, we shall return to the focal conditions V_r of Section 17. When $m > 1$ a focal point $x = b$ of V_r, following $x = a^1$ in the interval $\aleph$, may have a multiplicity which exceeds 1. The classical Sturm theorem is concerned with focal points of multiplicity 1. In order to

* The reader will find it interesting and profitable to study the treatment of comparison theorems and selfadjoint systems by William T. Reid [3]. Our "comparison matrix" $\|b_{hk}\|$ is not an object of study.

extend the classical Sturm theorem to the case $m > 1$ we shall replace each *ordinary** focal point $x = b$ of multiplicity $\nu > 1$, by a set of ν symbolically distinct points, each with the x-coordinate b, enumerating these ν points as follows.

Definition 21.1. *The algebraic enumeration of focal points of* $\mathrm{V_r}$, *following* $\mathrm{x} = \mathrm{a}^1$ *in* $\aleph$.

Let $x = b > a^1$ be an "ordinary" focal point of V_r in $\aleph$ of multiplicity $\nu > 0$. Let κ be the count of ordinary focal points of V_r in (a^1,b). We understand that the k-th algebraic focal point $x = e_k$ of V_r, following $x = a^1$ in $\aleph$, then exists and is *carried* by $x = b$, provided k is any one of the ν integers $\kappa + 1, \kappa + 2, \ldots, \kappa + \nu$.

The following lemma is an immediate consequence of Definition 21.1.

LEMMA 21.1. *If the "count" of ordinary focal points of* $\mathrm{V_r}$ *following* $\mathrm{x} = \mathrm{a}^1$ *in* $\aleph$ *exceeds a positive integer* μ, *there exists a sequence,*

$$\mathrm{e}_1 \leq \mathrm{e}_2 \leq \cdots \leq \mathrm{e}_\mu, \tag{21.3}$$

of algebraic focal points of $\mathrm{V_r}$ *in* $\aleph$ *such that for* $\mathrm{k} = 1, \ldots, \mu$, $\mathrm{e_k}$ *is the* k-th *algebraic focal point of* $\mathrm{V_r}$ *following* $\mathrm{x} = \mathrm{a}^1$ *in* $\aleph$.

Definition 21.2. *The comparison subinterval* $[\mathrm{a}^1,\mathrm{d}]$. The comparison of the algebraic focal points of V_r and $\hat{V}_\rho$ is restricted to focal points on a *fixed subinterval* $(a^1,d]$ of $\aleph$, where $x = a^1$ is the point at which the BC conditions of the focal conditions are defined, and $d > a^1$ is prescribed in $\aleph$.

The principal hypotheses in our comparison theorems depend on which of the three cases (21.0) is under consideration.

A First Hypothesis in Case I. In Case I, $r = \rho > 0$ and we assume that the conditions V''_r and $\hat{V}''_r$ are *identical* conditions, that is, that

$$c^1_{ih}u_h \equiv \hat{c}^1_{ih}u_h \qquad (i = 1, \ldots, m). \tag{21.4}$$

Apart from this first condition in Case I the hypotheses are in terms of a difference functional which we now define.

Definition 21.3. *The difference functional* $\mathrm{D^b_r}$ *in Case* I. Set

$$\Delta b_{hk} = b_{hk} - \hat{b}_{hk} \qquad (h, k = 1, \ldots, r) \tag{21.5$'$}$$

and for $x \in \aleph$ and arbitrary m-tuples η and ζ set

$$\Delta\omega(x,\eta,\zeta) = \omega(x,\eta,\zeta) - \hat{\omega}(x,\eta,\zeta). \tag{21.5$''$}$$

* Focal points $x = c$, introduced in Definition 17.5, are termed *ordinary*. Such focal points are to be distinguished from the *algebraic* focal points introduced in Definition 21.1 below.

For $a^1 < b \leq d$ let D_r^b have values

$$D_r^b(\eta) = \Delta b_{hk}u_h u_k + 2\int_{a^1}^{b} \Delta\omega(x,\eta(x),\eta'(x))\,dx \qquad (r > 0) \tag{21.6$'$}$$

for η any D^1-mapping of $[a^1,b]$ into R^m which satisfies the identical conditions V_r'' and $\hat{V}_r''$ with the r-tuple u and vanishes when $x = b$.

Definition 21.4. *The difference functional* D_0^b *in Cases* II *and* III. For any constant b such that $a^1 < b \leq d$, let D_0^b have values

$$D_0^b(\eta) = 2\int_{a^1}^{b} \Delta\omega(x,\eta(x),\eta'(x))\,dx \tag{21.6$''$}$$

for η any D^1-mapping of $[a^1,b]$ into R^m which vanishes when $x = a^1$ and $x = b$.

Definition 21.5(a). *Hypothesis* H(I). In Case I, $r = \rho > 0$, conditions V_r'' and $\hat{V}_r''$ are assumed identical and the difference functional D_r^d : (21.6)′ is assumed positive definite*.

Definition 21.5(b). *Hypotheses* H(II) *in Case* II *and* H(III) *in Case* III. In both cases the difference functional D_0^d of Definition 21.4 is assumed positive definite*.

Note that Hypothesis H(I) implies that D_r^b is positive definite for each $b \in (a^1,d]$. A similar remark is valid concerning Hypotheses H(II) and H(III).

Our first comparison theorem concerns Cases I and II.

COMPARISON THEOREM 21.1. *Under Hypothesis* H(I) *in Case* I *and Hypotheses* H(II) *in Case* II *if*

$$\mathrm{e}_1 \leq \mathrm{e}_2 \leq \cdots \leq \mathrm{e}_s \qquad (s > 0) \tag{21.7}$$

are the algebraic focal points in the subinterval $(a^1,d]$ *of* $\aleph$, *of focal conditions* V_r *with* $r > 0$ *in Case* I *and* $r = 0$ *in Case* II, *there will exist a sequence*

$$\hat{\mathrm{e}}_1 \leq \hat{\mathrm{e}}_2 \leq \cdots \leq \hat{\mathrm{e}}_s \tag{21.8}$$

of algebraic focal points of $\hat{V}_r$ *in* $(a^1,d]$, *such that* $\hat{\mathrm{e}}_k < \mathrm{e}_k$ *for* $k = 1, \ldots, s$ *and* $\hat{\mathrm{e}}_k$ *is the* k-th *algebraic focal point of* $\hat{V}_r$ *following* $x = a^1$.

Note. In general there will be algebraic focal points of $\hat{V}_r$ in $(a^1,d]$ following those enumerated in (21.8).

Notation for Proof. For each b such that $a^1 < b \leq d$, systems $\mathscr{W}_{r,b}$ and $\hat{\mathscr{W}}_{r,b}$ over the interval $[a^1,b]$ are introduced, based on the focal conditions

* See Note following the proof of Corollary 21.1.

V_r and $\hat{V}_r$, respectively (Definition 17.7). For each such b, 0-frame $\Lambda_{p,b}$ can be defined over (a^1,b) so as to be admissible, both relative to the DE of $\hat{V}_r$ and of V_r. The x-coordinates of the m-planes $\Pi_1^b, \ldots, \Pi_p^b$ of $\Lambda_{p,b}$ will necessarily depend on b, but can be chosen so that p is independent of $b \in (a^1,d]$. Set $\mu = r + mp$. For each $b \in (a^1,d]$ index forms (Definition 15.2)

$$Q_b^0\#(\mathscr{W}_{r,b},\Lambda_{p,b});\ Q_b^0\#(\hat{\mathscr{W}}_{r,b},\Lambda_{p,b}) \qquad \text{(cf. (19.3)')} \tag{21.9}$$

in μ-tuples z are well-defined and will be denoted, respectively, by

$$Q_{r,b},\ \hat{Q}_{r,b} \qquad (a^1 < b \leq d). \tag{21.10}$$

We shall verify the following lemma.

LEMMA 21.2. *Under the hypotheses of Comparison Theorem* 21.1 *in Cases* I *and* II, $\hat{Q}_{r,b}(z) < Q_{r,b}(z)$, *whenever the μ-tuple* z *is nonnull.*

Proof of Lemma 21.2 *when* $r > 0$. Let η and $\hat{\eta}$ be D^1-mappings of $[a^1,b]$ into R^m such that, in the sense of Definition 15.2, with $[a^1,b]$ replacing $[a^1,a^2]$,

$$\text{graph}\ \eta = h_b^{0,z}\#(\mathscr{W}_{r,b},\Lambda_{p,b}) \tag{21.11}$$

$$\text{graph}\ \hat{\eta} = h_b^{0,z}\#(\hat{\mathscr{W}}_{r,b},\Lambda_{p,b}). \tag{21.12}$$

According to (15.2)″ and (15.3), regarded as models when $[a^1,b] = [a^1,a^2]$,

$$Q_{r,b}(z) = b_{hk}u_hu_k + \int_{a^1}^{b} 2\omega(x,\eta(x),\eta'(x))\,dx \tag{21.13)$'$}$$

$$\hat{Q}_{r,b}(z) = \hat{b}_{hk}u_hu_k + \int_{a^1}^{b} 2\hat{\omega}(x,\hat{\eta}(x),\hat{\eta}'(x))\,dx \tag{21.13)$''$}$$

where the initial r-tuple of the μ-tuple z is the r-tuple u. By virtue of the minimizing* property of the $p + 1$ subarcs into which graph $\hat{\eta}$ is subdivided by the m-planes $\Pi_1^b, \ldots, \Pi_p^b$ of the frame $\Lambda_{p,b}$,

$$\hat{Q}_{r,b}(z) \leq^{\dagger} \hat{b}_{hk}u_hu_k + \int_{a^1}^{b} 2\hat{\omega}(x,\eta(x),\eta'(x))\,dx = P_{r,b}(z), \tag{21.14}$$

introducing $P_{r,b}(z)$.

It follows from the definition of the difference functional D_r^b in (21.6)′, that

$$P_{r,b}(z) = b_{hk}u_hu_k + \int_{a^1}^{b} 2\omega(x,\eta(x),\eta'(x))\,dx - D_r^b(\eta). \tag{21.15}$$

* Minimizing the integral in (21.13)″ in the fixed end point problem, by virtue of Corollary 7.1.

† The equality would hold here if $\eta(x)$ and $\eta'(x)$ were replaced by $\hat{\eta}(x)$ and $\hat{\eta}'(x)$.

For a nonnull μ-tuple z, η is nonnull on $[a^1,b]$. Since $D_r^b(\eta)$ is then positive by hypothesis, we infer that

$$P_{r,b}(z) < b_{hk}u_h u_k + \int_{a^1}^{b} 2\omega(x,\eta(x),\eta'(x))\,dx = Q_{r,b}(z). \tag{21.16}$$

Lemma 21.2 follows from (21.14) and (21.16) when $r > 0$.

Proof of Lemma 21.2 *when* r = 0. When $r = 0$, one deletes the above quadratic forms in the r-tuple u. The functional $D_0^b(\eta)$ of (21.6)″ replaces $D_r^b(\eta)$ in (21.15). So modified, the proof when $r > 0$, is valid when $r = 0$, and establishes Lemma 21.2 when $r = 0$.

The proof of Lemma 21.2 is complete.

The most important consequence of Lemma 21.2, all that we use in proving Theorem 21.1, is that, under the hypotheses of Theorem 21.1,

$$\text{index } \hat{Q}_{r,b} \geq (\text{index} + \text{nullity})Q_{r,b} \tag{21.17}$$

for each focal point $b \in (a^1,d]$ of focal conditions V_r : (17.6) or (17.7). Moreover, the relations (21.17)′ can be established under much hypotheses, termed *nuclear*, as we shall see.

Completion of proof of Theorem 21.1. Let e_n be chosen arbitrarily from the algebraic sequence (21.7) of Theorem 21.1. Set $b = e_n$ and

$$(\text{index} + \text{nullity})Q_{r,b} = t \qquad (r \geq 0). \tag{21.18}$$

The algebraic enumeration of focal points $e_1, \ldots, e_s$ of (21.7) is such that $n >$ index $Q_{r,b}$ and

$$t \geq n, \qquad e_t = e_n. \tag{21.19}$$

By (21.17)

$$\text{index } \hat{Q}_{r,b} \geq (\text{index} + \text{nullity})Q_{r,b} = t. \tag{21.20}$$

According to Focal Point Theorem 17.3 the count k of focal points of $\hat{V}_r$ in (a^1,b) is such that

$$k = \text{index } \hat{Q}_{r,b}. \tag{21.21}$$

The algebraic enumeration of the algebraic focal points of $\hat{V}_r$ is then such that $\hat{e}_k < b$. Since $b = e_n$ and $k \geq t \geq n$ we infer that

$$\hat{e}_n \leq \hat{e}_k < b = e_n. \tag{21.22}$$

This completes the proof of Comparison Theorem 21.1.

Comparison Theorem 21.1 can be extended to Case III in the form of Corollary 21.1 below. In Case III focal conditions V_0 and $\hat{V}_r$, $r > 0$, are compared.

COROLLARY 21.1. *Under Hypothesis* H(III) *in Case* III, *if*

$$e_1 \le e_2 \le \cdots \le e_s \qquad (s > 0) \tag{21.23}$$

are the algebraic focal points in the subinterval $(a^1,d]$ *of* $\aleph$ *of focal conditions* V_0 *at* $x = a^1$, *there then exists a sequence*

$$\hat{e}_1 \le \hat{e}_2 \le \cdots \le \hat{e}_s, \tag{21.24}$$

of algebraic focal points of $\hat{V}_r$ *such that for* $k = 1, \ldots, s$, $\hat{e}_k < e_k$ *and* $\hat{e}_k$ *is the* k*-th algebraic focal point of* $\hat{V}_r$ *following* $x = a^1$ *in* $\aleph$.

Let $\hat{V}_0$ be focal conditions of dimension 0 in which the DE are those of $\hat{V}_r$. The difference functional D_0^b : (21.6)″, associated with V_0 and $\hat{V}_0$, is identical with the difference functionals D_0^b associated with V_0 and $\hat{V}_r$, and so is positive definite by hypothesis. The validity of Comparison Theorem 21.1 in Case II then implies the following.

(A) *Corresponding to the algebraic focal points* (21.23) *of* V_0 *in* $(a^1,d]$ *there exists a sequence*

$$e'_1 \le e'_2 \le \cdots \le e'_s \tag{21.25}$$

of algebraic focal points of $\hat{V}_0$ *such that for* $k = 1, \ldots, s$, $e'_k < e_k$ *and* e'_k *is the* k*-th algebraic focal point of* $\hat{V}_0$ *following* $x = a^1$ *in* $\aleph$.

To complete the proof of Corollary 21.1 use will be made of Theorem 19.4 applied to $\hat{V}_0$ and $\hat{V}_r$, in place of V_0 and V_r of Theorem 19.4. This application is possible since the DE of $\hat{V}_0$ and of $\hat{V}_r$ are the same, by definition of $\hat{V}_0$.
It follows that if b is any value in $(a^1,d]$, then the count of ordinary focal points of $\hat{V}_r$ in $(a^1,b]$ is at least as great as the count of focal points of $\hat{V}_0$ in $(a^1,b]$. The existence of e'_k, as the k-th algebraic focal point of $\hat{V}_0$ following $x = a^1$, then implies the existence of a k-th algebraic focal point $\hat{e}_k$ of $\hat{V}_r$ following $x = a^1$ in $\aleph$ such that $\hat{e}_k \le e'_k$. Hence

$$\hat{e}_k \le e'_k < e_k \qquad (k = 1, \ldots, s). \tag{21.26}$$

This completes the proof of Corollary 21.1.

Conditions for the positive definiteness of D_r^d. Comparison Theorem 21.1 and its Corollary 21.1 will be supplemented by two theorems. In Theorem 21.2 the range of i, j is $1, \ldots, m$ and when $r > 0$ the range of h, k is $1, \ldots, r$. We refer to R_{ij}, Q_{ij}, P_{ij} of Appendix I.

THEOREM 21.2. *The condition of Theorem 21.1 and Corollary 21.1 that the functional D_r^d be positive definite is satisfied if the following is true:*

(a_1) $$Q_{ij}(x) = \hat{Q}_{ij}(x) \qquad (x \in [a^1,d])$$

(a_2) $$(R_{ij}(x) - \hat{R}_{ij}(x))\zeta_i\zeta_j \geq 0 \qquad (x \in [a^1,d],\ \zeta \in R^m)$$

(a_3) *The quadratic form with* m-*square matrix* $\|P_{ij}(x) - \hat{P}_{ij}(x)\|$ *is positive definite for values of* x *everywhere dense in* $[a^1,d]$,

(a_4) $$\Delta b_{hk}u_h u_k \geq 0 \qquad (u \in R^r, \text{ when } r > 0).$$

The Picone form of the Sturm Comparison Theorem when $m = 1$ is satisfied, as we shall see, as a consequence of Theorem 21.2.

The conditions of Theorem 21.2 of the Sturm-Picone type are by no means necessary when the functionals D_r^d of (21.6) are positive definite, as Theorem 21.3 will make clear.

Definition 21.6. *Focal conditions* ΔV_r, $r > 0$. In Theorem 21.3 we shall assume that

(21.27) $$(R_{ij}(x) - \hat{R}_{ij}(x))\zeta_i\zeta_j > 0 \qquad (\mathbf{0} \neq \zeta \in R^m).$$

Corresponding to the focal conditions V_r and $\hat{V}_\rho$, when $r = \rho > 0$, we shall set up focal conditions to be denoted by ΔV_r. The underlying quadratic form of focal conditions V_r : (17.6) has values $\omega(x,\eta,\zeta)$; the corresponding underlying quadratic form of focal conditions ΔV_r shall have the values $\Delta\omega(x,\eta,\zeta)$ of (21.5)″. For $x \in \aleph$ and arbitrary m-tuples η and ζ set

(21.28) $$\Phi(x,\eta,\zeta) = \Delta\omega(x,\eta,\zeta).$$

With Φ so defined, focal conditions ΔV_r shall have the form:

$$\begin{bmatrix} \frac{d}{dx}\Phi_{\zeta_i}(x,\eta,\eta') = \Phi_{\eta_i}(x,\eta,\eta') & \qquad (i = 1, \ldots, m) & (21.29)' \\ \eta_i(a^1) = c^1_{ih}u_h & \qquad (i = 1, \ldots, m) & (21.29)'' \\ c^1_{ih}\xi_i(a^1) = \Delta b_{hk}u_k & \qquad (h = 1, \ldots, r) & (21.29)''' \end{bmatrix}$$

where Δb_{hk} is given by (21.5)′ and the conditions (21.29)″ are common to focal conditions V_r and $\hat{V}_\rho$ in accord with (21.4).

The following theorem, under hypothesis (21.27), complements Theorem 21.2.

THEOREM 21.3(i). *When* $r > 0$, *a sufficient condition that the functional* D_r^d *of* (21.6)′ *be positive definite is that the focal conditions* ΔV_r : (21.29) *have no focal point in* $(a^1,d]$.

(ii) *When* $r = 0$ *a sufficient condition that the functional* D_0^d *of* (21.6)″ *be positive definite is that the point* $x = a^1$ *have no conjugate point relative to the* DE (21.29)′ *in the interval* $(a^1,d]$.

Theorem 21.3(i) follows from Theorem 19.5(ii) on replacing the interval $(a^1,a^2]$ by the interval $(a^1,d]$. Theorem 21.3(ii) similarly follows from Theorem 7.4. The conditions of Theorem 21.3 are necessary as well as sufficient, as one readily shows.

Nuclear Comparisons. We shall now compare two sets of focal conditions V_r and $\hat{V}_\rho$ of (21.0), under hypotheses termed *nuclear*, which are weaker than those of the preceding theorems but which lead to the same conclusions.

Nuclear hypotheses. Hypotheses K(I), K(II), K(III) which we shall now define in Cases I, II, III, respectively, are less restrictive than Hypotheses H(I), H(II), H(III), respectively. Nevertheless Theorem 21.1 remains valid if H(I) and H(II) in Comparison Theorem 21.1 are replaced by K(I) and K(II) respectively. Corollary 21.1 similarly remains valid if Hypothesis H(III), therein, is replaced by Hypothesis K(III). The merit of Hypotheses H(I), H(II), H(III) is their simplicity and the relative simplicity of the proof of Theorem 21.1 and Corollary 21.1 under these hypotheses. We prepare for the definition of K(I), K(II), K(III).

Notation Recalled. For simplicity of notation we suppose that the focal conditions V_r, $r > 0$ and V_0 have the same DE and that $\hat{V}_r$, $r > 0$, and $\hat{V}_0$, similarly, have the same DE. For each value b such that $a^1 < b \leq d$ let canonical systems $\mathscr{W}_{r,b}$ and $\hat{\mathscr{W}}_{r,b}$, based on the focal conditions V_r and $\hat{V}_r$, respectively, be introduced, as in Definition 17.7, both when $r > 0$ and $r = 0$. The functionals $I^0_{r,b}\#\mathscr{W}_{r,b}$ and $\hat{I}^0_{r,b}\#\hat{\mathscr{W}}_{r,b}$ are then defined when $r > 0$, by their respective values

$$I^0_{r,b}(\eta) = b_{hk}u_hu_k + 2\int_{a^1}^{b} \omega(x,\eta(x),\eta'(x))\,dx \qquad (\eta \in \{\mathscr{W}_{r,b}\}) \tag{21.30}$$

$$\hat{I}^0_{r,b}(\eta) = \hat{b}_{hk}u_hu_k + 2\int_{a^1}^{b} \hat{\omega}(x,\hat{\eta}(x),\hat{\eta}'(x))\,dx \qquad (\hat{\eta} \in \{\hat{\mathscr{W}}_{r,b}\}) \tag{21.31}$$

When $r = 0$ the forms $b_{hk}u_hu_k$ and $\hat{b}_{hk}u_hu_k$ are to be deleted. (Cf. Definition 15.1.)

The V_r*-Nuclear Space* $\{V_r\}_d$. There is given a set of focal conditions V_r of form (17.6) or (17.7) and a subinterval $[a^1,d]$ of $\aleph$ on which focal points are to be compared. A canonical system $\mathscr{W}_{r,d}$ has been adjoined to V_r in Definition 17.7. The functional $I^0_{r,d}\#\mathscr{W}_{r,d}$ is introduced in Definition 15.1. The V_r-nuclear vector subspace $\{V_r\}_d$ of the domain of $I^0_{r,d}\#\mathscr{W}_{r,d}$, as defined below, is similar but not identical with the V_r-nuclear subspace $\{V_r\}$ of the domain $\{\mathscr{W}_r\}$ as characterized in Definition 19.0.

The geometric meaning of $\{V_r\}_d$ can be made transparent by the use of two geometric terms. Let $x = b \in (a^1,d]$ be a focal point of V_r and v a non-null solution of V_r such that $v(b) = \mathbf{0}$; the restriction $\hat{v}$ of v to the interval $[a^1,b]$ will then be called a *truncated* solution of V_r. A mapping η^b of $[a^1,d]$ into R^m such that $\eta^b(x) = \hat{v}(x)$ for $a^1 \leq x \leq b$ and

$$\eta^b(x) = \mathbf{0} \qquad (i = 1, \ldots, m;\, b \leq x \leq d) \tag{21.32}$$

will be called an *axial extension* of $\hat{v}$ over the interval $[a^1,d]$. The following definition extends Definition 19.0.

Definition 21.7. *The nuclear space* $\{V_r\}_d$. Suppose that the count κ of focal points of V_r in $(a^1,d]$ is positive. By the nuclear space $\{V_r\}_d$ is then meant the vector subspace of the domain of $I^0_{r,d}\#\mathscr{W}_{r,d}$ generated by the axially extended truncated solutions of V_r with "corners" at the focal points of V_r in $(a^1,d]$. (Note that $\dim \{V_r\}_d = \kappa$.)

A basic theorem follows. It is called *nuclear* to distinguish it from the weaker Comparison Theorem 21.1. Its hypotheses follow.

Definition 21.8. Hypotheses K(I), K(II), and K(III) are defined as follows.

Hypothesis K(I). In comparing focal conditions V_r and $\hat{V}_r$, $r > 0$, in Case I of (21.0), it is assumed that conditions V_r'' and $\hat{V}_r''$ are identical and that $\hat{I}^0_{r,d}\#\hat{\mathscr{W}}_{r,d}$ is negative definite on the V_r-nuclear subspace $\{V_r\}_d$ of the domain of $I^0_{r,d}\#\mathscr{W}_{r,d}$.

Hypothesis K(II). In comparing V_0 and $\hat{V}_0$, it is assumed that $\hat{I}^0_{0,d}\#\hat{\mathscr{W}}_{0,d}$ is negative definite on the V_0-nuclear subspace of the domain of $I^0_{0,d}\#\mathscr{W}_{0,d}$.

Hypothesis K(III). In comparing V_0 and $\hat{V}_r$, $r > 0$, it is assumed that $\hat{I}^0_{r,d}\#\hat{\mathscr{W}}_{r,d}$ is negative definite on the V_0-nuclear subspace of the domain of $I^0_{0,d}\#\mathscr{W}_{0,d}$.

NUCLEAR COMPARISON THEOREM 21.4. *If Hypotheses* H(I) *and* H(II) *in Theorem* 21.1 *are replaced by the weaker Hypotheses* K(I) *and* K(II), *respectively, the resultant theorem is true.*

The proof of Theorem 21.4 follows the proof of Theorem 21.1 until Lemma 21.2 is reached. Lemma 21.2 is not in general true under the hypotheses of Theorem 21.4, as examples will show. A new lemma is needed to prove Theorem 21.4.

LEMMA 21.3. *Under the hypotheses of Theorem* 21.4, *the index forms* (21.10) *are such that*

$$\text{(21.33)} \qquad \textit{index } \hat{Q}_{r,b} \geq (\textit{index} + \textit{nullity})Q_{r,b} \qquad ((\text{cf. } 21.17))$$

for each focal point $x = b \in (a^1,d]$ *of focal conditions* V_r.

Let k be the count of focal points of V_r in the interval $(a^1,b]$. Let $\{V_r\}_b$ be the V_r-nuclear subspace of the domain of $I^0_{r,b}\#\mathscr{W}_{r,b}$. Then dim $\{V_r\}_b = k$. Lemma 21.3 will follow from statement (i).

(i). *Under the hypotheses of Theorem* 21.4, $\hat{I}^0_{r,b}\#\hat{\mathscr{W}}_{r,b}$ *is negative definite on* $\{V_r\}_b$, *so that index* $\hat{I}^0_{r,b} \geq k$ *in accord with Definition* 15.4.

Proof of (i). Let v be a nonnull mapping in $\{V_r\}_b$. The mapping v has an x-domain $[a^1,b]$ and an "axial extension" $\bar{v}$ over $[a^1,d]$. We see that

$$\text{(21.34)} \qquad \hat{I}^0_{r,b}(v) = \hat{I}^0_{r,d}(\bar{v}).$$

By hypothesis of Theorem 21.4 the right member of (21.34) is negative. Hence $\hat{I}^0_{r,b}$ is negative definite on $\{V_r\}_b$. Since dim $\{V_r\}_b = k$, (i) follows in accord with Definition 15.4.

Thus (i) *is true*.

It follows, respectively, from (i) and Theorem 15.3 that

$$\text{(21.35)} \qquad k \leq \text{index } \hat{I}^0_{r,b} = \text{index } \hat{Q}_{r,b}.$$

Since

$$\text{(21.36)} \qquad k = (\text{index} + \text{nullity})Q_{r,b},$$

by Theorems 17.3 and 15.2, the relation (21.33) follows.

Thus Lemma 21.3 *is true*.

Completion of Proof of Theorem 21.4. Granting (21.33), the proof of Theorem 21.4 is completed exactly as was the proof of Theorem 21.1, from (21.17) on.

Thus Nuclear Comparison Theorem 21.4 *is true and implies the conclusions of Comparison Theorem* 21.1.

We state the following corollary of Theorem 21.4.

COROLLARY 21.2. *If in Corollary* 21.1 *Hypothesis* H(III) *is replaced by Hypothesis* K(III), *the resultant theorem is true*.

The proof of Corollary 21.2 is identical with the proof of Corollary 21.1 except that statement (A) in the proof of Corollary 21.1 is now to be inferred from Theorem 21.4 instead of from Theorem 21.1

We close this section with a theorem which is the first step in showing that our nuclear hypotheses K(I), K(II), and K(III) of Definition 21.8 are less restrictive than the respective hypotheses H(I), H(II), and H(III) of Definitions 21.5(a) and 21.5(b). We begin by showing that K(I) is implied by H(I). This is the case in which focal conditions V_r, $r > 0$ and $\hat{V}_r$ are compared. We shall prove the following theorem.

THEOREM 21.5. *The satisfaction of hypothesis* H(I), *as defined for the interval* $[a^1,d]$, *implies the satisfaction of the nuclear hypothesis* K(I), *defined for the same interval* $[a^1,d]$.

Before coming to the proof proper of Theorem 21.5 we shall formulate a lemma which is a trivial extension of Lemma 19.2(ii). This lemma concerns the V_r-nuclear subspace $\{V_r\}_d$ (Definition 21.7) of the domain U of the quadratic functional $I^0_{r,d}\#\mathscr{W}_{r,d}$ associated in Definition 17.7 with the given focal conditions V_r. The proof of Lemma 19.2(ii) shows that the following is true.

(i) *The quadratic functional* $I^0_{r,d}\#\mathscr{W}_{r,d}$ *vanishes for each mapping* η *in the nuclear subspace* $\{V_r\}_d$ *of* U.

To prove Theorem 21.5, we refer to the difference functional D^d_r, $r > 0$, set up in (21.6)′ when the focal points of V_r and $\hat{V}_r$ are to be compared. According to Definition 21.3 the domain U of D^d_r is the vector space of D^1-mappings η of $[a^1,d]$ into R^m which satisfy the conditions $\eta^1_i = c^1_{ih}u_h$ of V_r and $\hat{V}_r$ with some r-tuple and the condition $\eta(d) = \mathbf{0}$. If $I^0_{r,d}$ and $\hat{I}^0_{r,d}$ are the quadratic functionals appearing in Definition 21.8 of K(I), we infer from (21.6)′ that

$$D^d_r(\eta) \equiv I^0_{r,d}(\eta) - \hat{I}^0_{r,d}(\eta) \qquad (\eta \in U). \tag{21.37}$$

Under hypothesis H(I) of Definition 21.5(a), D^d_r is positive definite over U, while (i) affirms that $I^0_{r,d}(\eta) = 0$ for $\eta \in \{V_r\}_d$. Relation (21.37) then implies that $\hat{I}^0_{r,d}$ is negative definite over $\{V_r\}_d$, or equivalently that K(I) is satisfied. Thus Theorem 21.5 is true.

One proves similarly that hypothesis H(II) implies nuclear hypothesis K(II). H(II) and K(II) are the hypotheses when one compares V_0 and $\hat{V}_0$. That H(II) implies K(II) follows from the proof of Theorem 21.5 on deleting all reference to the r-tuple u. One replaces $r > 0$ by $r = 0$ in (21.37), taking U as the vector space of mappings η of $[a^1,d]$ into R^m such that $\eta(a^1) = \eta(d) = \mathbf{0}$.

That H(III) implies K(III) follows from the fact that H(II) implies K(II).

That the hypotheses H(I), H(II), H(III) are actually more restrictive than the respective hypotheses K(I), K(II), and K(III), is readily shown by examples.

Historical Note. A comparison theorem of Leighton, given as Theorem 1.4 on page 4 of Swanson, comes under our Nuclear Comparison Theorem 21.4 when $m = 1$, when $s = 1$ in (21.7), and when Cases II or III of (21.0) occur. Moreover, Leighton has shown in Example 1 on page 6 of Swanson that his hypothesis (our nuclear hypothesis) implies the conclusion of his theorem in cases in which the corresponding Sturm-Picone Theorem fails to do so.

Exercises

Suppose that the Definitions 21.5 of Hypotheses H(I), H(II), H(III) are modified by replacing the condition of positive definiteness by positive semidefiniteness. Let the resultant weakened hypotheses be denoted by H^0(I), H^0(II), and H^0(III), respectively.

21.1. Show that Theorem 21.1 and Corollary 21.1 remain valid under Hypotheses H^0(I), H^0(II), H^0(III), provided the phrase "such that $\hat{e}_k < e_k$" therein is replaced by the phrase "such that $\hat{e}_k \leq e_k$."

A review of the original proofs of Theorem 21.1 and its corollary will suffice to verify the conclusion of Exercise 21.1. Changes to be made in the original proof include the following:

(1) The symbol $<$ appears in Lemma 21.2, in (21.16), (21.22), (A) and in (21.26), and should be replaced by $\leq$.

(2) In (21.17), (21.20), and (21.21) the word index should be replaced by (index + nullity).

(3) The count k appearing in (21.21), should be the count of focal points of $\hat{V}_r$ in $(a^1,b]$ rather than (a^1,b).

Suppose that the definitions of the Hypotheses K(I), K(II), K(III) are modified by replacing the phrase "negative definite" by negative semidefinite. Let the resultant weakened hypotheses be denoted, respectively, by K^0(I), K^0(II), K^0(III).

21.2. Show that the Nuclear Comparison Theorem 21.4 and its corollary hold under the Hypotheses K^0(I), K^0(II), K^0(III), provided the phrase "such that $\hat{e}_k < e_k$" is replaced by the phrase "such that $\hat{e}_k \leq e_k$."

22. The Sturm Comparisons When $m = 1$

The object of this section is to show how the Comparison Theorems of Section 21 imply and extend classical Sturm-like theorems. Leighton's fundamental theorem when $m = 1$, presented by Swanson as Theorem 1.4, is implied. Reid's basic theorem on first focal points, presented as Theorem 1.29 by Swanson is also implied in modified forms under weaker hypotheses. Reid's hypothesis that $(A/a)' \geq 0$ in the interval $[\alpha,\beta]$ is replaced by the hypothesis that A/a increase monotonically, enabling us to simplify the proof.

A first objective of Section 22 is to show how focal conditions enter in the classical cases and unify the theory.

On page 59, Bôcher [1] presents two sets of conditions whose solutions are to be compared. His interval $[a,b]$ will be replaced by our interval $[a^1,d]$. The Sturmian conditions whose solutions are to be compared have the respective forms,†

$$(22.0) \qquad \begin{bmatrix} \dfrac{d}{dx}(K\eta') + G\eta = 0 \\ \eta(a^1) = c \\ \eta'(a^1) = c', |c| + |c'| \neq 0 \end{bmatrix}$$

$$(22.1) \qquad \begin{bmatrix} \dfrac{d}{dx}(\hat{K}\eta') + \hat{G}\eta = 0 \\ \eta(a^1) = \hat{c} \\ \eta'(a^1) = \hat{c}', |\hat{c}| + |\hat{c}'| \neq 0 \end{bmatrix}.$$

As in the classical theory it is assumed that K, G, $\hat{K}$, $\hat{G}$ are continuous mappings of $[a^1,d]$ into R with $K(x) > 0$ and $\hat{K}(x) > 0$.

Comparable Conditions (22.0) *and* (22.1). In comparing the zeros of nonnull solutions of conditions (22.0) and (22.1) three cases are to be considered:

(22.2)

Case (i) $c = \hat{c} \neq 0$; Case (ii) $c = \hat{c} = 0$; Case (iii) $c = 0$, $\hat{c} \neq 0$.

These correspond, respectively, as we shall see, to Cases I, II, III, enumerated in (21.0). The reader will observe that in this enumeration of cases we have not included the general case $c \neq 0$, $\hat{c} \neq 0$. No generality is lost by this exclusion for the following reason. If $c \neq 0$ and $\hat{c} \neq 0$, then by multiplying $\hat{c}$ and $\hat{c}'$ by a suitable nonnull constant, conditions (22.1) can be replaced by conditions (22.1)* such that $\hat{c} = c$ and each nonnull solution η of (22.1) equals a nonnull solution of (22.1)* times a nonnull constant.

We shall term conditions (22.0) and (22.1) which come under one of the cases enumerated in (22.2), directly *comparable Sturmian conditions*.

Notation. To implement the results of Section 21 we shall set

$$(22.3)' \qquad \begin{aligned} 2\omega(x,\eta,\zeta) &= K(x)\zeta^2 - G(x)\eta^2 \qquad (x \in [a^1,d]) \\ 2\hat{\omega}(x,\eta,\zeta) &= \hat{K}(x)\zeta^2 - \hat{G}(x)\eta^2 \qquad (x \in [a^1,d]) \end{aligned}$$

† In Bôcher, G is replaced by $-G$.

for arbitrary η and ζ in R. The counterpart of the r-square matrix $\|b_{hk}\|$ and ρ-square matrix $\|\hat{b}_{hk}\|$ of Section 21 will be two matrices of single elements

$$\beta = K(a^1)cc', \qquad \hat{\beta} = \hat{K}(a^1)\hat{c}\hat{c}'. \tag{22.3)''$$

Focal conditions equivalent to Sturmian conditions. In Case (i) of (22.2) we associate with the Sturmian conditions (22.0) and the comparable conditions (22.1), "free focal conditions" V_1 and $\hat{V}_1$ of the respective forms (22.4)′ and (22.4)″.

$$(22.4)' \quad \left[\begin{array}{ll} V_1' : \dfrac{d}{dx}\,\omega_\zeta(x,\eta,\eta') = \omega_\eta(x,\eta,\eta') & \\ V_1'' : \eta(a^1) = cu, \qquad (c \neq 0) & \\ V_1''' : c\xi(a^1) = \beta u & \end{array}\right] \quad \begin{array}{l}(x \in [a^1,d]) \\ (\text{see } (22.3))\end{array}$$

$$(22.4)'' \quad \left[\begin{array}{ll} \hat{V}_1' : \dfrac{d}{dx}\,\hat{\omega}_\zeta(x,\eta,\eta') = \hat{\omega}_\eta(x,\eta,\eta') & \\ \hat{V}_1'' : \eta(a^1) = \hat{c}u \qquad (\hat{c} \neq 0) & \\ \hat{V}_1''' : \hat{c}\xi(a^1) = \hat{\beta}u & \end{array}\right] \quad \begin{array}{l}(x \in [a^1,d]) \\ (\text{see } (22.3))\end{array}$$

It is understood that in the conditions V_1

$$\xi(a^1) = \omega_\zeta(a^1,\eta(a^1),\eta'(a^1)) = K(a^1)\eta'(a^1)$$

and that in the conditions $\hat{V}_1$

$$\hat{\xi}(a^1) = \hat{\omega}_\zeta(a^1,\eta(a^1),\eta'(a^1)) = \hat{K}(a^1)\eta'(a^1).$$

With this understood, V_1''' and $\hat{V}_1'''$ can be written in the respective forms

$$\eta'(a^1) = c'u; \qquad \eta'(a^1) = \hat{c}'u. \tag{22.5}$$

Notation. Let $(22.0)_0$ and $(22.0)_1$ denote the Sturmian conditions (22.0) when $c = 0$ and $c \neq 0$, respectively. Let $(22.1)_{\hat{0}}$ and $(22.1)_{\hat{1}}$ similarly denote the Sturmian conditions (22.1) when $\hat{c} = 0$ and $\hat{c} \neq 0$ respectively. With this notation we state an equivalence lemma.

LEMMA 22.1. *The Sturmian conditions* $(22.0)_1$ *are equivalent* to the focal conditions* (22.4)′, *while the Sturmian conditions* $(22.1)_{\hat{1}}$, *are equivalent to the focal conditions* $\hat{V}_1$: (22.4)″.

* A set of Sturmian conditions X and a set of focal conditions Y, when $m = 1$, will be said to be *equivalent* if each nonnull solution of the conditions X {or Y} is a constant multiple of a solution of the conditions Y {or X}.

Proof. The conditions $(22.0)_1$, as well as the focal conditions V_1, are such that a nonnull solution η satisfies the condition,

$$\frac{\eta'(a^1)}{\eta(a^1)} = \frac{c'}{c}. \tag{22.6}$$

Hence the conditions $(22.0)_1$ and the conditions V_1 are equivalent. Conditions $(22.1)_{\hat{1}}$ and $\hat{V}_1$ are similarly equivalent.

Sturmian conditions $(22.0)_0$ and $(22.1)_{\hat{0}}$ are equivalent to focal conditions of dimension 0 as affirmed in the following lemma. (See Definition 17.4.)

LEMMA 22.2. *The Sturmian conditions* $(22.0)_0$ *are equivalent to focal conditions* V_0 *of the form*

$$\begin{bmatrix} V_0' : \dfrac{d}{dx}\, \omega_\zeta(x,\eta,\eta') = \omega_\eta(x,\eta,\eta') \\ \\ V_0'' : \eta(a^1) = 0 \end{bmatrix} \tag{22.7$'$}$$

while the Sturmian conditions $(22.1)_{\hat{0}}$ *are equivalent to focal conditions* $\hat{V}_0$ *of the form*

$$\begin{bmatrix} \hat{V}_0' : \dfrac{d}{dx}\, \hat{\omega}_\zeta(x,\eta,\eta') = \hat{\omega}_\eta(x,\eta,\eta') \\ \\ \hat{V}_0'' : \eta(a^1) = 0 \end{bmatrix}. \tag{22.7$''$}$$

The comparisons to be made in Theorem 22.1 are limited to the three cases enumerated in (22.2), or equivalently to the comparisons:

In Case (i) of V_1 and $\hat{V}_1$.
In Case (ii) of V_0 and $\hat{V}_0$.
In Case (iii) of V_0 and $\hat{V}_1$.

These comparisons of focal conditions come respectively under Cases I, II, and III of (21.0).

We shall interpret the Nuclear Comparison Theorem 21.4 and its corollary when $m = 1$.

The Nuclear Comparison Theorem when $m = 1$. Let

$$\mathscr{W}_{1,d},\ \hat{\mathscr{W}}_{1,d},\ \mathscr{W}_{0,d},\ \hat{\mathscr{W}}_{0,d}, \tag{22.8}$$

be the canonical systems "based" on the respective focal conditions,

$$V_1,\ \hat{V}_1,\ V_0,\ \hat{V}_0, \tag{22.9}$$

in the sense of Definition 17.7.

The functional $I^0_{1,d}\#\mathscr{W}_{1,d}$ has the values (Definition 15.1)

$$I^0_{1,d}(\eta) = \beta u^2 + \int_{a^1}^{d} (K(x)\eta'^2(x) - G(x)\eta^2(x))\, dx \tag{22.10}$$

for η a D^1-mapping of $[a^1,d]$ into R which satisfies the conditions V_1'' with a parameter u and with $\eta(d) = 0$. The functional $\hat{I}^0_{1,d}\#\hat{\mathscr{W}}_{1,d}$ is similarly represented with $\hat{\beta}$, $\hat{K}$, $\hat{G}$, $\hat{V}_1''$, respectively replacing β, K, G, V_1''. The functionals $I^0_{0,d}\#\mathscr{W}_{0,d}$ and $\hat{I}^0_{0,d}\#\hat{\mathscr{W}}_{0,d}$ are represented by deleting β, $\hat{\beta}$, and u, and requiring that $\eta(a^1) = \eta(d) = 0$.

If the V_1 and V_0-nuclear spaces with x-domains $[a^1,d]$ are defined as in Definition 21.7, the principal hypotheses in our Nuclear Comparison Theorem when $m = 1$ are as follows.

Hypothesis K(I). In comparing V_1 and $\hat{V}_1$ it is assumed that $c = \hat{c} \neq 0$ and that $\hat{I}^0_{1,d}\#\hat{\mathscr{W}}_{1,d}$ is negative definite on the V_1-nuclear subspace of the domain of $I^0_{1,d}\#\mathscr{W}_{1,d}$.

Hypothesis K(II). In comparing V_0 and $\hat{V}_0$ it is assumed that $\hat{I}^0_{0,d}\#\hat{\mathscr{W}}_{0,d}$ is negative definite on the V_0-nuclear subspace of the domain of $I^0_{0,d}\#\mathscr{W}_{0,d}$.

Hypothesis K(III). In comparing V_0 and $\hat{V}_1$ it is assumed that $\hat{I}^0_{1,d}\#\hat{\mathscr{W}}_{1,d}$ is negative definite on the V_0-nuclear subspace of the domain of $I^0_{0,d}\#\mathscr{W}_{0,d}$.

Theorem 21.4 yields the following theorem when $m = 1$. (Cf. Bôcher [1], p. 59.)

NUCLEAR COMPARISON THEOREM 22.1, m = 1. *Under Hypotheses* K(I), K(II), K(III) *in the respective Cases* (i), (ii), (iii), *if*

$$e_1 < e_2 < \cdots < e_s \qquad (s > 0) \tag{22.11}$$

are the zeros in $(a^1,d]$ *of a nonnull solution of Sturmian conditions* (22.0), *then any nonnull solution* $\hat{\eta}$ *of Sturmian conditions* (22.1) *directly comparable to the given conditions* (22.0) *has zeros,*

$$\hat{e}_1 < \hat{e}_2 < \cdots < \hat{e}_s, \tag{22.12}$$

such that $\hat{e}_k < e_k$, *for* $k = 1, \ldots, s$, *and* $\hat{e}_k$ *is the* k*-th zero of* $\hat{\eta}$ *following* $x = a^1$.

Granting the Equivalence Lemmas 22.1 and 22.2, Theorem 22.1 follows from Theorem 21.4 in Cases (i) and (ii) and from Corollary 21.2 in Case (iii).

Remark on s. A comparison theorem similar to the above in which $s = 1$ does not a priori imply Theorem 22.1 when $s > 1$. A proof is required. When $m > 1$ such an implication is ruled out altogether because successive focal points (including conjugate points as a special case) may have multiplicities which exceed 1. It is for these reasons that we have concerned ourselves with a sequence of $s \geq 1$ "algebraic" focal points, rather than with the special case $s = 1$.

Theorem 22.1 in weak form. If in the definition of the hypotheses *K(I)*, *K(II)*, *K(III)*, the condition of negative definiteness is replaced by the condition of negative semidefiniteness, the resultant conditions will be denoted, respectively, by $K^0(I)$, $K^0(II)$, $K^0(III)$. Exercise 21.2 indicates the truth of Theorem 22.1 in the following weak form.

WEAK NUCLEAR COMPARISON THEOREM 22.2. *Theorem* 22.1 *holds under the weakened Hypotheses* $K^0(I)$, $K^0(II)$, $K^0(III)$, *provided the phrase "such that* $\hat{e}_k < e_k$*" is replaced by the phrase "such that* $\hat{e}_k \leq e_k$.*"*

The theorem of Leighton, presented as Theorem 1.4 by Swanson, is implied by Theorem 22.1 when $m = 1$ and $s = 1$. It comes under Case (ii).

Modified Reid Theorems. Theorems 22.1 and 22.2, restricted to Case (i), imply modified forms of a comparison Theorem of Reid, presented by Swanson as Theorem 1.29.

In our notation Reid is concerned with the first focal points to the right or left of $x = a^1$, of our focal conditions V_1 and $\hat{V}_1$ in the case in which our constants β and $\hat{\beta}$ vanish. With Reid G and $\hat{G}$ are continuous on the interval $[a^1,d]$, K and $\hat{K}$ positive and of class C^1 (of class C^0 with us). We replace the hypothesis of Reid that $(K/\hat{K})' \geq 0$ by the hypothesis that $K/\hat{K}$ be monotonically increasing. It is sufficient to treat the case of focal points *following* $x = a^1$.

We begin with a *strong** variant of Reid's Theorem. A *weak* form will follow and imply Reid's Theorem.

THEOREM 22.3. *Suppose that* $K/\hat{K}$ *is monotonically increasing in the interval* $[a^1,d]$, *that* $\hat{G}(x) > 0$ *and that*

$$\frac{G(x)}{K(x)} \leq \frac{\hat{G}(x)}{\hat{K}(x)} \qquad (a^1 \leq x \leq d). \tag{22.13}$$

If a first focal point $x = b$ *of* V_1 *exists in* $[a^1,d]$ *when* $\beta = 0$ *and if, for at least one value of* $x \in (a^1,b]$ *the sign* $<$ *prevails in* (22.13), *then a first focal point* $x = \hat{b}$ *of* $\hat{V}_1$ *exists in* $[a^1,d]$ *when* $\hat{\beta} = 0$, *and is such that* $\hat{b} < b$.

For simplicity and without loss of generality in comparing the focal points of V_1 and $\hat{V}_1$ we can suppose that $c = \hat{c} = 1$.

* We term Theorems 22.3 and 22.4 "strong" and "weak," respectively, because of their conclusions $\hat{b} < b$, and $\hat{b} \leq b$.

The Solution v *of* $\hat{V}_1$. There exists a C^1-solution v of focal conditions $\hat{V}_1$ when $\hat{\beta} = 0$ such that $v(a^1) = 1$, and $v'(a^1) = 0$. The DE of $\hat{V}_1$ then shows that

$$(22.14) \qquad \frac{d}{dx}(\hat{K}(x)v'(x)) = -\hat{G}(x)v(x) < 0 \qquad (a^1 \leq x < h)$$

where h is the maximum real number such that $v(x) > 0$ for $a^1 < x < h \leq d$. From (22.14) we infer that $v'(x) < 0$ for $a^1 < x < h$.

The Solution z *of* V_1. By hypothesis there exists a solution z of focal conditions V_1 when $\beta = 0$ such that $z(a^1) = 1$, $z'(a^1) = 0$, $z(b) = 0$ and $z(x) > 0$ for $a^1 \leq x < b$. Let k be the minimum value on $[a^1,b]$ such that

$$(22.15) \qquad z'(x) \leq 0 \qquad [k \leq x \leq b].$$

The value k exists and may equal a^1 but cannot equal b. Note that $z'(k) = 0$.

An Auxiliary Comparison. We replace the comparison of V_1 and $\hat{V}_1$ under the conditions of Theorem 22.3 by a similar comparison in which the interval $[a^1,d]$ is replaced by the subinterval $[k,b]$ of (22.15). The focal conditions to be compared will be denoted by V_1^k and $\hat{V}_1^k$. The DE shall be those of V_1 and $\hat{V}_1$ respectively, restricted to the interval $[k,b]$. The constants β and $\hat{\beta}$ shall be null. The end conditions at $x = k$ shall be defined by constants $c = \hat{c} = 1$. The point $x = b$ is a first focal point of V_1^k, following $x = k$. Two cases must be distinguished, namely

$$\text{Case, } k = a^1, \qquad \text{Case, } k > a^1.$$

We shall apply Theorems 22.1 and 22.2, with $[a^1,d]$ replaced by $[k,b]$, and compare V_1^k and $\hat{V}_1^k$ in these two cases. The comparison is under Case (i) of (22.2). The following statement (α) will be proved. It implies Theorem 22.3. The brace indicates an alternative.

(α) *Under the conditions of Theorem* 22.3 *the first focal point* $\hat{b}^{(k)}$ *of* $\hat{V}_1^k$ *following* $x = k$, *exists in Case*, $k = a^1$, $\{k > a^1\}$, *together with the first focal point* $\hat{b}$ *of* $\hat{V}_1$ *and is such that*

$$(22.16) \qquad \hat{b} = \hat{b}^{(k)} < b, \ \{\hat{b} < \hat{b}^{(k)} \leq b\},$$

thereby establishing Theorem 22.3.

Proof of (α) *when* $k = a^1$. In this case $\hat{V}_1^k = \hat{V}_1$. It follows from Nuclear Theorem 22.1 that the focal point $\hat{b}$ of $\hat{V}_1$ exists in the interval $[a^1,b]$ and that $\hat{b} < b$, provided the integral

$$(22.17) \qquad L = \int_{a^1}^{b} (\hat{K}(x)z'^2(x) - \hat{G}(x)z^2(x))\, dx < 0.$$

In this case $\hat{b} = \hat{b}^{(k)} < b$.

Proof that L < 0 *when* k $= a^1$. Note that

$$L = \int_{a^1}^{b} \frac{\hat{K}}{K}(Kz'^2 - Gz^2)\,dx + \int_{a^1}^{b} \hat{K}z^2\left(\frac{G}{K} - \frac{\hat{G}}{\hat{K}}\right) dx. \tag{22.18}$$

The second integral in (22.18) is negative under the hypotheses of Theorem 22.3. Set $\hat{K}(x)/K(x) = \rho(x)$. Since $\rho(x)$ is monotonically decreasing, the second law of the mean for integrals implies that the first integral in (22.18), for some ξ such that $a^1 \leq \xi \leq b$, equals

$$\rho(a^1)\int_{a^1}^{\xi}(Kz'^2 - Gz^2)\,dx \tag{22.19}$$

An integration by parts shows that

$$\int_{a^1}^{\xi}(Kz'^2 - Gz^2)\,dx = [zKz']_{a^1}^{\xi} = [zKz']^{\xi} \leq 0 \tag{22.20}$$

by (22.15) when $k = a^1$. Thus the integral L : (22.17) is negative when $k = a^1$.

It follows from Theorem 22.1, applied in Case (i) with $[a^1,d]$ replaced by $[a^1,b]$, that the first focal point $\hat{b}$ of $\hat{V}_1$ exists when $k = a^1$, and $\hat{b} < b$. Thus (α) is true when $k = a^1$.

Proof of (α) *when* k $> a^1$. In this case the integral L of (22.17) is replaced by the integral

$$L_k = \int_{k}^{b} [\hat{K}(x)z'^2(x) - \hat{G}(x)z^2(x)]\,dx. \tag{22.21}$$

According to Theorem 22.2, applied in Case (i) with $[a^1,d]$ replaced by $[k,b]$, the first focal point $\hat{b}^{(k)}$ of $\hat{V}_1^k$ exists and is such that $\hat{b}^{(k)} \leq b$, provided $L_k \leq 0$.

Proof that $L_k \leq 0$ *when* k $> a^1$. As in the case when $k = a^1$,

$$L_k = \int_{k}^{b} \rho(x)(Kz'^2 - Gz^2)\,dx + \int_{k}^{b} \hat{K}z^2\left(\frac{G}{K} - \frac{\hat{G}}{\hat{K}}\right) dx. \tag{22.22}$$

The second integral in (22.22) is nonpositive but not necessarily negative when $k > a^1$. The first integral in (22.22) has the value (22.19), with a^1 replaced by k. Moreover (22.20) holds with a^1 replaced by k. Thus $L_k \leq 0$ with $k > a^1$.

It follows from Theorem 22.2 that the first focal point $\hat{b}^{(k)}$ of $\hat{V}_1^k$ exists and is such that $\hat{b}^{(k)} \leq b$. We continue by proving (i).

(i) *When* k $> a^1$, $\hat{b}$ *exists with* $\hat{b} < \hat{b}^{(k)}$.

Proof of (i). Let w be a solution of focal conditions $\hat{V}_1^k$ such that $w(k) = z(k)$ and $w'(k) = 0$. Then $w(\hat{b}^{(k)}) = 0$. If a first focal point $\hat{b}$ of $\hat{V}_1$ exists with $\hat{b} \leq k$, statement (*i*) is true when $k > a^1$. Otherwise $v(x) > 0$ for $a^1 \leq x \leq k$. Let $\bar{v}$ then be a positive multiple of v such that $\bar{v}(k) = z(k)$. Now $k < h$ of (22.14) since $v(x) > 0$ in $[a^1,k]$. Moreover

$$\bar{v}(k) = w(k) > 0,\ \bar{v}'(k) < 0,\ w'(k) = 0,\ w(\hat{b}^{(k)}) = 0. \tag{22.23}$$

It follows (with the aid of the classical Sturm Separation Theorem) that $\bar{v}(x)$ vanishes at a first point $\hat{b}$ following $x = k$ but preceding $x = \hat{b}^{(k)}$.

Statement (α) follows when $k > a^1$.

Thus (α) is true in all cases and Theorem 22.3 follows.

A review of the proof of Theorem 22.3 shows the truth of the following modification of Reid's theorem.

THEOREM 22.4. *Suppose that* $K/\hat{K}$ *is monotonically increasing in the interval* $[a^1,d]$, *that* $\hat{G} \geq 0$ *and that* $G(x)/K(x) \leq \hat{G}(x)/\hat{K}(x)$. *If then a first focal point* $x = b$ *of* V_1 *exists in* $[a^1,d]$ *when* $\beta = 0$, *a first focal point* $x = \hat{b}$ *of* $\hat{V}_1$ *exists in* $[a^1,d]$ *when* $\hat{\beta} = 0$ *and is such that* $\hat{b} \leq b$.

This theorem implies Reid's theorem but differs in that $\hat{G}$ is not required to be positive nor K and $\hat{K}$ of class C^1.

The proof of Theorem 22.4 includes the following modifications of the proof of Theorem 22.3. The solution v of $\hat{V}_1$ is such that $v'(x) \leq 0$ for $a^1 < x < h$. In (22.16) and (22.17) the symbol $<$ is to be replaced by $\leq$. The integrals in (22.18) and (22.22) are proved nonpositive. Statement (i) should read "When $k > a^1$, $\hat{b}$ exists with $\hat{b} \leq \hat{b}^{(k)}$." In (22.23) $\bar{v}'(k) \leq 0$.

Theorem 22.4 follows.

The conditions $\hat{\beta} = \beta = 0$ in Theorems 22.3 and 22.4 unnecessarily restrict the class of focal conditions V_1 and $\hat{V}_1$ for which these comparisons are valid. Theorems similar to Theorems 22.3 and 22.4 hold if proper conditions are put on β and $\hat{\beta}$.

Exercise

22.1. Show that the conclusions of the Nuclear Comparison Theorem 22.1 are valid if the nuclear hypotheses $K(I)$, $K(II)$, $K(III)$ are replaced by the more restrictive hypotheses formulated as follows in Cases (i), (ii), (iii), respectively.

In Case (i), $c = \hat{c} > 0, c' \geq \hat{c}'$; $K(x) \geq \hat{K}(x) > 0$; $G(x) \leq \hat{G}(x)$ with $G(x) \equiv \hat{G}(x)$ in no subinterval of $[a^1, d]$.

In Cases (ii) *and* (iii), $c = 0$ with K, $\hat{K}$, G, $\hat{G}$ conditioned as in the preceding paragraph.

Suggestions. In Case (i) the systems to be compared should be put in forms (22.4)′ and (22.4)″. Use can then be made of Theorems 21.1 and 21.2.

The condition that $c = \hat{c} > 0$ does not limit the generality of the comparison of systems (22.4)′ and (22.4)″, when $c \neq 0$ and $\hat{c} \neq 0$ since the zeros of a nonnull solution u of the JDE are the zeros of ku for each constant $k \neq 0$.

In Case (ii) use can be made of Theorems 21.1 and 21.2, supplemented by Corollary 21.1 in Case (iii).

The resultant comparison theorems imply the classical theorem as formulated by Bôcher [1] on p. 59, under the conditions above the theorem on p. 59. These theorems are weaker than the corresponding Nuclear Comparison Theorems. See Historical Note at end of Section 21.

23. The Variation of Conjugate Points (of $x = c$ with c)

For each $x \in \aleph$ there is given a quadratic form $\omega(x,\eta,\zeta)$, conditioned as in Appendix I. The corresponding JDE have the form

$$\frac{d}{dx}\,\omega_{\zeta_i}(x,\eta,\eta') = \omega_{\eta_i}(x,\eta,\eta') \qquad (i = 1, \ldots, m). \tag{23.1}$$

Conjugate points of a point $x = c$ in $\aleph$ relative to the DE : (23.1), taken in the sense of Section 3 of Appendix I, will be called *ordinary*. An ordinary conjugate point of $x = c$ in the interval $\aleph$ has a multiplicity ν. In order to formulate a theorem on the variation of conjugate points of $x = c$ with c, we shall adopt conventions by virtue of which an ordinary conjugate point of multiplicity ν is replaced by ν properly indexed points, termed *algebraic conjugate* points of $x = c$. When $m = 1$ no such convention is needed, since each ordinary conjugate point has a multiplicity 1.

The following definition covers two cases. Conditions in a brace { } refer to the second case.

Definition 23.1. *The* k-th *algebraic conjugate point of* x = c, *following* {*preceding*} x = c. The point $x = c$ is given in $\aleph$. Let $x = d$ in $\aleph$ be an ordinary conjugate point of $x = c$ of multiplicity ν, following {preceding} $x = c$. Let κ be the count of points in the interval (c,d) {interval (d,c)}. We understand then that the k-th *algebraic* conjugate point $c_k^+\{c_k^-\}$ of $x = c$, following {preceding} $x = c$, then exists and is carried by $x = d$, for k any one of the integers

$$\kappa + 1, \kappa + 2, \ldots, \kappa + \nu. \tag{23.2}$$

The following theorem is a consequence of Corollary 20.1.

THEOREM 23.1. *Let* a < b *be points in* $\aleph$. *A necessary and sufficient condition that* a_k^+ *exist and have the carrier* b *is that* b_k^- *exist and have the carrier* a.

The first of two theorems follows.

THEOREM 23.2(i). *Let* a *and* b *be two points in* $\aleph$ *whose* k-*th algebraic conjugate points* a_k^+ *and* b_k^+ *exist in* $\aleph$. *If* $\mathrm{a} < \mathrm{b}$ *then* $\mathrm{a}_\mathrm{k}^+ < \mathrm{b}_\mathrm{k}^+$.

(ii). *If* $\mathrm{a} < \mathrm{b}$ *and* b_k^+ *exists in* $\aleph$, *then* a_k^+ *exists in* $\aleph$.

The following definition is needed in proving Theorem 23.2.

Definition 23.2. *The focal count* X_c. Let X be a relatively compact subinterval of $\aleph$ and $x = c$ a point in $\aleph$. Let F_c be the von Escherich family of solutions of the DE (23.1) which vanish at $x = c$. Let X_c denote the count of focal points of F_c in X.

Proof of Theorem 23.2(i). Let d be the carrier of b_k^+. Note that the count $[b,d]_b \geq m + k$. It follows from this relation and Corollary 20.1 that

$$m + k \leq [b,d]_b = [b,d]_d. \tag{23.3)$'$ }$$

Hence $[b,d)_d \geq k$. Note that

$$(a,d)_d \geq [b,d)_d \geq k.$$

From Corollary 20.1 and this relation

$$(a,d)_a = (a,d)_d \geq k. \tag{23.3)$''$ }$$

Thus (i) of Theorem 23.2 is true.

Proof of Theorem 23.2(ii). Let d be the carrier of b_k^+ and set $\kappa = (b,d)_b$. Let ν be the multiplicity of d as a conjugate point of $x = b$. By virtue of the definition of κ and ν,

$$m + \kappa + \nu = [b,d]_b = [b,d]_d, \tag{23.4}$$

where the second equality follows from Corollary 20.1. By hypothesis $[a,d] \supset [b,d]$. From (23.4) and Corollary 20.1

$$m + \kappa + \nu \leq [a,d]_d = [a,d]_a. \tag{23.5}$$

Hence

$$\kappa + \nu \leq (a,d]_a. \tag{23.6}$$

Since $k \leq \kappa + \nu$, it follows from (23.6) and Definition 23.1 that a_k^+ exists in $(a,d]$.

The proof of Theorem 23.2 is complete.

Theorem 23.2 has the following *dual*, obtained by replacing algebraic conjugate points of $x = a$ following a, by algebraic conjugate points of $x = b$ preceding b. The symmetry of Definition 23.1 with respect to "preceding" and "following," the symmetry of Corollary 20.1 with respect to a and b, and the validity of the Separation Theorem, for all choices of the interval τ and von Escherich families, make it unnecessary to add a proof.

THEOREM 23.3(i). *Let* a *and* b *be points in* $\aleph$ *whose* k*-th algebraic conjugate points* a_k^- *and* b_k^- *exist. If* $a < b$ *then* $a_k^- < b_k^-$.

(ii). *Let* a *and* b *be given in* $\aleph$ *with* $a < b$. *If* a_k^- *exists in* $\aleph$ *then* b_k^- *exists in* $\aleph$.

The k-th **conjugate point homeomorphism.** The preceding theorems enable us to establish the following basic theorem. By Definition 23.1, c_k^+ is the k-th algebraic conjugate point of $x = c$ following c in $\aleph$, and d_k^- is the k-th algebraic conjugate point of $x = d$ preceding d in $\aleph$.

THEOREM 23.4. *If for at least one point* $x = a$ *in* $\aleph$, a_k^+ *exists, or equivalently, if for at least one point* $x = b$ *in* $\aleph$, b_k^- *exists, then the following is true.*

(i). *The subsets of* $\aleph$, *namely*

$$\aleph_k' = \{c \in \aleph \mid c_k^+ \text{ exists}\}, \qquad \aleph_k'' = \{d \in \aleph \mid d_k^- \text{ exists}\} \tag{23.7}$$

are, respectively, initial and terminal open subsets of $\aleph$ *(possibly* $\aleph$*).*

(ii). *The mappings*

$$c \to c_k^+ : \aleph_k' \to \aleph \tag{23.8$'$}$$

$$d \to d_k^- : \aleph_k'' \to \aleph \tag{23.8$''$}$$

are homeomorphism into $\aleph$ *which are inverses of one another, respectively, mapping* $\aleph_k'$ *onto* $\aleph_k''$ *and* $\aleph_k''$ *onto* $\aleph_k'$.

Proof of (i). We shall show that $\aleph_k'$ is an *initial open* subinterval of $\aleph$.

It follows from Theorem 23.2 (ii) that $\aleph_k'$ is an initial interval of $\aleph$ which is either open or contains a right end point $x = a$ in $\aleph$. In the latter case set $a_k^+ = c$ and let $d \in \aleph$ be such that $d > c$. Since $c_k^- = a$ by Theorem 23.1, d_k^- exists by Theorem 23.3(ii) with $d_k^- > c_k^- = a$. Set $d_k^- = b$. Then $b_k^+ = d$ by Theorem 23.1; since $a < b$, this is contrary to the hypothesis that a is the right end point of $\aleph_k'$.

We infer that $\aleph_k'$ is an initial open subinterval of $\aleph$. Similar $\aleph_k''$ is a terminal open subinterval of $\aleph$.

Proof of (ii). The mappings (23.8) are biunique and order preserving, by virtue of Theorems 23.2(i) and 23.3(i), respectively. They are inverses of one another in accord with Theorem 23.1. It follows from Theorems 23.2 and 23.3 that a prescribed compact subinterval N' of $\aleph_k'$ is mapped by (23.8)′ biuniquely *onto* a compact subinterval N'' of $\aleph_k''$. Hence the mapping (23.8)′ is continuous. The mapping (23.8)″ is continuous for similar reasons.

Hence (ii) is true.

The proof of Theorem 23.4 is complete.

Part IV

GENERAL BOUNDARY CONDITIONS

CHAPTER 7

The Oscillation Number, Index W_0

24. Conjugate Points and Characteristic Roots

Definition 24.0 (a). *The oscillation number of the* DE

$$\frac{d}{dx}\omega_{\zeta_i}(x,\eta,\eta') = \omega_{\eta_i}(x,\eta,\eta') \qquad (i = 1, \ldots, m) \quad (x \in \aleph) \tag{24.0}$$

over the open interval (a^1,a^2) will be taken as the count of conjugate points of $x = a^1$ in (a^1,a^2) of the DE (24.0).

The Comparison Theorems of Section 21 relate the above oscillation number to the oscillation numbers of similar DE (24.0). The Separation Theorem 20.1 relates this oscillation number to the count of focal points in (a^1,a^2) of an arbitrary von Escherich family of solutions of the DE (24.0). However, focal conditions involve boundary conditions at merely *one* end point of the interval $[a^1,a^2]$.

Let W_r, $r > 0$, be an arbitrary system of canonical conditions whose DE are the DE (24.0). These conditions involve boundary conditions at *both* end points of the interval $[a^1,a^2]$. What is lacking is a knowledge of how the above oscillation number of the DE (24.0) conditions solutions of these more general systems W_r. Our Oscillation Theorem 24.1 gives such conditions in a precise way. The following definition names the entities which appear in this Oscillation Theorem.

Definition 24.0 (b). *Index* W_r *and nullity* W_r, $r \geq 0$. *Index* W_r shall be the count of negative characteristic roots λ of the system W_r of Section 15. *Nullity* W_r shall be the multiplicity of $\lambda = 0$ as a characteristic root of the system W_r.

By Theorem 19.2, *index* W_0 *is the oscillation number over* (a^1,a^2) *of Definition* 24.0.

OBJECTIVE. *In Chapter 7 we seek to evaluate the difference*

$$\text{(24.1)*} \qquad \textit{index}\ \mathrm{W_r} - \textit{index}\ \mathrm{W_0} \qquad (\mathrm{r} > 0)$$

when $\lambda = 0$ *is not a characteristic root of* $\mathrm{W_r}$.

We are thus concerned with the difference between the following two numbers:

(1) *The count of negative characteristic roots of the system* W_r, i.e. index W_r;
(2) *The count of conjugate points of* $\mathrm{x} = \mathrm{a}^1$ *in* $(\mathrm{a}^1,\mathrm{a}^2)$, i.e. index W_0.

The conjugate points are relative to the DE

$$\text{(24.2)} \qquad \frac{d}{dx}\,\omega_{\zeta_i}(x,\eta,\eta') = \omega_{\eta_i}(x,\eta,\eta') \qquad (i = 1, \ldots, m).$$

The key to an exact evaluation of the difference (24.1) is the study of a vector space of solutions of the DE (24.2) which we shall now define.

Definition 24.1. *The end space* $((\mathrm{W_r}))$, $\mathrm{r} > 0$ *of* $\mathrm{W_r}$. The vector space over R of C^1-solutions of the DE (24.2) which satisfy the conditions (see (15.0))

$$\text{(24.3)} \qquad \eta_i^s = c_{ih}^s u_h \qquad (s = 1, 2;\ i = 1, \ldots, m)$$

with some r-tuple u (possibly null), will be termed the *end space* $((W_r))$ of W_r.

In Section 5 of Appendix I we have shown that

$$\text{(24.4)} \qquad \dim\,((W_r)) = r$$

when W_r is ND. As a consequence there exist infinitely many bases $\mathscr{B}$ of the end space $((W_r))$ of the form,

$$\text{(24.5)} \qquad \mathscr{B} = (\eta^{(1)}, \ldots, \eta^{(r)})$$

where for $h = 1, \ldots, r$, $\eta^{(h)}$ is a suitably chosen nonnull mapping in $((W_r))$. An arbitrary mapping $\eta \in ((W_r))$ has the form

$$\text{(24.6)} \qquad \eta = w_h \eta^{(h)}$$

for a choice of an r-tuple w of real numbers $w_1, \ldots, w_r$ uniquely determined by η.

Theorem 24.1, below, is the principal theorem of Chapter 7. As we shall see in Theorem 24.1, when W_r is ND,

$$\text{(24.7)} \qquad \text{index } W_r - \text{index } W_0 = \kappa + \nu,$$

where κ and ν are, respectively, the index κ and nullity ν of quadratic end forms $D_{\mathscr{B}}(w)$ of W_r which we now define.

* We say that W_r is ND (*nondegenerate*) when $\lambda = 0$ is not a characteristic root of W_r.

Definition 24.2. *The quadratic end forms* $D_{\mathscr{B}}(w)$ *of* $\mathrm{W_r}$, $\mathrm{r} > 0$. Corresponding to an arbitrary base $\mathscr{B}$: (24.5) of the end space $((W_r))$ of a ND W_r and to the functional $I_r^0 \# W_r$ based on W_r, defined in Section 15, we term the symmetric quadratic form

$$D_{\mathscr{B}}(w) = I_r^0(w_h \eta^{(h)}) = d_{hk} w_h w_k \tag{24.8}^*$$

the quadratic end form of $\mathrm{W_r}$ *over the base* $\mathscr{B}$ *of* $((\mathrm{W_r}))$.

In Section 5 of Appendix I, the definition (24.8) of $D_{\mathscr{B}}(w)$ is made more explicit. It is shown that the quadratic end forms $D_{\mathscr{B}}$ and $D_{\hat{\mathscr{B}}}$ over different bases $\mathscr{B}$ and $\hat{\mathscr{B}}$ of $((W_r))$ are *R-equivalent*, that is, obtainable, one from the other, by suitable nonsingular linear transformations of the variables of one form into the variables of the other.

The principal theorem, Theorem 24.1, evaluates the difference (24.1). It is understood that W_r and W_0 are systems of canonical conditions of form (15.0) and (15.1), respectively, with *common* DE.

OSCILLATION THEOREM 24.1. *If* $\mathrm{W_r}$, $\mathrm{r} > 0$ *is an* ND *system of free canonical conditions* (15.0) *and if* κ *and* ν *are, respectively, the index and nullity common to the* R*-equivalent end forms* $\mathrm{D}_{\mathscr{B}}(\mathrm{w})$ *of* $\mathrm{W_r}$, *then*

$$\textit{nullity}\ \mathrm{W_0} = \nu \tag{24.9}$$

and

$$\textit{index}\ \mathrm{W_r} - \textit{index}\ \mathrm{W_0} = \kappa + \nu \leq \mathrm{r}. \tag{24.10}$$

The proof of Oscillation Theorem 24.1 is based on an auxiliary theorem on subordinate quadratic forms, formulated in Section 25, proved in Appendix II and applied in Section 26 to prove Oscillation Theorem 24.1.

COROLLARY 24.1. *Under the conditions of Theorem* 24.1.

$$0 \leq \textit{index}\ \mathrm{W_r} - \textit{index}\ \mathrm{W_0} \leq \mathrm{r}. \tag{24.11}$$

In Section 26 we shall prove that the relations (24.11) are true even if the canonical conditions W_r are degenerate, that is even if $\lambda = 0$ is a characteristic root of W_r. The conclusions of Theorem 24.1 are not true when W_r is degenerate.

Theorem 24.1 is particularly revealing when applied in Section 27 to the system W_r associated in (11.20) with a periodic extremal.

* The coefficients d_{hk} are unique if we suppose that $d_{hk} = d_{kh}$, as we do.

Exercises

In case $m = 1$, with η and ζ 1-tuples, set

$$2\omega(x,\eta,\zeta) \equiv \zeta^2 - \eta^2.$$

Let the conditions $W_1(\lambda)$ of a canonical system W_1 based on an interval $[a^1,a^2] = [0,\tau]$ have the form

$$\begin{bmatrix} \eta'' + \eta + \lambda\eta = 0 \\ \eta(0) = \eta(\tau) = u \\ \xi(0) - \xi(\tau) = 0 \end{bmatrix} \qquad \begin{matrix} (\lambda \in R) \\ (u \in R) \end{matrix}$$

so that characteristic solutions have the period τ. Let a system W_0 be defined by the same differential equations and by end conditions $\eta(0) = \eta(\tau) = 0$.

24.1. If $\tau = 3\pi$ show that the system W_1 is ND, that

$$\begin{aligned} &\text{index } W_1 = 3, \qquad &\text{nullity } W_1 = 0 \\ &\text{index } W_0 = 2, \qquad &\text{nullity } W_0 = 1, \end{aligned}$$

that $x \to \sin x$ is a base $\mathscr{B}$ of the "end space" $((W_1))$ and that the corresponding end form $D_{\mathscr{B}}(w)$ vanishes identically so that $\kappa = 0$ and $\nu = 1$.

25. The Auxiliary Theorem on Quadratic Forms

In this section we shall formulate the auxiliary theorem on quadratic forms to be used in Section 26 to prove Oscillation Theorem 24.1. See Morse [7].

For α and β on the range $1, \ldots, \mu$, let

$$Q(z) = a_{\alpha\beta} z_\alpha z_\beta \tag{25.1}$$

be a real-valued, symmetric quadratic form in μ-tuples z. For integers r and n such that $0 < r < \mu$ and $\mu = n + r$ set

$$(z_1, \ldots, z_\mu) = (u_1, \ldots, u_r : s_1, \ldots, s_n) \tag{25.2}$$

and, subject to (25.2), set

$$Q(z) = P(u,s). \tag{25.3}$$

Let $\mathbf{0}$ be the r-tuple with null components. We shall term $P(\mathbf{0},s)$ a quadratic form *subordinate* to $Q(z)$ or $P(u,s)$. When either $Q(z)$ or $P(\mathbf{0},s)$ is nonsingular, we shall give a formula for the difference

$$\text{index } P(u,s) - \text{index } P(\mathbf{0},s) \tag{25.4}$$*

* In Section 26 we shall show the difference

$$\text{index } W_r - \text{index } W_0 \qquad \text{(of (24.1))}$$

with proper interpretation, takes the form (25.4).

in terms of the index and nullity of a class of R-equivalent quadratic forms termed *complementary* to $P(\mathbf{0},s)$ (cf. Definition 25.2).

Quadratic forms complementary to $P(\mathbf{0},s)$. If either $Q(z)$ or $P(\mathbf{0},s)$ is nonsingular, the matrix of coefficients of the $n = \mu - r$ linear equations

$$\frac{\partial Q}{\partial z_{r+1}}(z) = \cdots = \frac{\partial Q}{\partial z_\mu}(z) = 0 \tag{25.5}$$

in the μ-tuples z, will have rank n. In such a case the points z in R^μ which satisfy (25.5) will make up an r-plane

$$\pi_r = \left\{ z \in R^\mu \,\middle|\, \frac{\partial Q}{\partial z_{r+1}}(z) = \cdots = \frac{\partial Q}{\partial z_\mu}(z) = 0 \right\}. \tag{25.6}$$

Definition 25.1. *A base* B *for* π_r. Let

$$B = (z^{(1)}, \ldots, z^{(r)}) \tag{25.7}$$

be a *base for* π_r, that is, a maximal set of linearly independent μ-tuples in π_r.

Given the base B of π_r, an arbitrary point $z \in \pi_r$ can be represented as a μ-tuple $z = w_h z^{(h)}$ in which $w = (w_1, \ldots, w_r)$ is an r-tuple uniquely determined by z and B. The restriction $Q \mid \pi_r$ has a representation as a symmetric quadratic form

$$H_B(w) = Q(w_h z^{(h)}) = (a_{\alpha\beta} z_\alpha^{(h)} z_\beta^{(k)}) w_h w_k \tag{25.8}$$

where the range of h and k is $1, \ldots, r$, and the range of α and β is $1, \ldots, \mu$ (cf. (25.1)).

Definition 25.2. *We term* $H_B(w)$ *a quadratic form, over the base* B : (25.7), **complementary** *to the form* $P(\mathbf{0},s)$ *subordinate to* $P(u,s)$.

The auxiliary theorem on quadratic forms which enables us to prove the Oscillation Theorem 24.1 is as follows.

AUXILIARY THEOREM 25.1. *If the above quadratic form* $Q(z) = P(u,s)$ *is nonsingular, the following is true:*

(i). *For any two bases* B *of the* r*-plane* π_r : (25.6) *the quadratic forms* $H_B(w)$, *complementary to the form* $P(\mathbf{0},s)$, *are* R*-equivalent, and accordingly have a common index* κ *and nullity* ν. *Moreover*

(ii). *Nullity* $P(\mathbf{0},s) = \nu$

and

(iii). *Index* $P(u,s)$ − *Index* $P(\mathbf{0},s) = \kappa + \nu$.

The proof of Theorem 25.1 is given in Appendix II.

26. Proof of Oscillation Theorem 24.1

It follows from Theorem 15.2 (ii) that the difference, index W_r − index W_0, of Theorem 24.1, equals

$$\text{index } Q^0\#(W_r,\Lambda_p) - \text{index } Q^0\#(W_0,\Lambda_p) \qquad (r > 0) \tag{26.1}$$

for a suitable choice of a 0-frame Λ_p over (a^1,a^2). Set $n = mp$ and $\mu = n + r$. With a proper naming of its variables, $Q^0\#(W_r,\Lambda_p)$ is a symmetric quadratic form in μ variables

$$(z_1, \ldots, z_\mu) = (u_1, \ldots, u_r : s_1, \ldots, s_n) \tag{26.2}$$

and, subject to (26.2), will be denoted by

$$\hat{Q}(z) = \hat{P}(u,s). \tag{26.3}$$

With a proper naming of its variables $Q^0\#(W_0,\Lambda_p)$ is the form $\hat{P}(\mathbf{0},s)$ "subordinate" to $\hat{P}(u,s)$. This follows from the definitions in Section 15, culminating in (15.3). Moreover, the quadratic form $\hat{Q}(z)$ is *nonsingular* by Theorem 15.1 (i), since $\lambda = 0$ is not a characteristic root of W_r by hypothesis of Theorem 24.1. Auxiliary Theorem 25.1 on quadratic forms can accordingly be applied to $\hat{P}(u,s)$ and to $\hat{P}(\mathbf{0},s)$, as subordinate to $\hat{P}(u,s)$.

The Application of Theorem 25.1. Paralleling the definition in (25.6) of the r-plane π_r, we here set

$$\hat{\pi}_r = \left\{ z \in R^\mu \,\middle|\, \frac{\partial \hat{Q}}{\partial z_{r+1}}(z) = \cdots = \frac{\partial \hat{Q}}{\partial z_\mu}(z) = 0 \right\} \qquad (r > 0). \tag{26.4}$$

Since $\hat{Q}(z)$ is nonsingular, $\hat{\pi}_r$, like π_r, is an r-plane in R^μ. It follows from (15.8)′ that, for μ-tuples $z \in \hat{\pi}_r$, the broken secondary extremal $h^{0,z}\#(W_r,\Lambda_p)$ of Definition 15.2(b) is without corners, and is accordingly the graph of a mapping $\eta = \varphi(z)$, contained in the "end space" $((W_r))$ of W_r. We see that the mapping

$$z \to \varphi(z) : \hat{\pi}_r \to ((W_r)) \tag{26.5}$$

is linear, biunique and surjective.

By virtue of (26.5) there is a 1–1 correspondence between the bases B of $\hat{\pi}_r$ and the bases $\mathscr{B}$ of $((W_r))$ in which a base

$$B = (z^{(1)}, \ldots, z^{(r)}) \tag{26.6}$$

of $\hat{\pi}_r$ corresponds to a base,

$$\mathscr{B} = (\varphi(z^{(1)}), \ldots, \varphi(z^{(r)})) = \varphi(B) \qquad (\text{of } ((W_r))). \tag{26.7}$$

The mapping (26.5) has another basic property. If I_r^0 is "based" on W_r, that is if $I_r^0 = I_r^0 \# W_r$, then

$$\hat{Q}(z) = I_r^0(\varphi(z)) \qquad (z \in \hat{\pi}_r) \tag{26.8}$$

by virtue of (15.3). Corresponding to a base B of $\hat{\pi}_r$ let $\hat{H}_B(w)$ be a quadratic form, complementary to $\hat{P}(\mathbf{0},s)$ in the sense of Definition 25.2. It follows, respectively, from (25.8), from (26.8), and from Definition 24.2, of a quadratic end form $D_{\mathscr{B}}(w)$ of $((W_r))$, that

$$\hat{H}_B(w) \equiv \hat{Q}(w_h z^{(h)}) \equiv I_r^0(w_h \varphi(z^{(h)})) \equiv D_{\varphi(B)}(w) \qquad (w \in R^r). \tag{26.9}$$

Conclusion of Proof of Oscillation Theorem 24.1. It follows from (26.9) that the index κ and nullity ν of a quadratic form $\hat{H}_B(w)$, complementary to the form $\hat{P}(\mathbf{0},s)$ subordinate to $\hat{P}(u,s)$, equal the respective index and nullity of the *end form* $D_{\varphi(B)}(w)$ of W_r. On recalling Definition 24.0(b) of Index W_r and Nullity W_r, we find that

$$\begin{cases} \text{nullity } W_0 = \text{nullity } \hat{P}(\mathbf{0},s) \\ \text{index } W_r = \text{index } \hat{P}(u,s) \\ \text{index } W_0 = \text{index } \hat{P}(\mathbf{0},s) \end{cases} \tag{26.10}$$

by virtue of Theorem 15.2, with $\lambda = 0$ therein, and the definition of $\hat{Q}(z) = \hat{P}(u,s)$.

Oscillation Theorem 24.1 *follows from the Auxiliary Theorem* 25.1 *on quadratic forms, applied to* $\hat{P}$(u,s).

We return to the relation (24.11) of Section 24.

Another proof of (24.11). When W_r is ND, the relations (24.11) are a corollary of Theorem 24.1. However, (24.11) can be readily proved under the weaker hypotheses of the following theorem.

THEOREM 26.1. *If* $\mathrm{W_r}$ *is a system of canonical conditions* (15.0) *for which* $\mathrm{r} > 0$ *and if* $\mathrm{W_0}$ *is a system of dimension* 0 *with the same* DE, *then*

$$0 \le \text{index } W_r - \text{index } W_0 \le r. \tag{26.11}$$

Proof of Theorem 26.1. If the hypothesis that W_r is ND is dropped the exposition of Section 26 is valid up to and including the setting of $\hat{Q}(z) = \hat{P}(u,s)$ in (26.3). The three equalities (26.10) are also valid. To establish the relations (26.11) it is sufficient to show that

$$0 \le \text{index } \hat{P}(u,s) - \text{index } \hat{P}(\mathbf{0},s) \le r. \tag{26.12}$$

This is an elementary theorem on the quadratic form $\hat{P}(u,s)$ and its subordinate form $\hat{P}(\mathbf{0},s)$.

Proof of (26.12). The r-tuples u and n-tuples s combine to give a μ-tuple (u,s) where $\mu = r + n$. If index $\hat{P}(u,s) = \kappa$, there is a κ-plane π_κ meeting the origin in the space R^μ of the μ-tuples (u,s) on which the quadratic form $\hat{P}(u,s)$ is negative definite. In R^μ the κ-plane π_κ intersects the coordinate subspace R^n of the n-tuples $(s_1, \ldots, s_n)$ in a k-plane π_k for which $k \geq \kappa - r$. Since $\hat{P}(\mathbf{0},s)$ is negative definite on π_k, we infer that index $\hat{P}(\mathbf{0},s) \geq k \geq \kappa - r$. Hence

$$\text{index } \hat{P}(\mathbf{0},s) \geq \text{index } \hat{P}(u,s) - r. \tag{26.13}$$

It is a trivial consequence of the geometric definition of the index of a quadratic form that

$$\text{index } \hat{P}(u,s) \geq \text{index } \hat{P}(\mathbf{0},s). \tag{26.14}$$

From (26.14) and (26.13) we infer the truth of the relations (26.12).

The relations (26.11) follow from the relations (26.12) and the equalities (26.10).

Thus Theorem 26.1 is true.

27. Quadratic End Forms of the System W_r, $(r > 0)$

The quadratic end forms $D_{\mathscr{B}}(w)$ of W_r (Definition 24.2) have enabled us to evaluate the difference

$$\text{index } W_r - \text{index } W_0 \qquad (r > 0) \tag{27.1}$$

in the Oscillation Theorem 24.1, when W_r is ND. In this section we shall derive a simple formula for the coefficients of an end form $D_{\mathscr{B}}(w)$ of W_r. This formula is particularly useful when the conditions W_r are of the periodic type (11.20).

We are assuming that $\lambda = 0$ is not a characteristic root of W_r. By Lemma 5.1 of Appendix I, dim $((W_r))$ is then r. A base for $((W_r))$ has the form

$$\mathscr{B} = (\eta^{(1)}, \ldots, \eta^{(r)}) \tag{27.2}$$

of r linearly independent C^1-solutions of the DE

$$\frac{d}{dx}\,\omega_{\zeta_i}(x,\eta,\eta') = \omega_{\eta_i}(x,\eta,\eta') \qquad (i = 1, \ldots, m) \tag{27.3}$$

satisfying the BC : (15.0)″ with r-tuples denoted, respectively, by

$$u^{(1)}, \ldots, u^{(r)}. \tag{27.4}$$

Let the respective "canonical associates" ξ of the mappings η in $\mathscr{B}$: (27.2) be denoted by

$$\xi^{(1)}, \ldots, \xi^{(r)}. \tag{27.5}$$

The base $\mathscr{B}$ and associated sets (27.4) and (27.5) give the data needed in the following lemma.

LEMMA 27.1. *The end form* $D_{\mathscr{B}}(w)$ *of a system* W_r *of canonical conditions* (15.0) *is a symmetric quadratic form*

$$D_{\mathscr{B}}(w) = d_{hk} w_h w_k \tag{27.6}$$

in the r-*tuple* w *with coefficients*

$$d_{hk} = \Big[\eta_i^{(h)}(x)\xi_i^{(k)}(x)\Big]_{a^1}^{a^2} + b_{\mu\nu} u_\mu^{(h)} u_\nu^{(k)} \tag{27.7}$$

where h, k, μ, ν, *have the range* 1, . . . , r, *and* i *the range* 1, . . . , m. *The comparison matrix* $\|b_{\mu\nu}\|$ *is taken from conditions* (15.0).

Note first that the right member, say g_{hk}, of (27.7) is such that $g_{hk} = g_{kh}$ by virtue of (18.1) and the relation $b_{\mu\nu} = b_{\nu\mu}$.

Proof of (27.7). For η in the end space $((W_r))$ of W_r and I_r^0 based on W_r,

$$\begin{aligned} I_r^0(\eta) &= b_{\mu\nu}u_\mu u_\nu + \int_{a^1}^{a^2} 2\omega(x,\eta(x),\eta'(x))\,dx \qquad \text{(see (15.2)'')} \\ &= b_{\mu\nu}u_\mu u_\nu + \Big[\eta_i(x)\,\xi_i(x)\Big]_{a^1}^{a^2} \end{aligned} \tag{27.8}$$

where u is the r-tuple which satisfies (15.0)'' with η. By definition (24.8)

$$D_{\mathscr{B}}(w) = I_r^0(w_h \eta^{(h)}) \tag{27.9}$$

for each r-tuple w. The right member of (27.9) can be evaluated by means of the right member of (27.8). In this evaluation one sets

$$\eta_i = w_h \eta_i^{(h)}, \qquad \xi_i = w_k \xi_i^{(k)}, \qquad u_\mu = w_h u_\mu^{(h)}, \qquad u_\nu = w_k u_\nu^{(k)}. \tag{27.10}$$

Taking account of the fact that $d_{hk} = d_{kh}$ by hypothesis, relations (27.7) follow.

The periodic case. For each $x \in \aleph$ let $\omega(x,\eta,\zeta)$ be a quadratic form in the m-tuples η and ζ, conditioned as in Section 1 of Appendix I, with coefficients P_{ij}, Q_{ij}, R_{ij} which have a period $\tau > 0$ in x. The system of canonical conditions $W_m(\lambda)$ of form,

$$\left[\begin{aligned} &\frac{d}{dx}\omega_{\zeta_i}(x,\eta,\eta') - \omega_{\eta_i}(x,\eta,\eta') + \lambda\eta_i = 0 && (27.11)' \\ &\eta_i(0) = \eta_i(\tau) = u_i \qquad (i = 1, \ldots, m) && (27.11)'' \\ &\xi_i(0) - \xi_i(\tau) = 0, && (27.11)''' \end{aligned}\right] \qquad (u \in R^m)$$

defined for each $\lambda \in R$, will be denoted by PW_m (read the system W_m in the periodic case). The "derived" conditions (11.20) take the form (27.11) when $\omega \equiv \Omega$.

A characteristic solution (η, ξ) of the system PW_m has the period τ in x. The system PW_m is termed *nondegenerate* if conditions (27.11) have no nonnull solution when $\lambda = 0$.

The end space $((\mathrm{PW}_m))$ *of* PW_m. According to Definition 24.1, $((PW_m))$ is the vector space over R of solutions of the DE (27.3) such that $\eta(0) = \eta(\tau)$. When PW_m is ND, dim $((PW_m)) = m$, as Lemma 5.1 of Appendix I affirms.

Objective. To evaluate the index of a ND system $\mathrm{PW_m}$ *with the aid of Oscillation Theorem* 24.1, *first simplifying the evaluation of the index* κ *and nullity* ν *of the quadratic end forms of* $\mathrm{PW_m}$.

The nullity ν. The quadratic end forms of a ND system PW_m have a nullity which can be evaluated as follows. If

$$\mathscr{B} = (\eta^{(1)}, \ldots, \eta^{(m)}) \tag{27.12)$'$ (}$$

is a base of the end space $((PW_m))$, the nullity of the m-square matrix

$$\|\eta_i^{(h)}(0)\| \qquad (\text{range } i, h = 1, \ldots, m) \tag{27.12)$''$ (}$$

is the *multiplicity* ν of $x = \tau$ as a conjugate point of $x = 0$. By Theorem 24.1 this multiplicity is the nullity ν common to the end forms of PW_m. See Section 5 Appendix I.

Definition 27.1. *A reduced diagonal base for* $((\mathrm{PW}_m))$. If PW_m is ND and if ν is the nullity of the matrix (27.12)$''$, then for a suitable choice of a new base $\mathscr{B}$ for $((PW_m))$, the new matrix (27.12)$''$ will be a "diagonal" matrix all of whose elements vanish, except for the first $m - \nu$ elements in the matrix diagonal each of which can be taken as 1. Such a base will be termed a *reduced diagonal base* for $((PW_m))$.

Examples will show that the nullity ν of the end forms of a ND system PW_m can be any integer on the range $0, 1, \ldots, m$.

The index κ. We seek an evaluation of the index κ of the end forms of a ND system PW_m. In the special case in which the nullity ν of these end forms is m, the end forms vanish identically and so have a null index κ. The following theorem concerns the case in which $\nu < m$.

THEOREM 27.1. *Let* $\mathrm{PW_m}$ *be a ND system whose end forms have a nullity* $\nu < \mathrm{m}$. *If* $\mathscr{B}$ *is a "reduced diagonal base" of the end space* $((\mathrm{PW_m}))$ *of* $\mathrm{PW_m}$, *then the corresponding symmetric quadratic end form*

$$\mathrm{D}_{\mathscr{B}}(\mathrm{w}) = \mathrm{d_{ij} w_i w_j} \qquad (\text{Definition } 24.2) \tag{27.13}$$

in the m-*tuple* w *has a matrix whose elements are null except for an* (m − ν)-*square principal minor of elements*

$$d_{st} = \xi_t^{(s)}(\tau) - \xi_t^{(s)}(0) \qquad (s, t = 1, \ldots, m - \nu) \tag{27.14}$$

where $\xi^{(1)}, \ldots, \xi^{(m)}$ *are the "canonical associates" of the respective mappings* $\eta^{(1)}, \ldots, \eta^{(m)}$ *in the base* $\mathscr{B}$.

Proof. One evaluates the coefficients d_{ij}, $i, j = 1, \ldots, m$ of the form (27.13) with the aid of the formula (27.7). By hypothesis each coefficient $b_{ij} = 0$. Because $\mathscr{B}$ is a reduced diagonal base

$$\eta_i^{(k)}(0) = \eta_i^{(k)}(\tau) = \delta_i^k \qquad (i, k = 1, \ldots, m - \nu) \tag{27.15}$$

while the remaining elements in the m-square matrices

$$\|\eta_i^{(k)}(0)\| = \|\eta_i^{(k)}(\tau)\| \qquad (\text{range } i, k = 1, \ldots, m) \tag{27.16}$$

vanish. Theorem 27.1 follows from (27.7).

We summarize the facts concerning the system PW_m in Theorem 27.2. Recall first that the system PW_m is ND in the sense of Oscillation Theorem 24.1, if and only if the DE

$$\frac{d}{dx}\,\omega_{\zeta_i}(x,\eta,\eta') = \omega_{\eta_i}(x,\eta,\eta') \qquad (i = 1, \ldots, m) \tag{27.17}$$

have no nonnull periodic solution.

THEOREM 27.2. *If the system* $\mathrm{PW_m}$ *is ND the following is true:*

(i) *The nullity* ν *common to the end forms* (*Definition* 24.2) *of* $\mathrm{PW_m}$ *equals the multiplicity of* $x = \tau$ *as a conjugate point of* x = 0 *relative to the DE* (27.17). *If* ν = m *the index* κ *of these end forms vanishes. If* ν < m, κ *equals the count of negative characteristic roots of the* (m − ν)-*square matrix*

$$\|d_{st}\| = \|\xi_t^{(s)}(\tau) - \xi_t^{(s)}(0)\| \qquad (\textit{range } s, t = 1, \ldots, m - \nu) \tag{27.18}$$

of (27.14).

(ii) *If* K *is the count of conjugate points of* x = 0 *in* (0,τ] *then*

$$\textit{index } \mathrm{PW_m} = \mathrm{K} + \kappa. \tag{27.19}$$

The affirmations of (i) have been established. Let k be the count of conjugate points of $x = 0$ on the interval $(0,\tau)$. By Oscillation Theorem 24.1, ν is the multiplicity of $x = \tau$ as conjugate point of $x = 0$ and

$$\text{index } PW_m - k = \nu + \kappa.$$

Since $K = k + \nu$, (27.19) follows.

The system PW_m when $m = 1$. When $m = 1$ the JDE (27.17) reduces to a single DE

$$\frac{d}{dx}(r(x)\eta' + q(x)\eta) = q(x)\eta' + p(x)\eta \tag{27.20}$$

where p, q, r are continuous for $x \in R$ and have the period τ. By hypothesis $r(x) > 0$ for each x. When $m = 1$, the conditions $PW_m(\lambda)$: (27.11) in E_{m+1} become conditions $PW_1(\lambda)$ in E_2 and have the form

$$\begin{array}{ll} (27.21)' & \left[\frac{d}{dx}(r(x)\eta' + q(x)\eta) - (q(x)\eta' + p(x)\eta) + \lambda\eta = 0\right. \\ (27.21)'' & \quad \eta(0) = \eta(\tau) = u \qquad (u \in R) \\ (27.21)''' & \left. \quad \xi(0) - \xi(\tau) = 0 \right] \end{array}$$

The subscript i in (27.11) has just one value 1 when $m = 1$ and is omitted in (27.21).

Nondegeneracy. A system PW_1 will be said to be ND (nondegenerate) if there exists no solution of its JDE : (27.20) with a period τ, other than the null solution.

We shall define classes (L), (M), (N) of the above ND system PW_1. The intersection of any two of these classes will presently be shown to be empty. A ND PW_1 will be included in (L), (M), or (N), respectively, if its JDE has a solution η such that the corresponding condition

$$\begin{aligned} \mathrm{L} &: \eta(0) = \eta(\tau) = 1 : \eta'(\tau) > \eta'(0) \\ \mathrm{M} &: \eta(0) = \eta(\tau) = 1 : \eta'(\tau) < \eta'(0) \\ \mathrm{N} &: \eta(0) = \eta(\tau) = 0 : \eta'(\tau) \neq \eta'(0) \end{aligned}$$

is satisfied. Simple examples show that all three classes exist.

The following theorem is a corollary of Theorem 27.2.

THEOREM 27.3. *A ND system* PW_1 *is in one and only one of the above three classes.*

If μ *is the count of conjugate points of* $\mathrm{x} = 0$ *in* $(0,\tau)$ *relative to the JDE* : (27.20), *then the following is true. For ND systems* PW_1 *in classes* (*L*), (*M*), (*N*), *index* PW_1 *equals* μ, $\mu + 1$, $\mu + 1$ *respectively.*

By definition the end space $((PW_1))$ is the vector space of nonnull solutions η of the JDE : (27.20) such that $\eta(0) = \eta(\tau)$. Since $m = 1$, a base for $((PW_1))$ consists of just *one* nonnull solution η. Because of the existence of a nonnull solution η such that $\eta(0) = \eta(\tau)$, it is clear that a ND system belongs to at least one of the three classes (L), (M), (N).

The multiplicity ν of $x = \tau$ as a conjugate point of $x = 0$ is 1 or 0.

The case $\nu = 0$. If $\nu = 0$, a "reduced base" for $((PW_1))$: (Definition 27.1) consists of a solution η such that $\eta(0) = \eta(\tau) = 1$. If ξ is the canonical associate of η, the matrix (27.18) consists of just one element

$$\xi(\tau) - \xi(0) = r(0)(\eta'(\tau) - \eta'(0)) \neq 0 \qquad (\text{when } \nu = 0) \tag{27.22}$$

where $\eta'(\tau) - \eta'(0) \neq 0$ since the system PW_1 is ND by hypothesis. According to Theorem 27.1 the end form (27.13) of PW_1 defined by the above reduced base, is the quadratic form

$$r(0)(\eta'(\tau) - \eta'(0))w^2 \qquad (\text{by } (27.14))$$

and so has an index $\kappa = 0$ or 1 according as PW_1 belongs to (L) or (M).

The Case $\nu = 1$. In this case the end form (27.13) vanishes identically according to Theorem 27.2, since $\nu = m$, and so has an index $\kappa = 0$.

As in Theorem 27.2, let K be the count of conjugate points of $x = 0$ in $(0,\tau]$. The above analysis leads to the entries in the following table:

	ν	κ	K
Class (L)	0	0	μ
Class (M)	0	1	μ
Class (N)	1	0	$\mu + 1$

From this table it is clear that a ND system cannot belong to two of the classes (L), (M), (N), since ν, μ, κ, K are uniquely determined by the system PW_1. The table and the formula, index $PW_1 = K + \kappa$, of Theorem 27.2 confirm the conclusion of Theorem 27.3.

Note. The oscillation theorems which have been presented in this chapter may be called *comparative* oscillation theorems. They are to be distinguished from the oscillation theorems of Section 37.

28. A Minimizing Periodic Extremal g

This section is a continuation of Section 11. There is given a preintegrand f and extremal g which have the period τ in the sense of (8.9). Set $[a^1,a^2] = [0,\tau]$. The end conditions C_m satisfied by g have the form (11.19)′ and (11.19)″ and are denoted by $C_m(P)$ (read periodic conditions C_m).

The end function $\alpha \to \Theta(\alpha)$ associated in (8.18) with our end conditions is taken as a null function. Because $\Theta(\alpha) \equiv 0$ and because $g'(0) = g'(\tau)$ by hypothesis, the transversality conditions (9.1) based on $C_m(P)$ are satisfied by g. Thus g is a *critical extremal* of $(J,C_m(P))$ or more generally of $(J^\lambda,C_m(P))$ for each λ. See Definition 9.1.

By Definition 11.4 the index {nullity} of g as a critical extremal of $(J,C_m(P))$ is the count of negative {null} characteristic roots of the conditions (11.20), that is, the conditions

$$\left.\begin{array}{ll}(28.1)' & \frac{d}{dx}\Omega_{\zeta_i}(x,\eta,\eta') - \Omega_{\eta_i}(x,\eta,\eta') + \lambda\eta_i = 0 \\ (28.1)'' & \eta_i(0) = \eta_i(\tau) = u_i \\ (28.1)''' & \xi_i(0) - \xi_i(\tau) = 0\end{array}\right] \quad \begin{array}{l}(\lambda \in R) \\ (u \in R) \\ (i = 1, \ldots, m)\end{array}$$

where Ω is defined in (4.0). It is understood that the Legendre S-condition (7.9) holds along g.

For each x, Ω is a quadratic form in the m-tuples η and ζ. The coefficients in the form Ω are C^1-mappings of $[a^1,a^2]$ into R, unlike the quadratic forms $\omega(x,\eta,\zeta)$ of Section 1, Appendix I whose coefficients, R_{ij}, Q_{ij}, P_{ij} are required to be no more than continuous.

We shall give conditions under which a critical extremal g of $(J,C_m(P))$ affords a minimum to the integral **J** with values

$$(28.2) \qquad \mathbf{J}(\gamma) = \int_0^\tau f(x,\gamma(t),\gamma'(t))\,dt$$

on "$C_m(P)$-admissible" curves γ in a sufficiently small open neighborhood of g in E_{m+1}. Conditions for a minimum involve *conjugate* points which are taken relative to the JDE

$$(28.3) \qquad \frac{d}{dx}\Omega_{\zeta_i}(x,\eta,\eta') = \Omega_{\eta_i}(x,\eta,\eta') \qquad (i = 1, \ldots, m).$$

Weak minima are defined in Section 9 and *proper strong relative minima* in Definition 16.1.

Necessary Conditions. In order that an x-parameterized extremal g with period τ and x-domain $[0,\tau]$ afford a *weak* minimum to **J** relative to neighboring $C_m(P)$-admissible curves, it is necessary that the four following conditions hold.

(a_1) *The index of* g *as a critical extremal of* $(J,C_m(P))$ *must be zero*; *by Theorem* 16.1.

(a_2) *The integral*

$$\int_0^\tau \Omega(\mathrm{x},\eta(\mathrm{x}),\eta'(\mathrm{x}))\,\mathrm{dx} \geq 0$$

for each $C''_m(P)$-*admissible curve* η; *by Corollary* 16.1 (i).

(a_3) *The point* $\mathrm{x} = 0$ *is not conjugate to any point in* $(0,\tau)$; *by Theorem* 4.1.

(a_4) *The point* $\mathrm{x} = 0$ *is conjugate to the point* $\mathrm{x} = \tau$ only *if each solution* η

of the JDE (28.3) *for* $\mathrm{x} \in [0,\tau]$ *such that* $\eta(0) = \eta(\tau) = 0$, *admits an extension for all values of* x *which is a solution with the period* τ; *by Corollary* 16.1 (ii).

A Theorem of Poincaré when $m = 1$. According to Poincaré when $m = 1$ and the periodic extremal is minimizing in the above sense, the point $x = 0$ cannot be conjugate to the point $x = \tau$. See Hadamard [1], p. 432.

Proof. Were $x = 0$ conjugate to $x = \tau$ there would exist a nonnull solution of the JDE such that $\eta(0) = \eta(\tau) = 0$. According to (a_4), η would be a solution of the JDE with period τ, so that $\eta'(0) = \eta'(\tau)$. When $m = 1$, such a solution would vanish at least once on the interval $(0,\tau)$, contrary to (a_3).

Theorem 16.2m implies Theorem 28.1. In Theorem 28.1 the preintegrand f and extremal g have the peroid τ, in the sense of (8.9).

THEOREM 28.1. *Sufficient conditions that the* x-*parameterized extremal* g *with period* τ *afford a proper, strong minimum* **J**(g) *to the integral* **J** *with values* (28.2) *on* $\mathrm{C_m}$(P)-*admissible curves in some neighborhood of* g *in* X, *are that the Weierstrass and Legendre* S-*conditions hold relative to* f *and* g *and that the ordinary index and nullity of* g *be zero.*

Theorem 28.1 is supplemented as follows.

SUPPLEMENT TO THEOREM 28.1. *For the index and nullity of the periodic extremal* g *of Theorem* 28.1 *to be zero it is necessary and sufficient that the following two conditions be satisfied;*

(i) *For each nonnull solution* η *of the JDE* (28.3) *such that* $\eta(0) = \eta(\tau)$ *the integral*

$$\int_0^\tau 2\Omega(\mathrm{x},\eta(\mathrm{x}),\eta'(\mathrm{x}))\,d\mathrm{x} > 0. \tag{28.4}$$

(ii) *The point* $\mathrm{x} = 0$ *be conjugate to no point in* $(0,\tau]$.

Proof that the conditions are sufficient.

Let $P\hat{W}_m$ denote the canonical system defined for a prescribed λ by the conditions (28.1). The system $P\hat{W}_m$ is ND for the following reason. There is no nonnull solution $\bar{\eta}$ of the JDE (28.3) which has the period τ, since for such an $\bar{\eta}$

$$\int_0^\tau \Omega(x,\bar{\eta}(x),\bar{\eta}'(x))\,dx = 0 \tag{28.5}$$

contrary to the condition (28.4). Since $P\hat{W}_m$ is ND, Theorem 27.2 can be applied. By hypothesis the count K of conjugate points of $x = 0$ in $(0,\tau]$ is zero.

The relation (28.4) holds for each nonnull η in the end space $((\hat{W}_m))$ of $\hat{W}_m$. Hence the quadratic end forms of $\hat{W}_m$ are positive definite in accord with definition (24.8) and so have an index $\kappa = 0$. Thus K and κ both vanish in the formula (27.19) for index $P\hat{W}_m$, so that index $P\hat{W}_m = 0$. Nullity $P\hat{W}_m = 0$, since $P\hat{W}_m$ is ND. Finally by definition

$$\text{(28.6)} \qquad \text{index } g = \text{index } P\hat{W}_m; \text{ nullity } g = \text{nullity } P\hat{W}_m.$$

Thus the conditions of the theorem are sufficient.

Proof that the Conditions of the Theorem are Necessary. By hypothesis the elements in (28.6) vanish. Hence all characteristic roots σ of the system PW_m (with $\omega = \Omega$) are positive. The integral (28.4) gives the values of the quadratic functional $I_m^0\#PW_m$, and is accordingly positive definite by Theorem 16.4. It follows that $x = 0$ is conjugate to no point in $(0,\tau]$.

Thus the Supplement to Theorem 28.1 is true.

Historical Note. Hadamard, in [1] p. 432, has given a proof of the theorem of Poincaré. For related matter on periodic extremals see Carathéodory [1], p. 335, Hedlund [1], Morse and Pitcher [1]. For studies in the global theory of periodic extremals one may turn to Ljusternik [1]. References to later work in the global theory will be given in the book on Variational Topology to follow this book.

CHAPTER 8

Selfadjoint Boundary Conditions

29. General Mixed Boundary Conditions

Following Definition 15.0 we made the affirmation: arbitrary admissible selfadjoint BC, adjoined to the DE (15.0)′ are always *equivalent*, in a sense to be defined, to suitably chosen canonical BC of form (15.0)″ and (15.0)‴.

OBJECTIVE OF CHAPTER 8. *To complete the definition of terms in this affirmation, and to establish its validity.*

We are concerned with BC adjoined to DE of the form

$$(29.0)(\lambda) \qquad \frac{d}{dx}\,\omega^{\lambda}_{\xi_i}(x,\eta,\eta') = \omega^{\lambda}_{\eta_i}(x,\eta,\eta') \qquad (i = 1, \ldots, m)$$

conditioned as in Appendix I. In Section 15, W_r denoted a system of canonical conditions $W_r(\lambda)$, one for each value of λ, of form (15.0) when $r > 0$, and of form (15.1) when $r = 0$. The BC, when $r > 0$, have the form

$$\begin{aligned}&(29.1)'r \qquad && \eta^s_i = c^s_{ih}u_h \qquad (s = 1, 2;\ i = 1, \ldots, m)\\ &(29.1)''r && c^1_{ih}\xi^1_i - c^2_{ih}\xi^2_i = b_{hk}u_k \qquad (h = 1, \ldots, r)\end{aligned}$$

and when $r = 0$, the form (15.1)″, i.e., the form

$$(29.2) \qquad \eta^s_i = 0 \qquad (s = 1, 2;\ i = 1, \ldots, m).$$

The conditions (29.1) and (29.2) are on the $4m$ variables

$$(29.3) \qquad (\eta^s,\xi^s) = \begin{bmatrix} \eta^1_1, \ldots, \eta^1_m;\ \xi^1_1, \ldots, \xi^1_m \\ \eta^2_1, \ldots, \eta^2_m;\ \xi^2_1, \ldots, \xi^2_m \end{bmatrix}$$

with the $2m$ variables ξ^s_i unconditioned when $r = 0$. The conditions (29.1) and (29.2) are independent of λ.

In the classical theory the conditions on the $4m$-tuples (η^s, ξ^s) are ordinary, linear, and homogeneous conditions of the form

$$(29.4)^* \qquad g_\alpha(\eta^s, \xi^s) = 0 \qquad (\alpha = 1, \ldots, 2m)$$

where the forms $g_1, \ldots, g_{2m}$ are linearly independent. The conditions (29.1) lead to $2m$ linearly independent conditions of form (29.4) (see Ex. 11.2) on eliminating the r parameters $u_1, \ldots, u_r$. The conditions (29.2) have the form (29.4) a priori.

Definition 29.0. *The carrier of a set of BC*. By the *carrier* of a set of BC is meant the vector subspace of R^{4m} of the $4m$-tuples (η^s, ξ^s) which satisfy the given BC.

Definition 29.1. *Equivalent BC*. Two sets Z and $\hat{Z}$ of BC, adjoined to the DE (15.0)$'$, whether in parametric or nonparametric form, are termed *equivalent*, if their carriers are the same.

Definition 29.2. *Admissible BC*. We shall admit those BC which are defined by a finite set of linear homogeneous conditions on the $4m$-tuples (η^s, ξ^s) and a finite set (possibly empty) of auxiliary parameters. We require that the carrier of the $4m$-tuples (η^s, ξ^s), satisfying these conditions with a set of parameters (if any exist), be a $2m$-plane meeting the origin in the space of the $4m$ variables (η^s, ξ^s). Conditions (29.1) are an "admissible" set of BC with parameters $u_1, \ldots, u_r$.

Definition 29.3. *Adjoint boundary conditions*. Two admissible sets of BC, Z, and $\hat{Z}$ will be said to be *adjoint*, if the bilinear form

$$(29.5) \qquad \left[\eta_i^s \hat{\xi}_i^s - \hat{\eta}_i^s \xi_i^s\right]_{s=1}^{s=2} = P(\eta, \xi; \hat{\eta}, \hat{\xi})\dagger$$

in the $4m$ variables (η^s, ξ^s) and the $4m$ variables $(\hat{\eta}^s, \hat{\xi}^s)$ vanishes, whenever the $4m$-tuple (η^s, ξ^s) satisfies the conditions Z, and $(\hat{\eta}^s, \hat{\xi}^s)$ the conditions $\hat{Z}$. The conditions Z are termed *selfadjoint* if $P(\eta, \xi; \hat{\eta}, \hat{\xi}) = 0$ whenever the variables $(\hat{\eta}^s, \hat{\xi}^s)$, as well as the variables (η^s, ξ^s), satisfy the conditions Z.

Historical origin of the bilinear form (29.5). We refer to the operators $L_i(\eta)$, $i = 1, \ldots, m$, introduced in Appendix I. Let η and $\hat{\eta}$ be two C^1-mappings of the interval $\aleph$ into R^m whose canonical associates, ξ and $\hat{\xi}$ respectively, are of class C^1. Green's formula

$$(29.6) \qquad \int_{a^1}^{a^2} \big(\eta_i(x) L_i(\hat{\eta}(x)) - \hat{\eta}_i(x) L_i(\eta(x))\big)\, dx = P(\eta, \xi; \hat{\eta}, \hat{\xi})$$

* The conditions (29.4) are called *mixed* when they involve both the variables η_i^s and the variables ξ_i^s, as distinguished from the canonical conditions (29.1)$'$ or (29.2), which involve only the variables η_i^s.

† For simplicity this bilinear form is denoted by $P(\eta, \xi; \hat{\eta}, \hat{\xi})$ rather than by $P(\eta^s, \xi^s; \hat{\eta}^s, \hat{\xi}^s)$.

is readily verified and gives rise to the bilinear form (29.5). See Bôcher [1], p. 23.

The following theorem will be verified.

THEOREM 29.1. *Corresponding to any admissible set* Z *of BC, there exists at least one admissible set* $\hat{Z}$ *of BC adjoint to* Z. *Any other admissible set of BC adjoint to* Z *is equivalent to* $\hat{Z}$.

By hypothesis carrier Z is a vector subspace over R of dimension $2m$ in the vector space R^{4m} of $4m$-tuples (η^s, ξ^s). Let

$$(\eta^s_{(k)}, \xi^s_{(k)}) \qquad (k = 1, \ldots, 2m) \tag{29.7}$$

be the k-th $4m$-tuple in a base of carrier Z.

For a fixed base (29.7) the $2m$ linear conditions

$$P(\eta_{(k)}, \xi_{(k)}; \hat{\eta}, \hat{\xi}) = 0 \qquad (k = 1, \ldots, 2m) \tag{29.8}$$

on the $4m$-tuples $(\hat{\eta}^s, \hat{\xi}^s)$ are linearly independent and so define an admissible set $\hat{Z}$ of BC adjoint to the set Z. It is clear that if $\hat{Z}$ is adjoint to Z then carrier $\hat{Z}$ is unique. Hence any two admissible sets of BC adjoint to Z are "equivalent."

Thus Theorem 29.1 is true.

We continue with Lemmas 29.1 and 29.2. These lemmas prepare for the theorems of Section 30 and Section 31.

LEMMA 29.1. *The canonical sets of BC of systems* W_r *of Section* 15, *namely conditions* (29.1) *when* $r > 0$, *and conditions* (29.2) *when* $r = 0$, *are selfadjoint.*

The conditions (29.1). Let (η^s, ξ^s) be a $4m$-tuple satisfying the conditions (29.1) with an r-tuple u. Let $(\hat{\eta}^s, \hat{\xi}^s)$ be a second such $4m$-tuple, satisfying the conditions (29.1) with an r-tuple $\hat{u}$. For h, k on the range $1, \ldots, r$, i on the range $1, \ldots, m$ and s not summed,

$$\left[\eta^s_i \hat{\xi}^s_i\right]_{s=1}^{s=2} = \left[c^s_{ih} u_h \hat{\xi}^s_i\right]_{s=1}^{s=2} = -b_{hk} u_h \hat{u}_k \tag{29.9}$$

$$\left[\hat{\eta}^s_i \xi^s_i\right]_{s=1}^{s=2} = \left[c^s_{ih} \hat{u}_h \xi^s_i\right]_{s=1}^{s=2} = -b_{hk} \hat{u}_h u_k \tag{29.10}$$

in accord with (29.1). Since $b_{hk} = b_{kh}$, the bilinear form (29.9) equals the bilinear form (29.10).

The conditions (29.1) are accordingly selfadjoint. That the conditions (29.2) are selfadjoint is immediate.

Thus Lemma 29.1 is true.

LEMMA 29.2. *If two sets of canonical conditions* (29.1) *are equivalent and have common matrices* $\|c_{ih}^{s}\|$, *then their* r*-square symmetric matrices* $\|b_{hk}\|$ *are identical.*

Suppose that $\|b'_{hk}\|$ and $\|b''_{hk}\|$ are the r-square matrices of form $\|b_{hk}\|$ in the two given sets of conditions (29.1). Corresponding to a prescribed r-tuple u there exists a $4m$-tuple (η^s, ξ^s) which satisfies the first set of conditions (29.1) with the r-tuple u. By virtue of the hypothesis of equivalence, (η^s, ξ^s) satisfies the second set of conditions (29.1) as well, with an r-tuple $\hat{u}$ which must equal u because of the identity of the two matrices $\|c_{ih}^{s}\|$. From the conditions (29.1), it follows that

$$b'_{hk}u_h u_k \equiv b''_{hk}u_h u_k \qquad (u \in R^r). \tag{29.11}$$

Since the matrices $\|b'_{hk}\|$ and $\|b''_{hk}\|$ are symmetric, it follows from the identity (29.11) that $b'_{hk} = b''_{hk}$ for each pair h, k.

Thus Lemma 29.2 is true.

Exercises

29.1. When $m = 1$ the BC conditions

$$\begin{cases} a\eta^1 - b\xi^1 = 0 \\ c\eta^2 + d\xi^2 = 0 \end{cases} \tag{29.12}$$

are admissible, if $ad + bc \neq 0$. The range of (i) in (29.1) is $1, \ldots, m$. When $m = 1$, i is accordingly dropped as a subscript.

Show that the conditions (29.12) are selfadjoint.

29.2. When $m = 1$ let conditions

$$\begin{aligned} p_{11}\eta^1 + p_{12}\eta^2 &= q_{11}\xi^1 + q_{12}\xi^2 \\ p_{21}\eta^1 + p_{22}\eta^2 &= q_{21}\xi^1 + q_{22}\xi^2 \end{aligned} \tag{29.13}$$

be linearly independent. Introduce the matrices $\|p_{ij}\| = \mathbf{p}$, $\|q_{ij}\| = \mathbf{q}$. Show that the conditions (29.13) are *selfadjoint* if and only if the matrix $\mathbf{pq}^*$ is symmetric. See Theorem 30.1 (ii).

30. Conditions for Selfadjointness

We shall find necessary and sufficient conditions that an admissible set of BC given in matrix form be selfadjoint.

Let $\boldsymbol{\eta}$ and $\boldsymbol{\xi}$ be matrices, each consisting of one column containing, respectively, the elements

$$\eta_1^1, \ldots, \eta_m^1; \qquad \eta_1^2, \ldots, \eta_m^2 \tag{30.1}$$

$$\xi_1^1, \ldots, \xi_m^1; \qquad -\xi_1^2, \ldots, -\xi_m^2. \tag{30.2}$$

Let $\mathbf{p}$ and $\mathbf{q}$ be $2m$-square matrices of elements in R such that the matrix $\|\mathbf{p},\mathbf{q}\|$ of $2m$ rows and $4m$ columns has the rank $2m$. The matrix equality

$$\mathbf{p}\boldsymbol{\eta} = \mathbf{q}\boldsymbol{\xi}, \tag{30.3}$$

if written in the form, $\mathbf{p}\boldsymbol{\eta} - \mathbf{q}\boldsymbol{\xi} = \mathbf{0}$, is equivalent to equating to zero $2m$ linearly independent linear forms in the $4m$ variables (η^s, ξ^s). A set of admissible BC is accordingly defined by the conditions (30.3).

Conversely, an arbitrary admissible set of BC is equivalent, for a proper choice of the $2m$-square matrices $\mathbf{p}$ and $\mathbf{q}$, to conditions of the form (30.3).

The following theorem fulfills the mission of this section.

THEOREM 30.1 (i). *Conditions adjoint to the conditions* (30.3), *are equivalent to conditions on the* 4m-*tuples* $(\hat{\eta}^s, \hat{\xi}^s)$, *in matrix form*,

$$\hat{\boldsymbol{\eta}} = \mathbf{q}^*\mathbf{v}, \qquad \hat{\boldsymbol{\xi}} = \mathbf{p}^*\mathbf{v} \tag{30.4}$$†

where $\mathbf{v}$ *is a column of* 2m *parameters.*

(ii). *A necessary and sufficient condition that the conditions* (30.3) *be selfadjoint, is that the* 2m-*square matrix product* $\mathbf{p}\mathbf{q}^*$ *be symmetric.*

Proof of (i). The matrix

$$\mathbf{A} = \begin{Vmatrix} \mathbf{p}^* \\ \mathbf{q}^* \end{Vmatrix} \tag{30.5}$$

has $4m$ rows and $2m$ columns. Its rows and columns are the columns and rows, respectively, of $\|\mathbf{p},\mathbf{q}\|$. Hence rank $\mathbf{A} = 2m$. It follows that the conditions (30.4) on the $4m$-tuples $(\hat{\eta}^s, \hat{\xi}^s)$ are equivalent to $2m$ homogeneous independent linear conditions of form (29.4).

To prove that the conditions (30.4) are adjoint to the conditions (30.3), we have merely to show that the bilinear form (29.5) vanishes, subject to the conditions (30.3) and (30.4). To this end note that the bilinear form (29.5) is a matrix $\hat{\boldsymbol{\eta}}^*\boldsymbol{\xi} - \hat{\boldsymbol{\xi}}^*\boldsymbol{\eta}$ of one element. Subject to (30.4) this element equals

$$\mathbf{v}^*\mathbf{q}\boldsymbol{\xi} - \mathbf{v}^*\mathbf{p}\boldsymbol{\eta}, \tag{30.6}$$

and subject to (30.3), vanishes for arbitrary choices of the parameters $v_1, \ldots, v_{2m}$.

Thus (i) of Theorem 30.1 is true.

Proof of (ii). It follows from (i) that conditions (30.4) are adjoint to conditions (30.3). From Theorem 29.1 we infer that the conditions (30.3) are selfadjoint if and only if conditions (30.4) are equivalent to conditions (30.3), or equivalently, if and only if the matrix equation,

$$\mathbf{p}\hat{\boldsymbol{\eta}} = \mathbf{q}\hat{\boldsymbol{\xi}} \tag{30.7}$$

† $\mathbf{p}^*$ and $\mathbf{q}^*$ are matrices conjugate respectively to $\mathbf{p}$ and $\mathbf{q}$.

is satisfied by all columns $\hat{\eta}$, $\hat{\xi}$ which satisfy (30.4) for some $2m$-tuple v. For this it is necessary and sufficient that

$$\mathbf{pq}^*\mathbf{v} \equiv \mathbf{qp}^*\mathbf{v} \tag{30.8}$$

for arbitrary parameters $v_1, \ldots, v_{2m}$ or, equivalently, that $\mathbf{pq}^* = \mathbf{qp}^*$.

Thus Theorem 30.1 (ii) is true.

31. The Reduction of Selfadjoint *BC* to Canonical Form

We are concerned with selfadjoint BC adjoined to DE of the form (29.0). We begin with a definition.

Definition 31.1. *Accessory* r-*planes* χ_r. Let X be an admissible set of BC. Carrier X is then, by Definition 29.0, a $2m$-plane meeting the origin in the space R^{4m} of $4m$-tuples (η^s, ξ^s) : (29.3). The orthogonal projection of carrier X on the $2m$-dimensional coordinate subspace of the variables

$$\eta_1^1, \ldots, \eta_m^1 : \eta_1^2, \ldots, \eta_m^2 \tag{31.1}$$

will be an r-dimensional vector subspace over R of R^{4m}, to be termed the *accessory* r-*plane* χ_r of X. The integer r may be any one of the integers $0, 1, \ldots, 2m$, and will be called the *accessory dimension* r of X.

Example (1). When the boundary conditions are the special conditions

$$\eta_1^1 = \cdots = \eta_m^1 = \eta_1^2 = \cdots = \eta_m^2 = 0, \tag{31.2}$$

then $r = 0$, and the accessory r-plane χ_0 is (by convention) the origin.

Example (2). When the BC are the secondary conditions associated with a periodic extremal g, as in (11.20), the accessory dimension $r = m$ and the accessory m-plane χ_m is defined by the conditions,

$$\eta_i(a^1) = \eta_i(a^2) \qquad (i = 1, \ldots, m).$$

Example (3). When the BC are derived from end conditions C_{2m} of Section 8 which require the end points to rest on two m-dimensional manifolds cutting the extremal g transversally, then $r = 2m$. See Lemma 10.2.

Example (4). When the BC have the canonical form (29.1), the conditions (29.1)′ give a regular parametric representation of the accessory r-plane χ_r in terms of r-parameters $u_1, \ldots, u_r$.

The principal theorem of this section follows.

THEOREM 31.1. *An arbitrary selfadjoint set* X *of BC, adjoined to the DE* (29.0), *with an accessory* r-*plane* χ_r *of positive dimension* r, *is equivalent to canonical BC* (29.1) *in which*

(i) *the conditions* (29.1)′ *give a regular*† *linear representation of* χ_r *while*

(ii) *the conditions* (29.1)″ *involve an* r-*square symmetric matrix* $\|b_{hk}\|$ *uniquely determined by carrier* X *and the matrix* $\|c^s_{ih}\|$ *of* (29.1)′.

Proof of (i). We shall define the equivalence affirmed to exist in Theorem 31.1 as a resultant of four equivalences. If sets X and Y of selfadjoint BC are equivalent, we shall write $X \sim Y$.

A First Equivalence $X \sim Y$. The set X is admissible by hypothesis and is accordingly equivalent to a set Y of BC in matrix form,

$$\mathbf{p}\boldsymbol{\eta} = \mathbf{q}\boldsymbol{\xi}, \tag{31.3}$$

with $2m$-square matrices $\mathbf{p}$ and $\mathbf{q}$ such that the $\|\mathbf{p},\mathbf{q}\|$ has the rank $2m$.

A Second Equivalence $Y \sim \hat{Y}$. We shall modify the BC : (31.3) by *elementary* operations on the matrix $\|\mathbf{p},\mathbf{q}\|$. These operations are of two types: a row can be multiplied by an arbitrary nonnull constant, or a row can be added to another different row. If rank $\mathbf{q} = \rho$, a finite sequence of elementary operations will yield a matrix $\|\hat{\mathbf{p}},\hat{\mathbf{q}}\|$ in which the elements in the last $2m - \rho$ rows of $\hat{\mathbf{q}}$ vanish. One obtains thereby an equivalent set of BC

$$\hat{Y} : \hat{\mathbf{p}}\boldsymbol{\eta} = \hat{\mathbf{q}}\boldsymbol{\xi}. \tag{31.4}$$

The set $\hat{Y}$ will be termed *reduced.*

A Third Equivalence $\hat{Y} \sim \hat{Z}$. It follows from Theorem 30.1 (i) that the conditions $\hat{Y}$ are equivalent to matrix defined conditions,

$$\hat{Z} : \boldsymbol{\eta} = \hat{\mathbf{q}}^*\mathbf{v}, \qquad \boldsymbol{\xi} = \hat{\mathbf{p}}^*\mathbf{v} \tag{31.5}$$

where $\mathbf{v}$ is a column of $2m$ parameters $v_1, \ldots, v_{2m}$. If conditions (31.5) are satisfied by columns $\boldsymbol{\eta}$, $\boldsymbol{\xi}$, and $\mathbf{v}$ then

$$\hat{\mathbf{q}}\boldsymbol{\xi} = \hat{\mathbf{q}}\hat{\mathbf{p}}^*\mathbf{v}. \tag{31.6}$$

A Fourth Equivalence $\hat{Z} \sim Z$. Let $\mathbf{c}$ be the $2m \times \rho$ matrix given by the first ρ columns of $\hat{\mathbf{q}}^*$ and let $\mathbf{b}$ be the ρ-square, symmetric matrix given by the ρ-square principal minor of the matrix $\hat{\mathbf{q}}\hat{\mathbf{p}}^*$. The matrix $\hat{\mathbf{q}}\hat{\mathbf{p}}^*$, and hence

† A representation of χ_r of form (29.1)′ is called *regular*, since rank $\|c^s_{ih}\|$ equals the dimension r of χ_r.

the matrix $\mathbf{b}$, is symmetric by Theorem 30.1 (ii). If $\mathbf{u}$ is a column made up of the first ρ elements in $\mathbf{v}$, the BC

$$(31.7)\qquad Z\begin{cases}\boldsymbol{\eta} = \mathbf{cu} \\ \mathbf{c}^*\boldsymbol{\xi} = \mathbf{bu}\end{cases}\qquad (\text{rank } \mathbf{c} = \rho)$$

have the matrix form of canonical conditions (29.1) ρ. Carrier Z accordingly has the dimension $2m$. Moreover carrier $Z \supset$ carrier $\hat{Z}$, since a $4m$-tuple (η^s, ξ^s) which satisfies the conditions (31.5) with an m-tuple $\mathbf{v}$, satisfies conditions (31.7) with the initial r-tuple $\mathbf{u}$ of $\mathbf{v}$. However, carrier $Z =$ carrier $\hat{Z}$ since dim carrier $Z =$ dim carrier $\hat{Z} = 2m$. Hence $\hat{Z} \sim Z$.

The four equivalences $X \sim Y \sim \hat{Y} \sim \hat{Z} \sim Z$ imply the equivalence $X \sim Z$. The accessory dimensions r and ρ, respectively, of X and Z must then be equal. If one sets

$$(31.8)\qquad \|c_{ih}^s\| = \mathbf{c}, \qquad \|b_{hk}\| = \mathbf{b}$$

the conditions Z of (31.7) give a matrix representation of canonical conditions of form (29.1).

Thus (i) of Theorem 31.1 is true.

Proof of (ii). The matrix $\mathbf{b}$ in (31.7), and equivalently the matrix $\|b_{hk}\|$, in a representation (29.1) of conditions Z, is uniquely determined by carrier $X =$ carrier Z and the matrix $\mathbf{c} = \|c_{ih}^s\|$, in accord with Lemma 29.2.

Thus (ii) of Theorem 31.1 is true.

The equivalence class of each set X of selfadjoint BC, with accessory r-plane χ_r of positive dimension r, contains a set of canonical BC : Z, in matrix form (31.7). The following theorem tells when and how canonical BC are equivalent.

THEOREM 31.2 (i). *If canonical BC, Z of form* (31.7) *are given with* r > 0, *if* $\mathbf{a}$ *is any nonsingular* r*-square matrix and if one sets*

$$(31.9)\qquad \hat{\mathbf{c}} = \mathbf{ca}, \qquad \hat{\mathbf{b}} = \mathbf{a}^*\mathbf{ba}$$

and takes $\hat{\mathbf{u}}$ *as a column of parameters* $\hat{u}_1, \ldots, \hat{u}_r$, *then the matrix defined BC*

$$(31.10)\qquad \hat{Z}\begin{cases}\boldsymbol{\eta} = \hat{\mathbf{c}}\hat{\mathbf{u}} \\ \hat{\mathbf{c}}^*\boldsymbol{\xi} = \hat{\mathbf{b}}\hat{\mathbf{u}}\end{cases}$$

are equivalent to the conditions Z : (31.7).

(ii). *Conversely any selfadjoint BC which are canonical and equivalent to the BC*, Z : (31.7) *are of form* (31.10), *subject to* (31.9), *for a unique choice of a nonsingular* r*-square matrix* $\mathbf{a}$.

Proof of (i). Conditions (31.10) are satisfied by columns $\boldsymbol{\eta}$, $\boldsymbol{\xi}$, $\hat{\mathbf{u}}$ if and only if conditions (31.7) are satisfied by the columns $\boldsymbol{\eta}$, $\boldsymbol{\xi}$, and $\mathbf{u} = \mathbf{a}\hat{\mathbf{u}}$. The equivalence $Z \sim \hat{Z}$ follows.

Proof of (ii). Canonical BC : $\hat{Z}$ of the form (31.10) which are equivalent to the canonical BC : (31.7) must have the same accessory r-plane χ_r. The conditions $\boldsymbol{\eta} = \hat{\mathbf{c}}\hat{\mathbf{u}}$ and $\boldsymbol{\eta} = \mathbf{cu}$ must both afford a regular linear representation of χ_r. It follows that for some unique r-square nonsingular matrix $\mathbf{a}$, $\hat{\mathbf{c}} = \mathbf{ca}$. It remains to show that

$$\hat{\mathbf{b}} = \mathbf{a}^*\mathbf{ba}. \tag{31.11}$$

Were (31.11) not true there would exist *two* canonical sets $\hat{Z}$ of BC, both with $\hat{\mathbf{c}} = \mathbf{ca}$ and with carrier that of Z, but one for which (31.11) holds by (i), and one for which (31.11) does not hold. This is impossible by Lemma 29.2.

Hence (ii) is true.

We are lead to the following definition, and to a corollary of Theorem 31.2.

Definition 31.2. *The quadratic form*

$$b_{hk}u_h u_k \tag{31.12}$$

in the r-tuple u will be called the *accessory quadratic form* of the canonical selfadjoint BC : (29.1) or (31.7).

COROLLARY 31.1. *The accessory quadratic forms of two equivalent self-adjoint sets* Z : (31.7) *and* $\hat{\mathrm{Z}}$: (31.10) *of BC, are* R-*equivalent, with*

$$\mathrm{b}_{hk}\mathrm{u}_h\mathrm{u}_k = \hat{\mathrm{b}}_{hk}\hat{\mathrm{u}}_h\hat{\mathrm{u}}_k \tag{31.13}$$

if whenever the r-*tuples* u *and* $\hat{\mathrm{u}}$ *represents the same point on the accessory* r-*plane* χ_r *common to* Z *and* $\hat{\mathrm{Z}}$.

If $Z \sim \hat{Z}$, pairs (η^s, ξ^s) : (29.3) which satisfy the conditions (29.1) of Z and the corresponding conditions of $\hat{Z}$, are such that

$$\Big[\eta_i^s\xi_i^s\Big]_{s=2}^{s=1} = b_{hk}u_h u_k = \hat{b}_{hk}\hat{u}_h\hat{u}_k \tag{31.14}$$

whenever

$$\eta_i^s = c_{ih}^s u_h = \hat{c}_{ih}^s\hat{u}_h \qquad (s = 1, 2;\ i = 1, \ldots, m). \tag{31.15}$$

Part V

PRESTRUCTURES FOR CHARACTERISTIC ROOT THEORY

CHAPTER 9

Admissible Canonical Systems $\dot{\mathrm{W}}_r$

32. The Dependence on λ

In Section 15 the parameter λ of the *system* W_r of canonical conditions $W_r(\lambda)$ enters linearly in the *underlying* quadratic forms, in classical form

$$2\omega(x,\eta,\zeta) - \lambda \|\eta\|^2 = 2\omega^\lambda(x,\eta,\zeta). \tag{32.0}$$

Such forms are quadratic in the variables

$$(\eta_1, \ldots, \eta_m; \zeta_1, \ldots, \zeta_m) \tag{32.1}$$

for each $x \in \aleph$ and $\lambda \in R$. As conditioned in Appendix I these forms give rise to the DE

$$\frac{d}{dx}\,\omega_{\zeta_i}(x,\eta,\eta') - \omega_{\eta_i}(x,\eta,\eta') + \lambda\eta_i = 0 \qquad (i = 1, \ldots, m). \tag{32.2}$$

To these DE were adjoined canonical BC in (15.0) and (15.1), defined by special linear conditions whose coefficients were *independent of* λ.

This earlier prestructure for a theory of characteristic roots will be modified in two ways: (i) The parameter λ will enter in a more general way in the quadratic forms which will replace the form (32.0). (ii) The adjoined self-adjoint boundary conditions shall have coefficients which vary with λ subject to conditions to be imposed.

The modified DE. For each $x \in \aleph$ and each $\lambda \in R$ the *underlying* quadratic form shall be a form

$$2\dot{\omega}^\lambda(x,\eta,\zeta) = R_{ij}(x,\lambda)\zeta_i\zeta_j + 2Q_{ij}(x,\lambda)\zeta_i\eta_j + P_{ij}(x,\lambda)\eta_i\eta_j \tag{32.3*}$$

where i and j have the range $1, \ldots, m$ and the mappings of $\aleph \times R$ into R with values

$$R_{ij}(x,\lambda),\ Q_{ij}(x,\lambda),\ P_{ij}(x,\lambda) \tag{32.4}$$

* The dot over ω is a visible sign that the quadratic form $\dot{\omega}^\lambda(x,\eta,\zeta)$ is to be distinguished from the forms $\omega^\lambda(x,\eta,\zeta)$ *underlying* the canonical system W_r of Section 15.

shall be continuous. As previously, we require that for each $(x,\lambda) \in \aleph \times R$ the matrices $\|P_{ij}(x,\lambda)\|$ and $\|R_{ij}(x,\lambda)\|$ be symmetric and that the quadratic form with matrix $\|R_{ij}(x,\lambda)\|$ be positive definite. For each λ we associate with (32.3) the DE

$$(32.5) \qquad \frac{d}{dx}\,\overset{\lambda}{\omega}_{\zeta_i}(x,\eta,\eta') = \overset{\lambda}{\omega}_{\eta_i}(x,\eta,\eta') \qquad (i = 1, \ldots, m).$$

Selfadjoint Boundary Conditions X_λ. For each λ a set X_λ of boundary conditions is adjoined to the DE : (32.5), admissible and selfadjoint for each λ, in the senses of Section 29 and Section 30. The *canonical associate* ξ of a C^1-mapping η of the interval $\aleph$ into R^m is defined by setting

$$(32.6)' \qquad \xi_i = \overset{\lambda}{\omega}_{\zeta_i}(x,\eta,\eta') \qquad (i = 1, \ldots, m).$$

The BC are on the $4m$-tuples $(\eta(a^s),\eta'(a^s))_{(s=1,2)}$, or equivalently on the $4m$-tuples,

$$(\eta^1_1, \ldots, \eta^1_m : \xi^1_1, \ldots, \xi^1_m; \qquad \eta^2_1, \ldots, \eta^2_m : \xi^2_1, \ldots, \xi^2_m)$$

where

$$(32.6)'' \qquad \xi^s_i = \overset{\lambda}{\omega}_{\zeta_i}(x,\eta(a^s),\eta'(a^s)) \qquad (s = 1, 2;\ i = 1, \ldots, m).$$

We require that the accessory r-*plane* $\chi_r(\lambda)$ *of* X_λ (Definition 31.1) *be independent of* λ.

According to Theorem 31.1, for fixed λ, the BC X_λ, adjoined to the DE : (32.5), are equivalent to a set of canonical BC. Conditions on X_λ, beyond those already given, will be implied by our conditions on equivalent canonical BC now to be given. These conditions further limit the way X_λ varies with λ.

Canonical conditions equivalent to conditions X_λ. If for a fixed λ the accessory r-plane χ_r of the conditions X_λ has a positive dimension r, a regular linear representation

$$(32.7) \qquad \eta^s_i = c^s_{ih} u_h \qquad (s = 1, 2;\ i = 1, \ldots, m)$$

of χ_r can be chosen in many ways. Theorem 31.1 implies that the conditions X_λ are equivalent to the conditions (32.7), supplemented by r conditions of the form

$$(32.8) \qquad c^1_{ih}\xi^1_i - c^2_{ih}\xi^2_i = b_{hk}(\lambda)u_k \qquad (h = 1, \ldots, r)$$

where the matrix* $\|b_{hk}(\lambda)\|$ is r-square and symmetric.

If the accessory r-plane χ_r has the dimension 0, the BC X_λ are equivalent to the conditions

$$(32.9) \qquad \eta^1_1 = \cdots = \eta^1_m = \eta^2_1 = \cdots = \eta^2_m = 0.$$

* The matrix $\|b_{hk}(\lambda)\|$ will be called the *comparison matrix* of the system $\dot{W}_r$ of Definition 32.1 because of the fundamental role it plays in Comparison Theorem 35.1.

Canonical systems $\dot{W}_r$ of BC extending the canonical systems W_r of Section 15 are to be defined. The super dot distinguishes the systems $\dot{W}_r$ from the systems W_r defined in Section 15.

Definition 32.1, r > 0. **A canonical system*** $\dot{W}_r$, $r > 0$. For each $\lambda \in R$, the system $\dot{W}_r$ shall be defined by conditions $\dot{W}_r(\lambda)$ of the form,

$$\begin{bmatrix} \dfrac{d}{dx}\, \dot{\omega}^{\lambda}_{\zeta_i}(x,\eta,\eta') = \dot{\omega}^{\lambda}_{\eta_i}(x,\eta,\eta') & (i = 1, \ldots, m) & (32.10)' \\ \eta_i^s = c_{ih}^s u_h \quad (s = 1, 2;\ i = 1, \ldots, m) & & (32.10)'' \\ c_{ih}^1 \xi_i^1 - c_{ih}^2 \xi_i^2 = b_{hk}(\lambda) u_k & (h = 1, \ldots, r) & (32.10)''' \end{bmatrix}$$

where the $2m \times r$ matrix $\|c_{ih}^s\|$ has the rank r, the r-square matrix $\|b_{hk}(\lambda)\|$ is symmetric and the elements in this matrix vary *continuously* with λ. *We refer to these three conditions as conditions* (32.10)λ.

Definition 32.2, r = 0. **A canoncial system** $\dot{W}_0$. For each $\lambda \in R$, the system $\dot{W}_0$ shall be defined, by conditions $\dot{W}_0(\lambda)$: (32.11)λ of the form

$$\begin{bmatrix} \dfrac{d}{dx}\, \dot{\omega}^{\lambda}_{\zeta_i}(x,\eta,\eta') = \dot{\omega}^{\lambda}_{\eta_i}(x,\eta,\eta') & (i = 1, \ldots, m) & (32.11)' \\ \eta_1^1 = \cdots = \eta_m^1 = \eta_1^2 = \cdots = \eta_m^2 = 0 & & (32.11)'' \end{bmatrix}$$

To further condition a system $\dot{W}_r$, a quadratic functional $\dot{I}_r^{\lambda}$, *based* on $\dot{W}_r$, will now be defined.

Definition 32.3, r > 0. Following (15.2) as a model, the *functional* $\dot{I}_r^{\lambda} = I_r^{\lambda} \# \dot{W}_r$, *based* on $\dot{W}_r$, shall have values,†

$$\dot{I}_r^{\lambda}(\eta) = b_{hk}(\lambda) u_h u_k + \int_{a^1}^{a^2} 2\dot{\omega}^{\lambda}(x,\eta,\eta')\, dx \qquad (r > 0) \tag{32.12}$$

for each D^1-mapping η of $[a^1,a^2]$ into R^m which satisfies the conditions (32.10)″ with the r-tuple u, for the given λ.

Definition 32.3, r = 0. The *functional* $\dot{I}_0^{\lambda} = I_0^{\lambda} \# \dot{W}_0$, *based* on $\dot{W}_0$, shall have values,†

$$\dot{I}_0^{\lambda}(\eta) = \int_{a^1}^{a^2} 2\dot{\omega}^{\lambda}(x,\eta(x),\eta'(x))\, dx \qquad (r = 0) \tag{32.13}$$

* The system $\dot{W}_r$ of Definition 32.1 is termed *canonical* because its BC are canonical in form. Conditions $A_1, \ldots, A_4$, to be presently given, limit the way the parameter λ enters a system $\dot{W}_r$ (with superdot) as distinguished from a system W_r of Section 15 (without superdot). The BC of both W_r and $\dot{W}_r$ are canonical in form.

† The domain of $\dot{I}_r^{\lambda}$ is independent of λ and is denoted by $\{\dot{W}_r\}$. Cf. Definition 15.1.

for each D^1-mapping η of $[a^1,a^2]$ into R^m which satisfies the conditions $\eta(a^1) = \eta(a^2) = \mathbf{0}$.

As we shall see in Section 33, canonical systems $\dot{W}_r$, as conditioned above, have some but not *all* of the fundamental properties of the canonical systems W_r of Section 15. This leads us to distinguish between two levels of admissibility for systems $\dot{W}_r$.

Definition 32.4. *Discretely admissible and overall admissible canonical systems* $\dot{W}_r$ r ≥ 0.

(i) A system $\dot{W}_r$ of canonical conditions $\dot{W}_r(\lambda)$ as defined and conditioned above, will be said to be *discretely admissible*, that is admissible for each individual λ.

(ii) Such systems will be called *overall admissible* if the following four additional conditions are satisfied involving different values of λ.

Condition (A_1). For fixed (x,η,ζ) with $x \in \aleph$, η and ζ arbitrary m-tuples, $\dot{\omega}^\sigma(x,\eta,\zeta)$ shall decrease monotonically as σ increases.

Condition (A_2). When $r > 0$, the accessory quadratic form $b_{hk}(\sigma)u_h u_k$, associated in (32.10) with the conditions $\dot{W}_r$, shall decrease monotonically as σ increases, for fixed r-tuple u.

Condition (A_3). The functional $\eta \to \dot{I}_r^\sigma(\eta)$ of Definition 32.3 shall be positive definite for $-\sigma$ sufficiently large.

Condition (A_4).† $\dot{I}_r^\sigma(\eta) < 0$ whenever $\sigma > \lambda$ and η is a nonnull solution of conditions (32.10)λ when $r > 0$, or (32.11)λ when $r = 0$.

These four conditions are seen to be satisfied if $\dot{W}_r$ satisfies Conditions (A_1^*), (A_2), (A_3), where (A_1^*) is the following replacement of Condition (A_1).

Condition (A_1^*). For each fixed (x,η,ζ) with $x \in \aleph$, η a nonnull m-tuple and ζ an arbitrary m-tuple, $\dot{\omega}^\sigma(x,\eta,\zeta)$ shall strictly decrease as σ increases.

We are affirming that the satisfaction of Conditions (A_1), (A_2), (A_3), (A_4) is implied by the simpler conditions (A_1^*), (A_2), (A_3). The converse is clearly not true.

The following theorem shows that the class of "overall admissible" canonical systems $\dot{W}_r$ is not empty.

† One infers from (15.24) that $\dot{I}_r^\lambda(\eta) = 0$ for the mapping η given in Condition (A_4).

THEOREM 32.1. *Each system of canonical conditions* $\dot{\text{W}}_{\text{r}}$ *of Section* 15 *is "overall admissible" in the sense of Definition* 32.4, r $\geq$ 0.

It is clear that a system $\dot{W}_r$ of canonical conditions W_r (Definition 15.0) is "discretely admissible" in the sense of Definition 32.4. Such a system $\dot{W}_r$ is also overall admissible, since it satisfies Conditions (A_1^*), (A_2), (A_3), as we now verify.

Condition (A_1^*) is satisfied by $\dot{W}_r$, since for each $x \in \aleph$ and $\lambda \in R$ the underlying quadratic form of $\dot{W}_r$ is

$$\omega^\lambda(x,\eta,\zeta) = \omega(x,\eta,\zeta) - \frac{\lambda}{2}\|\eta\|^2 \qquad \text{(see Appendix I).} \tag{32.14}$$

Condition (A_3) is satisfied by $\dot{W}_r$, as the proof of Theorem 16.3 shows.

Condition (A_2) is satisfied by $\dot{W}_r$, since the coefficients b_{hk} of (15.0)″ are independent of λ.

The following definitions are variants of earlier definitions, applied to systems W_r of Section 15. The new definitions* are relative to $\dot{W}_r$.

Definition 32.5. *Characteristic solutions and roots* λ *of a system* $\dot{\text{W}}_{\text{r}}$. By a *characteristic solution* η of $\dot{W}_r$ with *root* λ is meant a nonnull C^1-mapping η of $[a^1,a^2]$ into R^m which satisfies the conditions $\dot{W}_r(\lambda)$: (32.10) or (32.11). By the *multiplicity* of a characteristic root λ of $\dot{W}_r$ is meant the number of linearly independent solutions of the conditions (32.10) with the parameter λ.

Definition 32.6. *σ-Conjugate points relative to* $\dot{\text{W}}_r$. Corresponding to a point $x = c$ in the interval $\aleph$, the conjugate points of $x = c$ in $\aleph$, relative to the DE

$$\frac{d}{dx}\dot{\omega}^\sigma_{\zeta_i}(x,\eta,\eta') = \dot{\omega}^\sigma_{\eta_i}(x,\eta,\eta') \qquad (i = 1, \ldots, m) \tag{32.15}$$

will be called the *σ-conjugate points of $x = c$ relative to* $\dot{\text{W}}_{\text{r}}$.

Definition 32.7. *A σ-frame* $\dot{\Lambda}_{\text{p}}$ *over* (a^1,a^2) *relative to* $\dot{\text{W}}_{\text{r}}$. Such a frame is defined as was a σ-frame Λ_p over (a^1,a^2) in Section 15, replacing σ-conjugate points, relative to W_r of Section 15, by σ-conjugate points, relative to $\dot{W}_r$.

Definition 32.8. *An index form* $\dot{\text{Q}}^\sigma\#(\dot{\text{W}}_{\text{r}},\dot{\Lambda}_{\text{p}})$. This quadratic form is defined as was the quadratic form $Q^\sigma\#(W_r,\Lambda_p)$, on making the replacements corresponding to a change from W_r to $\dot{W}_r$. One sets $\mu = r + mp$. A σ-frame $\dot{\Lambda}_p$ is defined for the given fixed σ. One makes use of σ-conjugate points and secondary σ-extremals relative to the DE (32.15) and for $z \in R^\mu$ defines the

* In these definitions $\dot{W}_r$ is assumed "discretely admissible."

broken secondary σ-extremal $\dot{h}^{\sigma,z}$ explicitly

$$h^{\sigma,z}\#(\dot{W}_r,\dot{\Lambda}_p) \qquad (\dot{W}_r,\dot{\Lambda}_p \text{ replacing } W_r,\Lambda_p) \tag{32.16}$$

as in Section 15. The definition (15.3) here takes the form

$$\dot{Q}^\sigma(z) = \dot{I}_r^\sigma(\eta) \tag{32.17}$$

when (graph η) $= \dot{h}^{\sigma,z}$.

The following lemma supplements Lemma 13.2.

LEMMA 32.1. *If* $\dot{\Lambda}_p$ *is a λ-frame over* (a^1,a^2), *relative to the discretely admissible system* $\dot{W}_r$, *and if* $\dot{W}_r$ *satisfies Condition* (A_1), *then* $\dot{\Lambda}_p$ *is a σ-frame over* (a^1,a^2) *for each* $\sigma \leq \lambda$.

The proof follows the proof of Lemma 13.1 as a model and is left to the reader. In this proof $2\dot{\omega}^\lambda(x,\eta,\zeta)$ replaces

$$2\Omega^\lambda(x,\eta,\zeta) = 2\Omega(x,\eta,\zeta) - \lambda\,\|\eta\|^2$$

of the earlier proof when $\lambda = \lambda_0$.

33. Canonical Systems $\dot{W}_r$

In this section we shall extend some of the theorems on the canonical systems W_r of Section 15, as theorems on the canonical systems $\dot{W}_r$ of Section 32. As in Definition 32.4 we distinguish between the "discretely admissible" and "overall admissible" systems $\dot{W}_r$. Theorems preceding the Index Theorem 33.5 concern discretely admissible systems $\dot{W}_r$. Index Theorem 33.5 and its corollaries concern "overall admissible" systems $\dot{W}_r$. The proof of Index Theorem 33.5 makes explicit use of the conditions (A_1), (A_2), (A_3), and (A_4) of Section 32.

Let a discretely admissible system $\dot{W}_r$ be given together with a prescribed fixed value σ. The system $\dot{W}_r$ and the value σ uniquely determine a system W_r of Section 15 as follows.

Definition 33.1. *The condition* $\dot{W}_r(\sigma) = W_r(\sigma)$, $r > 0$.* Let a discretely admissible system $\dot{W}_r$ be given, together with a prescribed and fixed value σ. For $r > 0$ there exists a unique canonical system W_r, as defined in Section 15, such that $\dot{W}_r(\sigma) = W_r(\sigma)$ in the following sense:

(1) For each $x \in \aleph$ and for the given σ, the relation

$$\omega^\sigma(x,\eta,\zeta) \equiv \dot{\omega}^\sigma(x,\eta,\zeta) \tag{33.0}$$

is an identity in the m-tuples η and ζ.

* When $r = 0$ Definition 33.1 holds on deleting $\|c_{ih}^s\|$, $\|b_{hk}\|$, and $\|b_{hk}(\sigma)\|$.

(2) The matrices $\|c_{ih}^s\|$ and $\|b_{hk}\|$ of W_r are identical with the matrices $\|c_{ih}^s\|$ and $\|b_{hk}(\sigma)\|$ of $\dot{W}_r(\sigma)$, evaluated for the prescribed σ.

The condition (33.0) determines $\omega^\lambda(x,\eta,\zeta)$ for all values of λ. In fact, for σ fixed and $x \in \aleph$,

$$\omega^\lambda(x,\eta,\zeta) \equiv \dot{\omega}^\sigma(x,\eta,\zeta) - \frac{\lambda - \sigma}{2}\|\eta\|^2 \qquad (\lambda \in R). \tag{33.1}$$

The following theorem is a corollary of Index Theorem 15.2 (ii).

THEOREM 33.1. *If the system* $\dot{\mathrm{W}}_\mathrm{r}$ *is "discretely admissible," then for a prescribed* σ, *the index of* $\dot{\mathrm{Q}}^\sigma\#(\dot{\mathrm{W}}_\mathrm{r},\dot{\Lambda}_\mathrm{p})$ *equals the count* $\kappa(\sigma)$ *of characteristic roots less than* σ, *of the canonical system* W_r *of Section* 15 *such that* $\dot{\mathrm{W}}_\mathrm{r}(\sigma) = \mathrm{W}_\mathrm{r}(\sigma)$ *for the given* σ.

Theorem 33.1 follows on noting that when $\dot{W}_r(\sigma) = W_r(\sigma)$

$$Q^\sigma\#(W_r,\Lambda_p) = \dot{Q}^\sigma\#(\dot{W}_r,\dot{\Lambda}_p), \tag{33.2}$$

if one takes $\Lambda_p = \dot{\Lambda}_p$, as is permissible, when (33.0) holds.

Theorem 19.3 implies the following.

THEOREM 33.2. *If the canonical system* $\dot{\mathrm{W}}_0$ *is discretely admissible, the following is true for each* σ:

(i) *The nullity of* $\dot{\mathrm{Q}}^\sigma\#(\dot{\mathrm{W}}_0,\dot{\Lambda}_\mathrm{p})$ *equals the multiplicity of* $\mathrm{x} = \mathrm{a}^2$ *as a conjugate point of* $\mathrm{x} = \mathrm{a}^1$ *relative to the DE* (32.15);

(ii) *The index of* $\dot{\mathrm{Q}}^\sigma\#(\dot{\mathrm{W}}_0,\dot{\Lambda}_\mathrm{p})$ *equals the count of conjugate points of* $\mathrm{x} = \mathrm{a}^1$ *in* $(\mathrm{a}^1,\mathrm{a}^2)$ *relative to the DE* (32.15).

Fixing σ, let W_0 be a canonical system of dimension 0, as defined in Section 15, such that for each $x \in \aleph$

$$\omega^\sigma(x,\eta,\zeta) \equiv \dot{\omega}^\sigma(x,\eta,\zeta). \tag{33.3}$$

Then

$$Q^\sigma\#(W_0,\Lambda_p) = \dot{Q}^\sigma\#(\dot{W}_0,\dot{\Lambda}_p) \tag{33.4}$$

provided one takes Λ_p as $\dot{\Lambda}_p$, as is permissible, since (33.3) holds by hypothesis. We infer from Theorem 19.3 that Theorem 33.2 holds for the given σ if the superdot is deleted from $\dot{Q}^\sigma\#(\dot{W}_0,\dot{\Lambda}_p)$ and the DE (32.15). Because of (33.3) and (33.4) Theorem 33.2 is thus true as stated.

Theorem 15.1 similarly implies the following:

THEOREM 33.3. *If the canonical system* $\dot{\mathrm{W}}_{\mathrm{r}}$ *is discretely admissible the following is true for each* σ:

(i) *A necessary and sufficient condition that the index form* $\dot{\mathrm{Q}}^{\sigma}\#(\dot{\mathrm{W}}_{\mathrm{r}},\dot{\Lambda}_{\mathrm{p}})$ *be singular, is that* σ *be a characteristic root of* $\dot{\mathrm{W}}_{\mathrm{r}}$.

(ii) *The nullity of* $\dot{\mathrm{Q}}^{\sigma}\#(\dot{\mathrm{W}}_{\mathrm{r}},\dot{\Lambda}_{\mathrm{p}})$ *equals the multiplicity* $\nu(\sigma)$ *of* σ *as a characteristic root of* $\dot{\mathrm{W}}_{\mathrm{r}}$.

The proof will be given when $r > 0$. The proof when $r = 0$ is similar.

The value σ is prescribed and fixed. Let W_r be the canonical system, as defined in Section 15, such that $W_r(\sigma) = \dot{W}_r(\sigma)$ in the sense of Definition 33.1. Then (33.2) holds, provided one takes Λ_p as $\dot{\Lambda}_p$. According to Theorem 15.1 Theorem 33.3 is valid if one deletes the superdots.

The quadratic forms in (33.2) are identical and so have the same nullity. By virtue of the relation $W_r(\sigma) = \dot{W}_r(\sigma)$, the multiplicity $\nu(\sigma)$ of σ as a characteristic root of $\dot{W}_r$, equals the multiplicity of σ as a characteristic root of W_r when $W_r(\sigma) = \dot{W}_r(\sigma)$. Thus Theorem 33.3 is true.

The domain $\{\dot{W}_r\}$ of a quadratic functional $\dot{I}_r^{\sigma}\#\dot{W}_r$ is independent of σ. With this understood the *index* and *nullity* of $\dot{I}_r^{\sigma}\#\dot{W}_r$ are defined as were the index and nullity of $I_r^{\sigma}\#W_r$ in Definition 15.4 adding appropriate superdots to the terms in Definition 15.4. The proof of Theorem 15.3 leads to a proof of the following analogous theorem.

THEOREM 33.4. (i) *The nullity of the quadratic functional* $\dot{\mathrm{I}}_{\mathrm{r}}^{\sigma}\#\dot{\mathrm{W}}_{\mathrm{r}}$ *equals the nullity of a corresponding index form* $\dot{\mathrm{Q}}^{\sigma}\#(\dot{\mathrm{W}}_{\mathrm{r}},\dot{\Lambda}_{\mathrm{p}})$ *for each admissible* σ-*frame* $\dot{\Lambda}_{\mathrm{p}}$.

(ii) *The index of* $\dot{\mathrm{I}}_{\mathrm{r}}^{\sigma}\#\dot{\mathrm{W}}_{\mathrm{r}}$ *exists and equals the index of each corresponding index form* $\dot{\mathrm{Q}}^{\sigma}\#(\dot{\mathrm{W}}_{\mathrm{r}},\dot{\Lambda}_{\mathrm{p}})$.

The remaining theorems of this section are no longer corollaries of theorems on systems W_r of Section 15.

Overall admissible canonical systems $\dot{W}_r$. There has been no reference in the preceding theorems of this section to characteristic roots of the system $\dot{W}_r$ with *different* values λ. However, our extension of Index Theorem 15.2 will concern such values. To make this extension the system $\dot{W}_r$ will be assumed "overall admissible." The extension of Index Theorem 15.2(ii) follows.

INDEX THEOREM 33.5. *If the canonical system* $\dot{\mathrm{W}}_{\mathrm{r}}$ *is overall admissible, then for a prescribed* γ, *index* $\dot{\mathrm{Q}}^{\gamma}\#(\dot{\mathrm{W}}_{\mathrm{r}},\dot{\Lambda}_{\mathrm{p}})$ *is the count* κ_{γ} *of characteristic roots of* $\dot{\mathrm{W}}_{\mathrm{r}}$ *less than* γ.

Index Theorem 15.2 was proved with the aid of model families F of quadratic forms satisfying Conditions I, II, III of Appendix III. Index Theorem 33.5 will be similarly proved with the aid of a model family G of quadratic forms satisfying Conditions I, II, IV, V of Appendix III. If $\dot{\Lambda}_p$ is a γ-frame over (a^1,a^2), the index form $\dot{Q}^\sigma\#(\dot{W}_r,\dot{\Lambda}_p)$ is well defined for $\sigma \leq \gamma$, in accord with Lemma 32.1. Condition I on the family of quadratic forms $\dot{Q}^\sigma \mid (\sigma \leq \gamma)$ is satisfied by virtue of the following extension of Lemmas 15.1 and 15.3.

LEMMA 33.1. *Let γ be fixed. Under the hypotheses of Index Theorem* 33.5, *the above quadratic forms* $\dot{Q}^\sigma\#(\dot{W}_r,\dot{\Lambda}_p)$ *are representable as symmetric quadratic forms*

$$\dot{Q}^\sigma(z) = a_{ij}(\sigma)z_i z_j \qquad (z \in R^\mu, \mu = r + mp) \tag{33.5)'$$

with coefficients which vary continuously with $\sigma \leq \gamma$.

The proof of this lemma can be modelled after the proof of Lemma 15.3. One makes use of a γ-frame

$$\dot{\Lambda}_p = (\Pi_1, \ldots, \Pi_p) \tag{33.5)''$$

independent of $\sigma \leq \gamma$.

Lemma 15.4 is replaced by the following lemma.

LEMMA 33.2. *Under the hypotheses of Index Theorem* 33.5, *the family of quadratic forms* $\dot{Q}^\sigma$ *of Lemma* 33.1, *defined for each $\sigma \leq \gamma$ and for* $z \in R^\mu$, *is a model family* G, *satisfying Conditions* I, II, IV, V *of Appendix* III.

Condition I of Appendix III is satisfied by virtue of Lemma 33.1.

Condition II of Appendix III requires that $\dot{Q}^\sigma$ be positive definite for $\sigma < 0$ and $|\sigma|$ sufficiently large. It is satisfied by virtue of the Condition (A_3) on $\dot{I}_r^\sigma\#\dot{W}_r$ and the equality

$$\dot{Q}^\sigma(z) = \dot{I}_r^\sigma(\eta) \qquad ((\text{graph } \eta) = \dot{h}^{\sigma,z}\#(\dot{W}_r,\dot{\Lambda}_p)) \tag{33.6}$$

of (32.17). Cf. (15.3). In (33.6), $\sigma \leq \gamma$ and $\dot{\Lambda}_p$ is the γ-frame (33.5)″.

Condition IV of Appendix III requires that $\dot{Q}^\sigma(z)$ shall decrease monotonically as σ increases for fixed $z \in R^\mu$. It is satisfied by virtue of (33.6) and Conditions (A_1) and (A_2) on $\dot{W}_r$.

Condition V of Appendix III requires the following: if y is a nonnull critical μ-tuple of Q^τ with $\tau < \gamma$, then

$$Q^\lambda(y) < 0 \qquad (\text{whenever } \tau < \lambda \leq \gamma). \tag{33.7}$$

Proof of (33.7). If y is a nonnull critical μ-tuple of $\dot{Q}^\tau$, a mapping $\bar{\eta} \in \{\dot{W}_r\}$ such that (graph $\bar{\eta}$) $= \dot{h}^{\tau,y}\#(\dot{W}_r,\dot{\Lambda}_p)$, as in (33.6), is a characteristic solution of $\dot{W}_r$ with root τ. Conditions (A_4) of $\dot{W}_r$ then implies that

$$\dot{I}_r^\lambda(\bar{\eta}) < 0 \qquad (\text{when } \tau < \lambda \leq \gamma). \tag{33.8)$'$}$$

If $\hat{\eta}$ is a mapping whose graph is the BSλ-extremal $\dot{h}^{\lambda,y}\#(\dot{W}_r,\dot{\Lambda}_p)$, then

$$\dot{I}_r^\lambda(\hat{\eta}) \leq \dot{I}_r^\lambda(\bar{\eta}) \tag{33.8)$''$}$$

by virtue of Corollary 7.1, applied to the secondary λ-extremals of graph $\hat{\eta}$ into which graph $\hat{\eta}$ is subdivided by the γ-frame $\dot{\Lambda}_p$. It follows from (33.6) that $\dot{Q}^\lambda(y) = \dot{I}^\lambda(\hat{\eta})$ and hence from (33.8), that (33.7) holds.

Lemma 33.2 is thus true.

Completion of Proof of Index Theorem 33.5. By virtue of Theorem 3.1 of Appendix III the index of $\dot{Q}^\gamma\#(\dot{W}_r,\dot{\Lambda}_p)$ equals the count of characteristic roots $\sigma < \gamma$ of the model family G of Lemma 33.2, or equivalently by Theorem 33.3, the count of characteristic roots $\sigma < \gamma$ of $\dot{W}_r$.

Thus Index Theorem 33.5 is true.

Corollaries of the Index Theorems 15.2 and 33.5. For a canonical system W_r, as defined in Section 15, Definition 24.0 introduced (index W_r) and (nullity W_r). Similar definitions will now be formulated for $\dot{W}_r$.

Definition 33.2. *Index* $\dot{\mathrm{W}}_\mathrm{r}$ *and nullity* $\dot{\mathrm{W}}_\mathrm{r}$. We suppose that the system $\dot{W}_r$ *is* overall admissible in the sense of Definition 32.4. *Index* $\dot{W}_r$ shall then denote the count of negative characteristic roots of $\dot{W}_r$, and *nullity* $\dot{W}_r$ the multiplicity of $\lambda = 0$ as a characteristic root of $\dot{W}_r$.

A first remarkable corollary of the Index Theorems 15.2 and 33.5 follows.

COROLLARY 33.1. *Let* $\dot{\mathrm{W}}_\mathrm{r}$ *and* W_r *be canonical systems, as defined in Section* 32 *and Section* 15, *respectively. If* $\dot{\mathrm{W}}_\mathrm{r}(0) = \mathrm{W}_\mathrm{r}(0)$ *in the sense of Definition* 33.1, *and if* $\dot{\mathrm{W}}_\mathrm{r}$ *is overall admissible, then*

$$\textit{index } \dot{\mathrm{W}}_\mathrm{r} = \textit{index } \mathrm{W}_\mathrm{r}; \qquad \textit{nullity } \dot{\mathrm{W}}_\mathrm{r} = \textit{nullity } \mathrm{W}_\mathrm{r}. \tag{33.9}$$

That nullity $\dot{W}_r =$ nullity W_r follows trivially from the condition $\dot{W}_r(0) = W_r(0)$. This relation also implies that

$$\dot{Q}^0\#(\dot{W}_r\#\dot{\Lambda}_p) = Q^0\#(W_r\#\Lambda_p), \tag{33.10}$$

if one takes the 0-frames $\dot{\Lambda}_p$ and Λ_p, as the same, as is possible when $\dot{W}_r(0) = W_r(0)$. The indices of the forms (33.10) are then equal and according to the Index Theorems 33.5 and 15.2, respectively, equal (index $\dot{W}_r$) and (index W_r).

Thus Corollary 33.1 is true.

Theorem 26.1 can be extended as follows with the aid of Index Theorem 33.5.

COROLLARY 33.2. *If* $\dot{\mathrm{W}}_{\mathrm{r}}$, $\mathrm{r} > 0$, *and* $\dot{\mathrm{W}}_0$ *are overall admissible canonical systems with a common underlying quadratic form* $\dot{\omega}^0(\mathrm{x},\eta,\zeta)$ *when* $\lambda = 0$, *then*

$$0 \leq \textit{index } \dot{\mathrm{W}}_{\mathrm{r}} - \textit{index } \dot{\mathrm{W}}_0 \leq \mathrm{r}. \tag{33.11}$$

Corollary 33.2 follows from Corollary 33.1 and Theorem 26.1. To see this let W_r and W_0 be canonical systems, as defined in Section 15, such that in the sense of Definition 33.1,

$$\dot{W}_r(0) = W_r(0), \qquad \dot{W}_0(0) = W_0(0). \tag{33.12$'$}$$

Then by Corollary 33.1

$$\text{index } \dot{W}_r = \text{index } W_r; \qquad \text{index } \dot{W}_0 = \text{index } W_0. \tag{33.12$''$}$$

However, the systems W_r and W_0 have in common the DE,

$$\frac{d}{dx}\,\omega_{\zeta_i}(x,\eta,\eta') = \omega_{\eta_i}(x,\eta,\eta') \qquad (i = 1, \ldots, m),$$

where $\omega(x,\eta,\zeta) \equiv \dot{\omega}^0(x,\eta,\zeta)$. Hence by Theorem 26.1

$$0 \leq \text{index } W_r - \text{index } W_0 \leq r. \tag{33.13}$$

The relations (33.11) follow from (33.13) and (33.12)$''$.

The preceding corollary has an extension to be used in Section 34.

COROLLARY 33.3. *Let a value* σ *be prescribed and fixed. If* $\dot{\mathrm{W}}_r$, $\mathrm{r} > 0$, *and* $\dot{\mathrm{W}}_0$ *are overall admissible systems whose underlying quadratic forms, defined for each* $\lambda \in \mathrm{R}$, *include a common form when* $\lambda = \sigma$, *then*

$$0 \leq \kappa_\sigma - \mathrm{k}_\sigma \leq \mathrm{r} \tag{33.14}$$

where κ_σ *and* k_σ *are, respectively, the counts of characteristic roots of* $\dot{\mathrm{W}}_r$ *and* $\dot{\mathrm{W}}_0$ *which are less than* σ.

The underlying quadratic forms $\dot{\omega}^\lambda(x,\eta,\zeta)$ and matrices $\|c^s_{ih}\|$ and $\|b_{hk}(\lambda)\|$ of $\dot{W}_r$, $r > 0$, will be called its *defining elements*. For $\dot{W}_0$ the underlying quadratic form, say $\dot{\omega}^\lambda(x,\eta,\zeta)$, is termed its *defining element*. Corollary 33.3 may be inferred from Corollary 33.2 as follows. The value σ is fixed.

Let overall admissible canonical systems $\dot{\mathbf{W}}_r$ and $\dot{\mathbf{W}}_0$ be introduced whose defining elements for each $\lambda \in R$ are the defining elements, respectively, of $\dot{W}_r$ and $\dot{W}_0$, evaluated at $\lambda + \sigma$. So defined, the systems $\dot{\mathbf{W}}_r$ and $\dot{\mathbf{W}}_0$ have in common an underlying quadratic form when $\lambda = 0$. Corollary 33.2 is then applicable to $\dot{\mathrm{W}}_r$ and $\dot{\mathrm{W}}_0$. By Definition 33.2

$$\text{(33.15)} \qquad \text{index } \dot{\mathrm{W}}_r = \kappa_\sigma, \qquad \text{index } \dot{\mathrm{W}}_0 = k_\sigma.$$

By Corollary 33.2

$$\text{(33.16)} \qquad 0 \leq \text{index } \dot{\mathrm{W}}_r - \text{index } \dot{\mathrm{W}}_0 \leq r$$

so that (33.14) follows.

A final corollary is needed in Section 34.

COROLLARY 33.4. *If* $\dot{\mathrm{W}}_0$ *is an overall admissible canonical system, the count of characteristic roots of* $\dot{\mathrm{W}}_0$ *less than a prescribed value* σ *equals the count of conjugate points of* $\mathrm{x} = \mathrm{a}^1$ *in* $(\mathrm{a}^1,\mathrm{a}^2)$ *relative to the DE* (32.15), *or equivalently the count of* σ*-conjugate points of* $\mathrm{x} = \mathrm{a}^1$ *in* $(\mathrm{a}^1,\mathrm{a}^2)$ *relative to* $\dot{\mathrm{W}}_0$. (*Definition* 32.6.)

The two counts in Corollary 33.4 both equal index $\dot{Q}^\sigma\#(\dot{W}_0,\dot{\Lambda}_p)$ by virtue, respectively, of Theorem 33.5 and Theorem 33.2 (ii).

The following theorem is useful in Section 36 in comparing characteristic roots of two overall admissible systems $\dot{W}_r$. The brace indicates an alternative.

THEOREM 33.6. *If the quadratic functional* $\dot{\mathrm{I}}^\sigma_\mathrm{r}\#\dot{\mathrm{W}}_\mathrm{r}$ *is negative definite* {*nonpositive*} *on a vector subspace* of* $\{\dot{W}_\mathrm{r}\}$ *of dimension* k, *then the count of characteristic roots of* $\dot{\mathrm{W}}_\mathrm{r}$ *less than* σ {*at most* σ} *is at least* k.

Under the hypotheses of the theorem

$$\text{(33.17)}' \qquad \text{index } \dot{I}^\sigma_r \geq k; \qquad \{\text{index} + \text{nullity } \dot{I}^\sigma_r \geq k\}.$$

Hence by Theorem 33.4, for a corresponding index form $\dot{Q}^\sigma\#(\dot{W}_r,\dot{\Lambda}_p)$,

$$\text{(33.17)}'' \qquad \text{index } \dot{Q}^\sigma \geq k; \qquad \{\text{index} + \text{nullity } \dot{Q}^\sigma \geq k\}.$$

From (33.17)″ and Index Theorem 33.5 {Index Theorem 33.5 and Theorem 33.3} we infer the truth of Theorem 33.6.

* $\{\dot{W}_r\}$ is the domain of $\dot{I}^\sigma_r\#\dot{W}_r$.

CHAPTER 10

The Existence of Characteristic Roots

34. Selfconjugate Limit Values

This section is concerned with the characteristic roots of an "overall" admissible system $\dot{W}_r$ of canonical conditions $\dot{W}_r(\lambda)$, defined for each $\lambda \in R$. See Definition 32.4.

It follows from Index Theorem 33.5 that the characteristic roots of the system $\dot{W}_r$ are isolated and bounded below. The multiplicity of a characteristic root λ of $\dot{W}_r$ is at most $2m$, since the vector space over R of solutions of the DE (32.5) for the given λ has a dimension $2m$. Examples show that characteristic roots λ, with multiplicities prescribed between 1 and $2m$ inclusive, exist for a proper choice of $\dot{W}_r(\lambda)$. For proper choice of an overall admissible system $\dot{W}_r$, the count of characteristic roots of $\dot{W}_r$ may be any number between 0 and $+\infty$ inclusive, as examples will show.

The problem of formulating existence theorems for characteristic roots of canonical systems $\dot{W}_r$ is simplified by the following definition and theorem.

Definition 34.1. *Differentiably compatible systems* $\dot{\mathrm{W}}_r$. Two overall admissible systems will be said to be *differentiably compatible* if they have the same underlying quadratic form $\dot{\omega}^{\lambda}(x,\eta,\zeta)$ for each $x \in \aleph$ and $\lambda \in R$. It is understood that the two systems can have different dimensions r.

THEOREM 34.1. *A necessary and sufficient condition that an overall admissible system* $\dot{\mathrm{W}}_r$, $\mathrm{r} > 0$ *have infinitely many characteristic roots is that the differentiably compatible system* $\dot{\mathrm{W}}_0$ *have infinitely many characteristic roots.*

Theorem 34.1 is an immediate consequence of Corollary 33.3.

Preparation for Existence Theorem 34.2. A lemma on σ-conjugate points, relative to $\dot{W}_0$, is needed. See Definition 32.6. Given $c \in \aleph$ the σ-conjugate points of $x = c$, relative to $\dot{W}_0$, are the "ordinary" conjugate points of $x = c$ relative to the DE

$$\frac{d}{dx}\,\dot{\omega}^{\sigma}_{\zeta_i}(x,\eta,\eta') = \dot{\omega}^{\sigma}_{\eta_i}(x,\eta,\eta') \qquad (i = 1, \ldots, m). \tag{34.0}$$

With this understood, let the k-th *algebraic σ-conjugate* points of $x = c$ following {preceding} $x = c$ be defined with Definition 23.1 as a model, and be denoted, respectively, by $c_k^+(\sigma)\{c_k^-(\sigma)\}$, if such points exist in $\aleph$. Concerning these points the following lemma is essential. The brace { } indicates an alternative.

LEMMA 34.1. *Let a value* c *be given in* $\aleph$, *together with values* $\hat{\sigma}$ *and* σ *such that* $\hat{\sigma} < \sigma$. *If a* k*-th algebraic* $\hat{\sigma}$*-conjugate point* $\mathrm{c}_{\mathrm{k}}^+(\hat{\sigma})\{\mathrm{c}_{\mathrm{k}}^-(\hat{\sigma})\}$ *of* $\mathrm{x} = \mathrm{c}$ *exists in* $\aleph$, *relative to* $\dot{\mathrm{W}}_0$, *following* c *{preceding* c,*} then a* k*-th algebraic* σ*-conjugate point* $\mathrm{c}_{\mathrm{k}}^+(\sigma)\{\mathrm{c}_{\mathrm{k}}^-(\sigma)\}$ *of* $\mathrm{x} = \mathrm{c}$ *exists in* $\aleph$, *relative to* $\dot{\mathrm{W}}_0$, *with**

$$\mathrm{c}_{\mathrm{k}}^+(\sigma) < \mathrm{c}_{\mathrm{k}}^+(\hat{\sigma}), \qquad \{\mathrm{c}_{\mathrm{k}}^-(\sigma) > \mathrm{c}_{\mathrm{k}}^-(\hat{\sigma})\}. \tag{34.1}$$

We shall prove the lemma when $c_k^+(\hat{\sigma})$ is given. The proof of the alternative lemma is similar.

Let a value $\gamma > \sigma$ be prescribed and fixed. For $\sigma \leq \gamma$ let index forms $\dot{Q}^{\sigma}\#(\dot{W}_0,\dot{\Lambda}_p)$ be set up with $[a^1,a^2]$ replaced by $[c,c_k^+(\hat{\sigma})]$ and with a γ-frame $\dot{\Lambda}_p$. The variables $z_1, \ldots, z_\mu$ of the forms $\dot{Q}^{\sigma}$ are $\mu = mp$ in number and the forms $\dot{Q}^{\sigma}$ define a model family G in the sense of Appendix III, as affirmed in Lemma 33.2.

By hypothesis $c_k^+(\hat{\sigma})$ exists and has been taken as the right end point of the interval $[c,c_k^+(\hat{\sigma})]$ over the interior of which the frame $\dot{\Lambda}_p$ has been defined. It follows from Theorem 33.2 and the definition of $c_k^+(\hat{\sigma})$ that

$$\text{index } \dot{Q}^{\hat{\sigma}} < k \leq (\text{index} + \text{nullity})\dot{Q}^{\hat{\sigma}}. \tag{34.2$'$}$$

It is a corollary of Theorem 3.1 of Appendix III that

$$\text{index } \dot{Q}^{\sigma} \geq (\text{index} + \text{nullity})\dot{Q}^{\hat{\sigma}} \tag{34.2$''$}$$

since $\hat{\sigma} < \sigma \leq \gamma$. We infer from (34.2)$''$ and (34.2)$'$ that index $\dot{Q}^{\sigma} \geq k >$ index $\dot{Q}^{\hat{\sigma}}$. It follows from Theorem 33.2 that the count of σ-conjugate points of $x = c$ in the interval $(c,c_k^+(\hat{\sigma}))$ is at least k. From the definition of the k-th algebraic σ-conjugate point of c we infer that $c_k^+(\sigma)$ exists when $\hat{\sigma} < \sigma \leq \gamma$ and

$$c_k^+(\sigma) < c_k^+(\hat{\sigma}). \tag{34.3}$$

We infer the truth of Lemma 34.1.

* The mapping $\sigma \to c_k^+(\sigma)\{\sigma \to c_k^-(\sigma)\}$ is continuous. For proof see *Note* at end of Section 34.

Selfconjugate limit values. From Lemma 34.1 the following can be inferred. If $c_1^+(\sigma)$ exists in $\aleph$ for some $c \in \aleph$ and some σ, then for $t > \sigma$, $c_1^+(t)$ exists and decreases as t increases, so that $\lim_{t\uparrow\infty} c_1^+(t)$ exists and has a value,

$$(34.4)' \qquad \lim_{t\uparrow\infty} c_1^+(t) \geq c.$$

Similarly, if $c_1^-(\sigma) < c$ exists in $\aleph$, then for $t > \sigma$, $c_1^-(t)$ exists and increases as t increases, so that $\lim_{t\uparrow\infty} c_1^-(t)$ exists and has a value,

$$(34.4)'' \qquad \lim_{t\uparrow\infty} c_1^-(t) \leq c.$$

The limits in (34.4)′ and (34.4)″ may or may not be c, as examples will show. We are led to a definition.

Definition 34.2. If in (34.4)′ {(34.4)″} the limit exists and equals c, c will be called a *selfconjugate right* {*left*} limit value, relative to $\dot{W}_0$.

The principal theorem of this section follows.

EXISTENCE THEOREM 34.2. *A necessary and sufficient condition that an "overall admissible" system* $\dot{\mathrm{W}}_0$ *of canonical system* $\dot{\mathrm{W}}_0(\lambda)$ *have infinitely many characteristic roots is that at least one of the two following cases occurs.*
Case 1. *Some value* $\mathrm{c} \in [\mathrm{a}^1,\mathrm{a}^2)$ *is a selfconjugate right limit value.*
Case 2. *Some value* $\mathrm{c} \in (\mathrm{a}^1,\mathrm{a}^2]$ *is a selfconjugate left limit value.*

Proof that the Condition is Necessary. Were the condition not necessary, then with each value $\alpha \in [a^1,a^2]$ there could be associated an open connected neighborhood $N(\alpha)$ of $x = \alpha$ relative to $[a^1,a^2]$, such that $N(\alpha)$ is free of λ-conjugate points of $x = \alpha$ preceding or following $x = \alpha$ for each λ. The interval $[a^1,a^2]$ could then be covered by a finite set of open neighborhoods $N(\alpha_k)$, $k = 1, \ldots, n$, of points $x = \alpha_k \in [a^1,a^2]$ such that $N(\alpha_k)$ contains a λ-conjugate point of $x = \alpha_k$ for no value of λ.

Let α and λ be prescribed in $[a^1,a^2]$ and R, respectively. Let F_α^λ be the von Escherich family of solutions of the DE

$$(34.5) \qquad \frac{d}{dx}\,\dot{\omega}^\lambda_{\zeta_i}(x,\eta,\eta') = \dot{\omega}^\lambda_{\eta_i}(x,\eta,\eta') \qquad (i = 1, \ldots, m)$$

which vanish when $x = \alpha$. The count of focal points of $F^\lambda_{\alpha_k}$ in $N(\alpha_k)$ is exactly m. By our Separation Theorem 20.1, the count of focal points of $F^\lambda_{a^1}$ in $N(\alpha_k)$ is then at most $2m$. Thus there are at most $n(2m)$ λ-conjugate points of $x = a^1$ in (a^1,a^2). Note that this count, $2nm$, is independent of the choice of λ.

This is impossible as we now show.

By hypothesis there exist infinitely many characteristic roots of the system $\dot{W}_0$. By Corollary 33.4 the count, say $n(\lambda)$, of characteristic roots of $\dot{W}_0$ less than λ, is the count of λ-conjugate points of $x = a^1$ in (a^1,a^2), relative to $\dot{W}_0$. By hypothesis $n(\lambda)$ is unbounded for $\lambda \in R$, contrary to the conclusion of the preceding paragraph.

The condition of Theorem 34.2 is accordingly necessary.

Proof that the Condition is Sufficient. Suppose first that Case 2 occurs, with $c \in (a^1,a^2]$. We shall then prove the following:

(μ) *If Case* 2 *occurs the count of λ-conjugate points of* $\mathrm{x} = \mathrm{a}^1$ *in* $(\mathrm{a}^1,\mathrm{a}^2)$, *relative to* $\dot{\mathrm{W}}_0$, *is not bounded as* $\lambda \uparrow \infty$.

Proof of (μ). We shall apply the Separation Theorem 20.1 to the von Escherich families $F_{a^1}^{\lambda}$ and F_c^{λ} of solutions of the DE (34.5) which vanish when $x = a^1$ and $x = c$, respectively. For λ sufficiently large $c_1^-(\lambda)$ exists in (a^1,a^2) by hypothesis. $F_{a^1}^{\lambda}$ must then have at least one focal point in the interval $[c_1^-(\lambda),c]$, since the count of focal points of F_c^{λ} in this interval exceeds m.

For some integer $k > 0$ this focal point of $F_{a^1}^{\lambda}$ is the carrier of the k-th algebraic λ-conjugate point $(a^1)_k^+(\lambda)$ of $x = a^1$. As λ increases $(a^1)_k^+(\lambda)$ will decrease in value, while $c_1^-(\lambda)$ will increase in value, in accord with Lemma 34.1. Moreover $\lim_{\lambda \uparrow \infty} c_1^-(\lambda) = c$ by hypothesis. Hence for λ sufficiently large

$$(a^1)_k^+(\lambda) < c_1^-(\lambda) \tag{34.6}$$

so that by the Separation Theorem there must be a k_1-th algebraic λ-conjugate point of $x = a^1$ in $[c_1^-(\lambda),c]$ for which $k_1 > k$.

Let n be a prescribed positive integer. Reasoning as in the preceding paragraph, we see that if λ is sufficiently large the count of λ-conjugate points of $x = a^1$ preceding $x = c$ must exceed n.

Statement (μ) follows.

When Case 2 occurs it follows from (μ) and Corollary 33.4 that, if λ is sufficiently large, the count of characteristic roots of $\dot{W}_0$ less than λ exceeds a prescribed integer n. Thus $\dot{W}_0$ has infinitely many characteristic roots if Case 2 occurs. That $\dot{W}_0$ has infinitely many characteristic roots, if Case 1 occurs, is established similarly, on interchanging the roles of a^1 and a^2. Cf. Corollary 20.1.

Thus Theorem 34.2 is true.

Test Sequences. The Existence Theorem 34.2 can be given an equivalent form involving Test Sequences which we shall now define and which are somewhat simpler to apply than the criteria of Theorem 34.2.

Definition 34.3 (i). A *Test Sequence* $\{\varphi_{c,\mu}\}$, $\mu = 1, 2, \ldots$. Corresponding to a prescribed value $c \in [a^1,a^2)$, let $e_1 > e_2 > \cdots$ be a sequence $\{e_\mu\}$ of positive numbers such that $e_\mu \downarrow 0$ as $\mu \uparrow \infty$. Let $\varphi_{c,\mu}$ be a nonnull D^1-mapping of $[c,c + e_\mu]$ into R^m which vanishes when $x = c$ and $x = c + e_\mu$.

Definition 34.3 (ii). A *Test Sequence* $\{\psi_{c,\mu}\}$, $\mu = 1, 2, \ldots$. Corresponding to a prescribed value $c \in (a^1,a^2]$ and a sequence $\{e_\mu\}$, as in Definition 34.3 (i), let $\psi_{c,\mu}$ be a nonnull D^1-mapping of $[c - e_\mu, c]$ into R^m which vanishes when $x = c$ and $c - e_\mu$.

THEOREM 34.3. *Under the hypotheses of Theorem* 34.2 *the following is true:*

(i) *A necessary and sufficient condition that Case* 1 *of Theorem* 34.2 *occur is that for some choice of* $c \in [a^1,a^2)$ *and of a Test Sequence* $\{\varphi_{c,\mu}\}$

$$\int_{c}^{c+e_\mu} \dot{\omega}^\mu(x,\varphi_{c,\mu}(x),\varphi'_{c,\mu}(x))\,dx \leq 0 \tag{34.7$'$}$$

for all sufficiently large integers μ.

(ii) *A necessary and sufficient condition that Case* 2 *of Theorem* 34.2 *occur is that for some choice of* $c \in (a^1,a^2]$ *and of a Test Sequence* $\{\psi_{c,\mu}\}$

$$\int_{c-e_\mu}^{c} \dot{\omega}^\mu(x,\psi_{c,\mu}(x),\psi'_{c,\mu}(x))\,dx \leq 0 \tag{34.7$''$}$$

for all sufficiently large integers μ.

Proof that the Condition of (i) *is Necessary.* We are assuming that Case 1 occurs in Theorem 34.2. That is, for some value $c \in [a^1,a^2)$, for some sufficiently large fixed value σ and for $t > \sigma$, 1-st algebraic conjugate values $c_1^+(t)$ of $x = c$ then exist in (a^1,a^2), decrease as t increases and have the limit c as $t \uparrow \infty$. For integers $\mu > \sigma$, set $c_1^+(\mu) = c + e_\mu$. For $\mu > \sigma$, there exists a nonnull solution $\varphi_{c,\mu}$ of the DE

$$\frac{d}{dx}\,\dot{\omega}^\mu_{\zeta_i}(x,\eta,\eta') = \dot{\omega}^\mu_{\eta_i}(x,\eta,\eta') \qquad (i = 1, \ldots, m) \tag{34.8}$$

which vanishes when $x = c$ and $x = c + e_\mu$. Then as in Lemma 4.2

$$\int_{c}^{c+e_\mu} \dot{\omega}^\mu(x,\varphi_{c,\mu}(x),\varphi'_{c,\mu}(x))\,dx = 0 \tag{34.9}$$

for all integers $\mu > \sigma$.

The condition of (i) is thus necessary.

Proof that the Condition of (i) *is Sufficient.* We are assuming that (34.9) holds for some Test Sequence $\{\varphi_{c,\mu}\}$ for all sufficiently large integers μ. It follows from Theorem 7.4 that Case I of Theorem 34.2 occurs for the given c.

The condition of (i) is thus sufficient.

That the condition of (ii) is necessary and sufficient is proved similarly.

Of the many corollaries of the theorems of this section, one of the most important follows.

COROLLARY 34.1. *The number of characteristic roots of a canonical system* W_r, *as defined in Section* 15, *is infinite.*

According to Theorem 34.1 the corollary is true if and only if it is true for a canonical system W_0 with the same DE as W_r. The underlying quadratic forms of W_0 have the form

$$\omega(x,\eta,\zeta) - \frac{\lambda}{2} \|\eta\|^2 \qquad (x \in \aleph,\ \lambda \in R) \tag{34.10}$$

as conditioned in Appendix I. Because of the way in which this form depends on λ the condition of Theorem 34.3 (i) is readily satisfied for each $c \in [a^1,a^2)$. Corollary 34.1 follows.

Note. The condition of Theorem 34.3 (ii) is also satisfied for each $c \in (a^1,a^2]$.

A Major Extension of Corollary 34.1. Let $\dot{W}_r$ be a system of conditions $\dot{W}_r(\lambda)$ defined for each λ, of form (32.10), if $r > 0$, and of form (32.11), if $r = 0$. We suppose $\dot{W}_r$ conditioned as follows:

Condition (α). For each $\lambda \in R$, for $x \in \aleph$ and for arbitrary m-tuples η and ζ

$$\dot{\omega}^\lambda(x,\eta,\zeta) = \omega(x,\eta,\zeta) - \frac{\lambda}{2} a_{ij}(x)\eta_i\eta_j, \tag{34.11}$$

where $\omega(x,\eta,\zeta)$ is conditioned as in Appendix I. For i, j on the range $1, \ldots, m$, the mapping $x \to a_{ij}(x)$ of $\aleph$ into R is supposed continuous and such that the quadratic form $a_{ij}(x)\eta_i\eta_j$ is positive or zero for each $x \in \aleph$, and for at least one value $c \in [a^1,a^2]$ is positive definite.

Condition (β). If $r > 0$ we require that the quadratic form $b_{hk}(\lambda)u_h u_k$ of (32.10)‴ decrease monotonically as λ increases, for each r-tuple u. The mappings $\lambda \to b_{hk}(\lambda)$ are assumed continuous in every case.

Condition (γ). Suppose finally that the associated quadratic functional $\dot{I}_r^\lambda \# \dot{W}_r$ of (32.12) is positive definite on its domain $\{\dot{W}_r\}$ when $\lambda = a$ for some value $a < 0$.

Corollary 34.1 is a special case of the following theorem:

THEOREM 34.4. *Under the above conditions* (α), (β), (γ), *on the system* $\dot{W}_r$, $\dot{W}_r$ *is "overall admissible" and has infinitely many positive characteristic roots.*

Condition (A_1) on $\dot{W}_r$ is satisfied by virtue of the form (34.11) of $\dot{\omega}^\lambda(x,\eta,\zeta)$.

Condition (A_2) on $\dot{W}_r$ is satisfied because of Condition (β).

Condition (A_3) on $\dot{W}_r$ is satisfied, since Condition (γ), supplemented by Conditions (β) and (α), implies that $\dot{I}_r^\lambda$ is positive definite on $\{\dot{W}_r\}$ for $\lambda \leq a$.

Condition (A_4) on $\dot{W}_r$ is satisfied, when $r > 0$ since a nonnull solution η of $\dot{W}_r$ with root τ, satisfying conditions (32.10)″ with an r-tuple u is such that for arbitrary σ and $\tau > \sigma$

$$\dot{I}_r^\sigma(\eta) - \dot{I}_r^\tau(\eta) = (b_{hk}(\sigma) - b_{hk}(\tau))u_h u_k + (\tau - \sigma)\int_{a^1}^{a^2} a_{ij}(x)\eta_i(x)\eta_j(x)\,dx > 0$$

When $r = 0$, Condition (A_4) is also satisfied, as one readily sees.

Thus $\dot{W}_r$ is overall admissible.

If c is a value in $[a^1,a^2)\{(a^1,a^2]\}$ for which the quadratic form $a_{ij}(c)\eta_i\eta_j$ is positive definite, an application of Theorem 34.3 shows that c is a selfconjugate right limit value {a selfconjugate left limit value}. According to Theorem 34.2 the overall admissible system $\dot{W}_0$, differentiably compatible to $\dot{W}_r$, then has infinitely many positive characteristic roots. By Theorem 34.1, $\dot{W}_r$ likewise has infinitely many positive characteristic roots.

Thus Theorem 34.4 is true.

Note. Verification of footnote to Lemma 34.1. We shall prove that the mapping $\sigma \to c_k^+(\sigma)$ of Lemma 34.1 is continuous at a prescribed value σ_0 at which $c_k^+(\sigma_0)$ exists. A proof of the continuity of the mapping $\sigma \to c_k^-(\sigma)$ at σ_0 is similar, provided $c_k^-(\sigma_0)$ exists.

Notation. Set $x_0 = c_k^+(\sigma_0)$. Let a and b be values prescribed in the interval $\aleph$ subject to the condition that $c < a < x_0 < b$ and that there is no σ_0-conjugate point of $x = c$ in the interval $[a,b]$, other than x_0. Let $\dot{Q}_a^\sigma$, $\dot{Q}_{x_0}^\sigma$, $\dot{Q}_b^\sigma$ be index forms set up as in Definition 32.8 with $r = 0$ and with the interval $[a^1,a^2]$ replaced by $[c,a]$, $[c,x_0]$, $[c,b]$, respectively. Let σ be restricted to an open subinterval I of σ_0.

If the intervals $[a,b]$ and I are sufficiently small, a σ_0-frame for $\dot{Q}_{x_0}^{\sigma_0}$ can serve* as a σ-frame for $\dot{Q}_{x_0}^\sigma$, $\dot{Q}_a^\sigma$, $\dot{Q}_b^\sigma$ for $\sigma \in I$, and the coefficients in these quadratic forms will vary continuously with σ. The forms $\dot{Q}_a^{\sigma_0}$ and $\dot{Q}_b^{\sigma_0}$ are nonsingular by virtue of our choice of a and b, and will remain nonsingular for $\sigma \in I$, if I is sufficiently small. We then establish the continuity of the mapping $\sigma \to c_k^+(\sigma)$ at σ_0 by showing that for I conditioned as above

$$a < c_k^+(\sigma) < b \qquad (\text{for } \sigma \in I). \tag{34.12}$$

Proof of (34.12). Let ρ and ν be, respectively, the counts of σ_0-conjugate points of $x = c$ preceding $x = x_0$ and at $x = x_0$. Then by Theorem 33.2

$$\rho = \text{index } \dot{Q}_a^{\sigma_0}, \qquad \rho + \nu = \text{index } \dot{Q}_b^{\sigma_0}. \tag{34.13}$$

* By Theorem 4.2 of Appendix I.

Because $\dot{Q}_a^\sigma$ and $\dot{Q}_b^\sigma$ are nonsingular for $\sigma \in I$ and because $\rho < k \leq \rho + \nu$

(34.14) $\quad \rho = \text{index } \dot{Q}_a^\sigma < k \leq \rho + \nu = \text{index } \dot{Q}_b^\sigma \qquad (\text{for } \sigma \in I).$

The relation $k \leq \text{index } \dot{Q}_b^\sigma$ implies that $c_k^+(\sigma)$ exists and is less than b for $\sigma \in I$. The relation index $\dot{Q}_a^\sigma < k$ implies that, if $c_k^+(\sigma)$ exists, $a < c_k^+(\sigma)$. Thus (34.12) is true.

Thus the footnote to Lemma 34.1 is true.

Exercise 34.1. Let $x \to C(x)$ be a continuous mapping of $\aleph$ into R such that $C(x) \geq 0$ and $C(x) \not\equiv 0$ for $x \in [a^1,a^2]$. Show that the system defined by the DE

$$\eta'' + \lambda C(x)\eta = 0$$

for each λ, with end conditions $\eta'(a^1) = \eta(a^2) = 0$, has infinitely many positive characteristic roots λ.

Suggestion. Set up an overall admissible system $\dot{W}_1$ of form (32.10), equivalent to the above differential equations and end conditions when $m = 1$, and apply Theorem 34.4.

35. Comparison of Characteristic Roots

The comparison of free focal conditions in Section 21 required for data conditions such as (21.1)″ and (21.1)‴ at one end point $x = a^1$. It was natural to compare the focal points of the corresponding von Escherich families.

In contrast canonical systems such as W_r in Section 15 or $\dot{W}_r$ in Section 32, with nontrivial BC at *both* end points of the interval $[a^1,a^2]$, cannot be naturally compared in terms of the focal points. A comparison of characteristic roots is simpler and more revealing.

Our principal theorem will compare two "overall" admissible canonical systems $\dot{W}_r$ and $\dot{W}_r^*$, as defined in Section 32. We suppose first that $r > 0$ and that $\dot{W}_r$ and $\dot{W}_r^*$ are *comparable* in that they are given with a common accessory end plane independent of λ with a regular representation

(35.1) $\qquad \eta_i^s = c_{ih}^s u_h \qquad (s = 1, 2;\ i = 1, \ldots, m).$

The "defining elements" of a canonical system $\dot{W}_r$ have been introduced in the proof of Corollary 33.3, and, in addition to the matrices $\|c_{ih}^s\|$ include underlying forms and matrices

(35.2) $\qquad \dot{\omega}^\lambda(x,\eta,\zeta),\ \|b_{hk}(\lambda)\| \qquad (\lambda \in R,\ x \in \aleph).$

The corresponding defining elements of $\dot{W}_r^*$ include the matrices $\|c_{ih}^s\|$ and underlying forms and matrices denoted respectively by

(35.3) $\qquad \dot{\omega}_*^\lambda(x,\eta,\zeta),\ \|b_{hk}^*(\lambda)\|\ .$

To formulate our comparison theorem we shall introduce the difference forms

$$(35.4) \qquad \Delta\dot{\omega}^{\lambda}(x,\eta,\zeta) = \dot{\omega}^{\lambda}(x,\eta,\zeta) - \dot{\omega}^{\lambda}_{*}(x,\eta,\zeta)$$

for each $x \in \aleph$ and $\lambda \in R$, and, for each $\lambda \in R$, r-square difference matrices,

$$(35.5) \qquad \|\Delta b_{hk}(\lambda)\| = \|b_{hk}(\lambda) - b^{*}_{hk}(\lambda)\| .$$

Definition 35.1. **The difference functionals** d_r^{λ}. For each $\lambda \in R$ set (cf. $D_r^{\lambda}(\eta)$ of (21.6)′)

$$(35.6) \quad d_r^{\lambda}(\eta) = \Delta b_{hk}(\lambda) u_h u_k + 2\int_{a^1}^{a^2} \Delta\dot{\omega}^{\lambda}(x,\eta(x),\eta'(x))\, dx \qquad (r > 0)$$

for each D^1-mapping η of $[a^1,a^2]$ into R^m which satisfies the conditions (35.1) with an r-tuple u.

We shall enumerate the characteristic roots of $\dot{W}_r$ in an *algebraic* sequence

$$(35.7) \qquad \sigma_1 \leq \sigma_2 \leq \sigma_3 \leq \cdots$$

which may be empty, finite or infinite, and in which each root appears a number of times equal to its multiplicity. Let the characteristic roots of $\dot{W}_r^{*}$ be enumerated in a similar sequence:

$$(35.8) \qquad \sigma_1^{*} \leq \sigma_2^{*} \leq \sigma_3^{*} \leq \cdots .$$

A basic comparison theorem follows. In this theorem alternative conditions are enclosed in a pair of braces { }. The theorem is concerned with the "overall" admissible comparable canonical systems $\dot{W}_r$ and $\dot{W}_r^{*}$.

COMPARISON THEOREM 35.1. *Suppose that an* n*-th characteristic root* σ_n *exists in the algebraic sequence of roots of* $\dot{W}_r$, $r > 0$. *If then the difference functional* $d_r^{\sigma_n}$: (35.6) *is positive definite {nonnegative}, a characteristic root* σ_n^{*} *exists in the algebraic sequence* (35.8) *for* $\dot{W}_r^{*}$, *and is such that*

$$(35.9) \qquad \sigma_n^{*} < \sigma_n \ \{\sigma_n^{*} \leq \sigma_n\}.$$

For arbitrary σ the index forms

$$(35.10) \qquad \dot{Q}^{\sigma}\#(\dot{W}_r,\dot{\Lambda}_p), \qquad \dot{Q}^{\sigma}_{*}\#(\dot{W}_r^{*},\dot{\Lambda}_p)$$

will be introduced with a common σ-frame $\dot{\Lambda}_p$. Set $\mu = r + mp$. We shall verify Lemma 35.1.

LEMMA 35.1. *Let* σ *be prescribed. When the difference functional* d_r^{σ} : (35.6) *is positive definite {nonnegative} the quadratic form*

$$(35.11) \qquad \dot{Q}^{\sigma}(z) - \dot{Q}^{\sigma}_{*}(z) \qquad (z \in R^{\mu})$$

is positive definite {nonnegative}.

Proof. Let z be a prescribed μ-tuple. Let $\bar{\eta}$ and η^* be D^1-mappings of $[a^1,a^2]$ into R^m such that, in the sense of Definition 15.2

$$\text{graph } \bar{\eta} = h^{\sigma,z}\#(\dot{W}_r,\dot{\Lambda}_p) \tag{35.12$'$}$$

$$\text{graph } \eta^* = h^{\sigma,z}\#(\dot{W}_r^*,\dot{\Lambda}_p). \tag{35.12$''$}$$

By virtue of Definition 32.8

$$\dot{Q}^\sigma(z) = b_{hk}(\sigma)u_h u_k + 2\int_{a^1}^{a^2} \dot{\omega}^\sigma(x,\bar{\eta}(x),\bar{\eta}'(x))\, dx \tag{35.13$'$}$$

$$\dot{Q}_*^\sigma(z) = b_{hk}^*(\sigma)u_h u_k + 2\int_{a^1}^{a^2} \dot{\omega}_*^\sigma(x,\eta^*(x),\eta^{*\prime}(x))\, dx \tag{35.13$''$}$$

where u is the initial r-tuple of the μ-tuple z. By virtue of the minimizing property (Corollary 7.1) of the $p + 1$ subarcs into which graph η^* is divided by the m-planes of the frame $\dot{\Lambda}_p$

$$\dot{Q}_*^\sigma(z) \leq b_{hk}^*(\sigma)u_h u_k + 2\int_{a^1}^{a^2} \dot{\omega}_*^\sigma(x,\bar{\eta}(x),\bar{\eta}'(x))\, dx = P^\sigma(z), \tag{35.14}$$

introducing $P^\sigma(z)$.

It follows from (35.14) and (35.6) that

$$P^\sigma(z) = b_{hk}(\sigma)u_h u_k + 2\int_{a^1}^{a^2} \dot{\omega}^\sigma(x,\bar{\eta}(x),\bar{\eta}'(x))\, dx - d_r^\sigma(\bar{\eta}). \tag{35.15}$$

Given σ and a nonnull μ-tuple z, the corresponding mapping $\bar{\eta}$ is nonnull. Since $d_r^\sigma(\bar{\eta})$ is then positive {nonnegative} by hypothesis, we infer from (35.15) and (35.13) that for $z \neq \mathbf{0}$

$$P^\sigma(z) < \dot{Q}^\sigma(z) \qquad \{P^\sigma(z) \leq \dot{Q}^\sigma(z)\}. \tag{35.16}$$

Lemma 35.1 follows from (35.14) and (35.16).

Completion of Proof of Theorem 35.1. When σ_n exists and $d_r^{\sigma_n}$ is positive definite {nonnegative}, Lemma 35.1 implies that for each nonnull μ-tuple z,

$$\dot{Q}_*^{\sigma_n}(z) < \dot{Q}^{\sigma_n}(z) \qquad \{\dot{Q}_*^{\sigma_n}(z) \leq \dot{Q}^{\sigma_n}(z)\}. \tag{35.17}$$

It follows from (35.17) that

$$\text{index } \dot{Q}_*^{\sigma_n} \geq (\text{index} + \text{nullity})\dot{Q}^{\sigma_n} = s, \tag{35.18$'$}$$

introducing s, and in the alternative case

$$\{(\text{index} + \text{nullity})\dot{Q}_*^{\sigma_n} \geq (\text{index} + \text{nullity})\dot{Q}^{\sigma_n}\}. \tag{35.18$''$}$$

The equality in (35.18)$'$ implies that in the algebraic sequence (35.7), $\sigma_s = \sigma_n$ and $s \geq n$. Set $k = \text{index } \dot{Q}_*^{\sigma_n}$. Theorem 33.5 implies that the count of characteristic roots less than σ_n is k. Our enumeration of characteristic roots of $\dot{W}_r^*$ is then such that $\sigma_k^* < \sigma_n$. Since $k \geq s \geq n$ we infer that $\sigma_n^* \leq \sigma_k^* < \sigma_n$, thereby completing the proof of Theorem 35.1 in case $d_r^{\sigma_n}$ is positive definite.

The proof that $\sigma_n^* \leq \sigma_n$ in the alternative case is similar, replacing all signs $<$ by $\leq$ and making use of Theorem 33.3 as well as of Theorem 33.5.

Theorem 35.1 has the following corollary:

COROLLARY 35.1. *If the canonical system* $\dot{W}_r$, $r > 0$ *has an infinite algebraic sequence* (35.7) *of characteristic roots and if the difference functional* d_r^σ, *associated in* (35.6) *with* $\dot{W}_r$ *and a second canonical comparable system* $\dot{W}_r^*$, *is positive definite* {*nonnegative*} *for each* σ, *then* $\dot{W}_r^*$ *has an infinite algebraic sequence* (35.8) *of characteristic roots such that*

$$\sigma_1^* < \sigma_1, \qquad \sigma_2^* < \sigma_2, \qquad \sigma_3^* < \sigma_3, \cdots$$
$$\{\sigma_1^* \leq \sigma_1, \qquad \sigma_2^* \leq \sigma_2, \qquad \sigma_3^* \leq \sigma_3, \cdots\}.$$

Comparisons when $r = 0$. In this case two overall admissible systems $\dot{W}_0$ and $\dot{W}_0^*$ are given with common end point conditions,

$$\eta(a^1) = \eta(a^2) = \mathbf{0} \tag{35.19}$$

A difference functional d_0^λ is defined by (35.6) for each λ; one deletes the quadratic form in the r-tuples u in (35.6) and restricts η to D^1-mappings of $[a^1,a^2]$ into R^m such that (35.19) holds. Comparison Theorem 35.1 then holds on setting $r = 0$ therein.

Definition 35.2. *Comparable systems* $\dot{W}_r$ *and* $\dot{W}_r^*$. Two overall admissible systems $\dot{W}_r$ and $\dot{W}_r^*$, with a common accessory end plane χ_r, with a common regular representation (35.1) when $r > 0$, or with common end conditions (35.19) when $r = 0$, will be said to be *comparable*.

The comparisons in Section 35 when $r \geq 0$ are restricted to overall admissible systems $\dot{W}_r$ and $\dot{W}_r^*$, "comparable" in the sense of Definition 35.2. The functionals $\dot{I}_r^\lambda$ based on comparable systems $\dot{W}_r$ and $\dot{W}_r^*$ have common domains $\{\dot{W}_r\} = \{\dot{W}_r^*\}$.

36. The $\dot{W}_r$-Nuclear Comparison Test

In Section 21, in comparing the focal points of two different sets of "comparable" focal conditions V_r and $\hat{V}_\rho$, we first introduced comparison conditions H(I), H(II), H(III) in the respective cases I, II, III of (21.0). These a priori conditions were then replaced by equally effective but less restrictive conditions K(I), K(II), K(III), defined in terms of the V_r-"nuclear" subspace of the domain of $I^0_{r,d}\#\mathcal{W}_{r,d}$.

In comparing characteristic roots rather than focal points we shall proceed similarly. In Theorem 35.1 we have given tests whose satisfaction guarantees that the n-th* characteristic root σ_n (if it exists) of an overall admissible system $\dot{W}_r$ is greater {at least as great as} the n-th characteristic root of a "comparable" overall admissible system $\dot{W}_r^*$. We term these tests *a priori σ_n-tests.* We shall replace these a priori σ_n-tests by less restrictive but equally effective tests on $\dot{W}_r$ and $\dot{W}_r^*$. These new tests will be called *nuclear tests* and will be defined in terms of the following vector space.

Definition 36.1. *The nuclear subspace* $\mathrm{N}_{\mathrm{r}}^\sigma$ *of* $\{\dot{\mathrm{W}}_{\mathrm{r}}\}$. For each characteristic root σ of $\dot{W}_r$ let N_r^σ denote the vector subspace over R of $\{\dot{W}_r\}$ generated by the characteristic solutions of $\dot{W}_r$ with roots at most σ. We term N_r^σ the σ-th *nuclear subspace* of $\{\dot{W}_r\}$.

Before stating our nuclear comparison theorem we shall derive some of the basic properties of N_r^σ.

The Bilinear Functional $\dot{\mathrm{B}}_{\mathrm{r}}^\sigma$ *of* $\dot{\mathrm{W}}_{\mathrm{r}}$. To define $\dot{B}_r^\sigma$ we fix the characteristic root σ and let γ be any characteristic solution of $\dot{W}_r$ with root σ. When $r > 0$, conditions (32.10) are satisfied by γ with an r-tuple v (in place of u). Let η be an arbitrary D^1-mapping of $[a^1,a^2]$ into R^m which satisfies the conditions (32.10)″ with an r-tuple u. Following the definition in Section 15 of the bilinear functional B_r^σ by its values (15.23), we here set

$$\dot{B}_r^\sigma(\gamma,\eta) = b_{hk}(\sigma)u_h v_k + \int_{a^1}^{a^2} (\eta_i \dot{\omega}_{\eta_i}^\sigma(x,\gamma,\gamma') + \eta_i' \dot{\omega}_{\zeta_i}^\sigma(x,\gamma,\gamma'))\, dx \tag{36.1}$$

and note that $\dot{B}_r^\sigma(\gamma,\eta) = 0$ as a consequence of the conditions imposed on γ and η. To verify this one makes the usual integration by parts.

When $r = 0$, $\dot{B}_0^\sigma(\gamma,\eta)$ is defined similarly by (36.1), deleting all reference to the r-tuples u and v, and requiring that η, as well as γ, vanish when $x = a^1$ and a^2.

There is thereby defined a bilinear functional $\dot{B}_r^\sigma$ whose domain is the space of pairs (γ,η) admitted above. When $r > 0$ it should be noted that the r-tuples v and u appearing on the right of (36.1) are uniquely determined, respectively by γ and η, depending linearly on γ and η for γ and η in the domain of $\dot{B}_r^\sigma$.

The bilinear functional $\dot{B}_r^\sigma$ is an aid in proving the following lemma.

LEMMA 36.1. *Let σ and λ be characteristic roots of* $\dot{\mathrm{W}}_{\mathrm{r}}$ *with $\lambda < \sigma$. If γ is a characteristic solution of* $\dot{\mathrm{W}}_{\mathrm{r}}$ *with root σ, then* $\dot{\mathrm{I}}_{\mathrm{r}}^\sigma(\gamma) = 0$ *and if η is any element of* $\mathrm{N}_{\mathrm{r}}^\sigma$, *then, when* $\dot{\mathrm{W}}_{\mathrm{r}}$ *is overall admissible,*

$$\dot{\mathrm{I}}_{\mathrm{r}}^\sigma(\gamma + \eta) = \dot{\mathrm{I}}_{\mathrm{r}}^\sigma(\eta) \leq \dot{\mathrm{I}}_{\mathrm{r}}^\lambda(\eta). \tag{36.2}$$

* Counting roots with their multiplicities.

As a matter of formal algebra,

$$\dot{I}_r^\sigma(\gamma) = \dot{B}_r^\sigma(\gamma,\gamma) = 0 \tag{36.3}$$

$$\dot{I}^\sigma(\gamma + \eta) = \dot{I}_r^\sigma(\gamma) + 2\dot{B}_r^\sigma(\gamma,\eta) + \dot{I}_r^\sigma(\eta) = \dot{I}_r^\sigma(\eta), \tag{36.4}$$

where the first two terms on the right of (36.4) vanish, as already noted. $\dot{W}_r$ is assumed "overall" admissible, so that Conditions (A_1), (A_2), (A_3), (A_4) of Definition 32.4 are satisfied. The final inequality in (36.2) then follows, from Conditions (A_1) and (A_2) since $\lambda < \sigma$, by hypothesis.

Thus Lemma 36.1 is true.

We continue with the following lemma.

LEMMA 36.2(a). *If σ is a characteristic root of $\dot{\mathrm{W}}_r$, then $\dot{\mathrm{I}}_r^\sigma(\chi) < 0$ for each nonnull mapping $\chi \in \mathrm{N}_r^\sigma$ other than a characteristic solution of $\dot{\mathrm{W}}_r$ with root σ.*

(b). *Dim N_r^σ is the count $\kappa(\sigma)$ of characteristic roots of $\dot{\mathrm{W}}_r$ at most σ.*

Proof of (a). The number of characteristic roots of $\dot{W}_r$ at most σ is at least 1. Let

$$\lambda_1 < \lambda_2 < \cdots < \lambda_n \qquad (n \geq 1) \tag{36.5$'$}$$

be n characteristic roots of $\dot{W}_r$ at most σ and let

$$\eta^{(1)}, \eta^{(2)}, \ldots, \eta^{(n)} \tag{36.5$''$}$$

be characteristic solutions of $\dot{W}_r$ with the respective roots $\lambda_1, \ldots, \lambda_n$. Set

$$\chi = \eta^{(1)} + \eta^{(2)} + \cdots + \eta^{(n)}. \tag{36.6}$$

An arbitrary nonnull mapping $\chi \in N_r^\sigma$ can be represented as a sum (36.6) for a suitable choice of the roots (36.5)$'$ and characteristic solutions (36.5)$''$.

If $n = 1$ and χ : (36.6) is not a characteristic solution with root σ, then $\lambda_1 < \sigma$ in (36.5)$'$ and χ is a characteristic solution with root λ_1. Hence $I_r^{\lambda_1}(\chi) = 0$ by (36.3). By Hypothesis A_4 of Section 32, $I_r^\sigma(\chi) < 0$. Suppose then that $n > 1$. When $n > 1$ the following n relations are consequences of Lemma 36.1 and can be successively verified.

$$\dot{I}_r^{\lambda_1}(\eta^{(1)}) = 0 \qquad \text{(by Lemma 36.1)} \tag{36.7$^{(1)}$}$$

$$\dot{I}_r^{\lambda_2}(\eta^{(1)} + \eta^{(2)}) = \dot{I}_r^{\lambda_2}(\eta^{(1)}) < 0 \qquad \text{(by (36.2) and } (A_4)) \tag{36.7$^{(2)}$}$$

$$\dot{I}_r^{\lambda_3}(\eta^{(1)} + \eta^{(2)} + \eta^{(3)}) \leq \dot{I}_r^{\lambda_2}(\eta^{(1)} + \eta^{(2)}) < 0 \tag{36.7$^{(3)}$}$$

(by (36.2) and $(36.7)^{(2)}$)

$\cdot \qquad \cdot$

$\cdot \qquad \cdot$

$\cdot \qquad \cdot$

$$\dot{I}_r^{\lambda_n}(\eta^{(1)} + \cdots + \eta^{(n)}) \leq \dot{I}_r^{\lambda_{n-1}}(\eta^{(1)} + \cdots + \eta^{(n-1)}) < 0 \tag{36.7$^{(n)}$}$$

(by (36.2) and, reasoning inductively, by $(36.7)^{(n-1)}$).

If $\lambda_n = \sigma$ the proof that $\dot{I}_r^\sigma(\chi) < 0$ is complete. If $\lambda_n \leq \sigma$, and if χ is given by (36.6), then by Conditions (A_1) and (A_2),

$$\dot{I}_r^\sigma(\chi) \leq \dot{I}_r^{\lambda_n}(\chi) < 0 \qquad (n > 1) \tag{36.8}$$

This completes the proof of Lemma 36.2(a).

Proof of (b). *A subset* X_r^σ *of* N_r^σ *generating* N_r^σ. A subset X_r^σ of N_r^σ which is a generating set of N_r^σ can be defined as follows. The mappings in X_r^σ shall be characteristic solutions of $\dot{W}_r$. Corresponding to each characteristic root λ of $\dot{W}_r$, at most σ, X_r^σ shall include a base for characteristic solutions of $\dot{W}_r$ with root λ and no other characteristic solutions with root λ. The number of mappings in X_r^σ will then equal the count $\kappa(\sigma)$ of characteristic roots of $\dot{W}_r$ at most σ.

To prove that $\dim N_r^\sigma = \kappa(\sigma)$ it is sufficient to prove that the $\kappa(\sigma)$ mappings in X_r^σ are linearly independent.

This is trivially true in case there are no characteristic roots of $\dot{W}_r$ less than σ. Suppose then that there are characteristic roots of $\dot{W}_r$ less than σ. Were there mappings of X_r^σ which are linearly dependent, there would then exist a sum χ of form (36.6), with $n > 1$, and χ identically zero. This, however, is impossible, since $\dot{I}_r^\sigma(\chi) < 0$ in accord with (36.8).

Thus the generating set X_r^σ of N_r^σ is a base for N_r^σ. Hence $\dim N_r^\sigma = \kappa(\sigma)$, as affirmed in Lemma 36.2(b).

Thus Lemma 36.2 is true.

Our nuclear comparison theorem is stated as follows. The brace { } indicates an alternative. Set

$$\dot{I}_r^{*\sigma} = I_r^\sigma \# \dot{W}_r^*. \tag{36.9}$$

NUCLEAR COMPARISON THEOREM 36.1. *Let* $\dot{W}_r$ *and* $\dot{W}_r^*$ *be comparable*† *overall*‡ *admissible canonical systems and* σ_n *a characteristic root of* $\dot{W}_r$ *in an algebraic enumeration of characteristic roots of* $\dot{W}_r$. *If the quadratic functional* $\dot{I}_r^{*\sigma_n}$ *is negative definite {nonpositive} when evaluated on the* σ_n*-th nuclear subspace* $N_r^{\sigma_n}$ *of* $\{\dot{W}_r\}$, *then an* n*-th algebraic characteristic root* σ_n^* *of* $\dot{W}_n^*$ *exists and* $\sigma_n^* < \sigma_n$ $\{\sigma_n^* \leq \sigma_n\}$.

By Lemma 36.2(b) $\dim N_r^{\sigma_n}$ equals the count k of characteristic roots of $\dot{W}_r$ at most σ_n. We shall apply Theorem 33.6, with σ therein replaced by σ_n and $\dot{W}_r$ replaced by $\dot{W}_r^*$. It follows from Theorem 33.6, so applied, that the count of characteristic roots of $\dot{W}_r^*$ less than σ_n {at most σ_n} is at least k. The conclusion of Theorem 36.1 follows.

† See Definition 35.2.
‡ See Definition 32.4.

The following theorem and examples show that the nuclear comparison tests of Theorem 36.1 are less restrictive than the corresponding "a priori σ_n-tests" of Comparison Theorem 35.1.

THEOREM 36.2. *Let comparable overall admissible systems* $\dot{\mathrm{W}}_{\mathrm{r}}$ *and* $\dot{\mathrm{W}}_{\mathrm{r}}^*$ *be given together with a characteristic root* σ *of* $\dot{\mathrm{W}}_{\mathrm{r}}$. *The conditions of Theorem* 36.1 *that the quadratic functional* $\dot{\mathrm{I}}_{\mathrm{r}}^{*\sigma}$ *be negative definite* {*nonpositive*} *on the nuclear subspace* $\mathrm{N}_{\mathrm{r}}^{\sigma}$ *of* $\{\dot{\mathrm{W}}_{\mathrm{r}}\}$ *are implied by the respective conditions of Theorem* 35.1 *that the difference functional* $\mathrm{d}_{\mathrm{r}}^{\sigma}$ *be positive definite* {*nonnegative*} *on* $\{\dot{\mathrm{W}}_{\mathrm{r}}\}$.

According to Definition 35.1, for each $\eta \in \{\dot{W}_r\}$,

$$d_r^{\sigma}(\eta) = \dot{I}_r^{\sigma}(\eta) - \dot{I}_r^{*\sigma}(\eta). \tag{36.10}$$

By Lemma 36.2(a), $\dot{I}_r^{\sigma}$ is nonpositive on the subspace N_r^{σ} of $\{\dot{W}_r\}$, while the test of Theorem 35.1 is satisfied if d_r^{σ} is positive definite {nonnegative} on $\{\dot{W}_r\}$. When this test is satisfied we infer from (36.10) that $\dot{I}_r^{*\sigma}$ is negative definite {nonpositive} on N_r^{σ}, since $\{\dot{W}_r\} \supset N_r^{\sigma}$.

Thus Theorem 36.2 is true.

One can give examples of systems W_r and W_r^*, in which the parameter λ enters linearly as in Section 15, and for which the first nuclear test of Theorem 36.1 is satisfied for a given characteristic root σ of W_r while the corresponding a priori test of Theorem 35.1 is not satisfied. Leighton's Example 1, as presented by Swanson on page 6, although intended for another purpose, will serve as an example for our purposes. In Leighton's example $m = 1$, $\sigma = 0$, $r = 0$, and $[a^1, a^2] = [0, \pi]$.

In Leighton's example, but in our notation,

$$I_0^{\lambda}(\eta) = \int_0^{\pi} (\eta'^2 - \eta^2 - \lambda\eta^2)\, dx \qquad (\lambda \in R) \tag{36.11}$$

$$I_0^{*\lambda}(\eta) = \int_0^{\pi} (\eta'^2 - C(x)\eta^2 - \lambda\eta^2)\, dx \tag{36.12}$$

where $C(x) = x + 1 - k$ with $0 < k < \pi/2$. The smallest characteristic root of the system W_0 with quadratic functional (36.11) is $\lambda = 0$. The corresponding nuclear subspace N_0^0 is generated by the characteristic solution $\sin x$. The choice of k is such that

$$I_0^{*0}(\sin x) = \int_0^{\pi} (k - x) \sin^2 x\, dx < 0, \tag{36.13}$$

so that the first nuclear test of Theorem 36.1 is satisfied. However, the difference functional of Theorem 35.1 has values

$$d_0^0(\eta) = \int_0^{\pi} (C(x) - 1)\eta^2\, dx = \int_0^{\pi} (x - k)\eta^2\, dx \tag{36.14}$$

and is not positive definite for $\eta \in \{W_0\}$, as one readily verifies. Thus the first test of Theorem 35.1 is not satisfied.

Other examples of this type can be readily set up.

37. Oscillation Criteria

From the variational point of view, oscillation theory is concerned with the "count" of conjugate points relative to a selfadjoint system of DE

$$\frac{d}{dx}\,\omega_{\zeta_i}(x,\eta,\eta') = \omega_{\eta_i}(x,\eta,\eta') \qquad (i = 1, \ldots, m) \tag{37.1}$$

defined for x in an open subinterval $\aleph$ of R and conditioned as in Appendix I. In the preceding sections the count of conjugate points has been limited to conjugate points of $x = a^1$ in a finite relatively compact subinterval of $\aleph$. In this section we suppose that $\aleph$ is the interval $(0,\infty)$ and study the count of conjugate points of points $x = a > 0$ in the corresponding open subinterval (a,∞) of $\aleph$.

When $m = 1$, conjugate points relative to the DE

$$\eta'' + c(x)\eta = 0 \qquad (c(x) > 0,\ x > 0) \tag{37.2}$$

have been studied in great detail. Hille, among others, has used nonvariational methods. Leighton, Wintner, Nehari, Howard and others have used variational methods based on the integral

$$\int_a^b (\eta'^2 - c(x)\eta^2)\, dx. \tag{37.3}$$

The DE (37.2) is the Euler equation of this integral. See Swanson, Chapter 2.

It is believed that the methods and theorems of this book will lead to a far-reaching oscillation theory when $m > 1$. Such a theory can be developed in *many* ways. Because of this we shall limit ourselves in this section to a formulation without proof of some of the theorems upon which such an oscillation theory can be built and shall return to a more extended development in a separately published paper.

Preliminary definitions are required. These definitions extend definitions given when $m = 1$. See Swanson, p. 44. In the following definitions a brace indicates an alternative.

Definition 37.1. *DE,* n-*oscillatory* $\{\infty$-*oscillatory*$\}$ *on* (a,∞). For $n \geq 0$ an integer and for $a > 0$, the system of m DE (37.1) will be said to be n-*oscillatory* $\{\infty$-*oscillatory*$\}$ on (a,∞), if the count of conjugate points of $x = a$ on (a,∞) is n {alternately ∞}.

Definition 37.2. *Oscillatory* {*nonoscillatory*} *DE.* The system of DE (37.1) will be termed *oscillatory* if, for each $a \in \aleph$, the system (37.1) is ∞-oscillatory on (a,∞). The system (37.1) will be said to be *nonoscillatory* if for each $a \in \aleph$ for which a is sufficiently large, the system is 0-oscillatory on (a,∞).

Our Separation Theorem 20.1 implies the following: if the system (37.1) is ∞-oscillatory on some *one* subinterval (a,∞), $a > 0$, of $\aleph$, it is ∞-oscillatory on *each* such subinterval of $\aleph$. The system of DE (37.1) is accordingly either "oscillatory" or "nonoscillatory" in the sense of Definition 37.2.

In the following theorem the integral

$$\int_a^b \omega(x,\eta(x),\eta'(x))\,dx \qquad (0 < a < b < \infty) \tag{37.4}$$

will be termed *positive definite*, subject to the conditions

$$\eta(a) = \eta(b) = \mathbf{0}, \tag{37.5}$$

if the integral is positive for each nonnull D^1-mapping of $[a,b]$ into R^m such that (37.5) holds.

THEOREM 37.1. *The system of DE* (37.1) *is nonoscillatory if and only if the integral* (37.4) *is positive definite, subject to* (37.5), *for each subinterval* [a,b] *of* $\aleph$ *for which* a *is sufficiently large.*

It should be recalled that the coefficients in the quadratic form

$$2\omega(x,\eta,\zeta) = R_{ij}(x)\zeta_i\zeta_j + 2Q_{ij}(x)\zeta_i\eta_j + P_{ij}(x)\eta_i\eta_j \tag{37.6}$$

are conditioned for $x \in \aleph$ as in Appendix I. With $\omega(x,\eta,\zeta)$ so conditioned Theorem 37.1 follows from Theorems 7.4 and 7.5.

Theorems 7.4, 7.5 and our Separation Theorem imply the following.

THEOREM 37.2. *The system of DE* (37.1) *is oscillatory if and only if corresponding to each value* $a \in \aleph$ *there exists a value* $b > a$ *so large that the integral* (37.4) *fails to be positive definite, subject to* (37.5).

We shall obtain other sufficient conditions that the system of DE (37.1) be oscillatory as a corollary of Theorem 37.2. To that end mappings termed *threads* are now defined.

Definition 37.3. *Threads.* A D^1-mapping* $x \to \gamma(x)$ of $[1,\infty)$ into R^m will be called a *thread*. There is no condition of boundedness on $\|\gamma(x)\|$ or on the initial value $\|\gamma(1)\|$.

* The mapping γ of Definition 37.3 will be termed a D^1-mapping if for each $a > 1$ the restriction $\gamma \mid [1,a]$ is a D^1-mapping.

Definition 37.4. *The Thread Condition.* It is a consequence of Theorem 37.2 that if the system of DE (37.1) is oscillatory there then exists a thread γ such that

$$\liminf_{x \uparrow \infty} \int_1^x \omega(x,\gamma(x),\gamma'(x))\, dx = -\infty. \tag{37.7}$$

The condition that (37.7) hold for some thread will be called the Thread Condition.

The condition of Leighton and Wintner (see Swanson, p. 45) that

$$\int_1^\infty c(x)\, dx = \infty \tag{37.8}$$

is a Thread Condition relative to the integrand of (37.3). The thread is the mapping $x \to \gamma(x)$ of $[1,\infty)$ into R such that $\gamma(x) \equiv 1$. According to Leighton and Wintner the condition (37.8) is sufficient that the DE (37.2) be oscillatory.

It is natural to ask for what systems of DE (37.1) the Thread Condition is sufficient that the system be oscillatory. A first theorem extending the above theorem of Leighton and Wintner follows.

THEOREM 37.3. *A DE* $\eta'' + \mathrm{c}(\mathrm{x})\eta = 0$ *for which* $\mathrm{c}(\mathrm{x}) \geq 0$ *for* $\mathrm{x} \geq 1$, *is oscillatory, if and only if the Thread Condition is satisfied by some thread* γ.

An analogous theorem concerns the following class of DE (37.1).

Class A. This is the class of systems of DE (37.1) for which the coefficients $R_{ij}(x)$ are bounded in absolute value, the coefficients $Q_{ij}(x)$ vanish identically, and the quadratic form $P_{ij}(x)\eta_i\eta_j$ is negative semidefinite, all for $x \geq 1$.

The following theorem extends Theorem 37.3.

THEOREM 37.4. *A system of DE* (37.1) *in Class* A *is oscillatory if and only if the corresponding Thread Condition is satisfied by some thread* γ.

The following class of DE (37.1) includes systems (37.1) for which the quadratic form $P_{ij}(x)\eta_i\eta_j$ is not required to be negative semidefinite.

Class B. This is the class of systems of DE (37.1) for each of which the coefficients $R_{ij}(x)$, $Q_{ij}(x)$, $P_{ij}(x)$ are bounded in absolute value for $x \geq 1$.

THEOREM 37.5. *A system of DE* (37.1) *in Class* B *is oscillatory if the corresponding Thread Condition is satisfied by some thread* γ *such that* $\|\gamma(\mathrm{x})\|$ *is bounded for* $\mathrm{x} \geq 1$.

A complete analysis of oscillation properties of systems of DE (37.1) requires a systematic relating of oscillation properties of the DE (37.1) to limiting properties of characteristic roots of conditions $W_0(\lambda)$ of (15.1). The basic interval $[a^1,a^2]$ is a subinterval of $[1,\infty]$. With $a^1 \geq 1$ prescribed and fixed, a^2 is allowed to increase without limit. Characteristic roots, as defined in Section 15 for each interval $[a^1,a^2]$, will tend decreasing to characteristic limits, possibly $-\infty$, that serve to define oscillation criteria. The author hopes to present the relevant details in the near future.

Example 37.1. The differential equation (see Swanson, p. 45)

$$\eta'' + \frac{k}{(x+1)^2}\eta = 0 \qquad (k \text{ a constant, } x > 0) \tag{37.9}$$

was shown by Kneser [2], [3] to be oscillatory when $k > \frac{1}{4}$. Theorem 37.5 implies this, since the Thread Condition is satisfied when

$$c(x) = \frac{k}{(x+1)^2};\ \gamma(x) = (x+1)^{1/2}, \qquad k > 1/4 \qquad (1 \leq x < \infty). \tag{37.10}$$

That the DE (37.9) is nonoscillatory when $k = \frac{1}{4}$ follows from the fact that when $k = \frac{1}{4}$ and $x > 0$, (37.9) has the nonvanishing solution $(x+1)^{1/2}$. That the DE (37.9) is nonoscillatory when $k < \frac{1}{4}$ follows on comparing the DE (37.9) when $k = \frac{1}{4}$ with the DE when $k < \frac{1}{4}$.

APPENDIX I*

Free Linear Conditions

1. Free Second-Order Differential Equations

For each value of x in an arbitrary open interval $\aleph$ and for i, j on the range $1, \ldots, m$, let

$$(1.0)' \qquad 2\omega(x,\eta,\zeta) = R_{ij}(x)\zeta_i\zeta_j + 2Q_{ij}(x)\zeta_i\eta_j + P_{ij}(x)\eta_i\eta_j$$

be a quadratic form in the m-tuples η and ζ such that

$$(1.0)'' \qquad R_{ij}(x) = R_{ji}(x), \qquad P_{ij}(x) = P_{ji}(x) \qquad (x \in \aleph)$$

and the mappings R_{ij}, Q_{ij}, P_{ij} of $\aleph$ into R are continuous. The form ω will be called *free* to distinguish it from the form Ω introduced in (4.0) of Chapter 1 and termed *derived.* The form Ω is derived in that it depends upon an antecedent preintegrand f and an extremal g of the Euler equations based on g. A "free" form ω requires no antecedent f and extremal g. To the form $2\omega(x,\eta,\zeta)$ we adjoin the *accessory* form

$$(1.1) \qquad 2\omega^\lambda(x,\eta,\zeta) = 2\omega(x,\eta,\zeta) - \lambda\eta_i\eta_i$$

for each real number λ. We suppose that

$$(1.2) \qquad R_{ij}(x)\zeta_i\zeta_j > 0 \qquad (x \in \aleph,\ \zeta \neq \mathbf{0}).$$

Canonical associates ξ. With each C^1-mapping η of $\aleph$ into R^m there is adjoined the *canonical associate* ξ of η relative to the quadratic form (1.0)′. The components of ξ by definition are

$$(1.3) \qquad \xi_i(x) = \omega_{\zeta_i}(x,\eta(x),\eta'(x)) \qquad (i = 1, \ldots, m).$$

* *References.* None of the three appendices has more than five sections. Each reference in an appendix is to an equation, lemma, or theorem in that appendix, unless otherwise indicated. Thus equation (3.1) in Appendix I refers to an equation in Section 3 of Appendix I. Equation 15.1 means an equation in Section 15 of the book and is uniquely defined since there is but one Section 15 in the book.

When a C^1-mapping η of $\aleph$ into R^m has a "canonical associate" ξ of class C^1, the m operators

$$\eta \to L_i(\eta) = \frac{d}{dx}(R_{ij}\eta'_j + Q_{ij}\eta_j) - (Q_{ji}\eta'_j + P_{ij}\eta_j) \tag{1.4}$$

are well-defined. We term the conditions

$$\frac{d}{dx}(R_{ij}(x)\eta'_j + Q_{ij}(x)\eta_j) = Q_{ji}(x)\eta'_j + P_{ij}(x)\eta_j \qquad (i = 1, \ldots, m) \tag{1.5}$$

the JDE with *underlying* form (1.0)′. If η and $\bar{\eta}$ are two C^1-mappings of the interval $\aleph$ into R^m whose canonical associates ξ and $\bar{\xi}$ are of class C^1, the Lagrange identity.

$$\eta_i L_i(\bar{\eta}) - \bar{\eta}_i L_i(\eta) \equiv \frac{d}{dx}(\eta_i\bar{\xi}_i - \bar{\eta}_i\xi_i) \tag{1.6}$$

is readily verified. Such operators L_i, $i = 1, \ldots, m$ are called *selfadjoint*. Cf. Bôcher [1].

A convention on D^1*-mappings.* D^1-mappings are defined in Chapter 1. If φ and ψ are D^1-mappings of $[a,b]$ into R, an affirmation that $\varphi'(x) = \psi'(x)$ is understood as holding only for those values of $x \in [a,b]$ for which φ' and ψ' exist.

A canonical associate ξ of a D^1-mapping η of $\aleph$ into R^m can be defined by the conditions (1.3) imposed at each point x at which $\eta'(x)$ exists. The canonical associate ξ of η will then have components of at least class D^0.

A classical definition of $\omega^\lambda(x,\eta,\zeta)$. As far as the theorems of this book are concerned the definition (1.1) of $\omega^\lambda(x,\eta,\zeta)$ could equally well be replaced by the definition

$$2\omega^\lambda(x,\eta,\zeta) = 2\omega(x,\eta,\zeta) - \lambda C_{ij}(x)\eta_i\eta_j \qquad (x \in \aleph) \tag{1.7}$$

where for each $x \in \aleph$, $C_{ij}(x)\eta_i\eta_j$ is a positive definite, symmetric, quadratic form in the m-tuple η, with coefficients $C_{ij}(x)$ which vary continuously with $x \in \aleph$. The corresponding Euler equations should then be written in the form

$$\frac{d}{dx}\omega_{\zeta_i}(x,\eta,\eta') = \omega_{\eta_i}(x,\eta,\eta') - \lambda C_{ij}(x)\eta_j \qquad (i = 1, \ldots, m) \tag{1.8}$$

rather than in the form in which $C_{ij}(x) = \delta^i_j$. There is no commitment to either definition of ω^λ if the Euler equations are written in the form

$$\frac{d}{dx}\omega^\lambda_{\zeta_i}(x,\eta,\eta') = \omega^\lambda_{\eta_i}(x,\eta,\eta') \qquad (i = 1, \ldots, m). \tag{1.9}$$

The form (1.1) will ordinarily be used.

If one prefers the definition (1.7) to the definition (1.1), then in the periodic case, the coefficients $C_{ij}(x)$ should have the period τ in x. When the definition

(1.7) of $\omega^\lambda(x,\eta,\zeta)$ is preferred, the definitions of the entities in Section 15, such as the canonical systems W_r, the free quadratic functionals $I_r^\lambda \# W_r$, the frames Λ_p and the index forms $Q^\lambda \#(W_r,\Lambda_p)$, remain verbally unchanged from the definitions when $C_{ij}(x) \equiv \delta_i^j$. Secondary λ-extremals and λ-conjugate points are then extremals and conjugate points relative to the DE (1.8).

2. Solution of the Differential Equations (1.5)

To solve the DE (1.5) we make a linear transformation from the unknown m-tuples η and η' to m-tuples η and ξ by setting

$$\xi_i = R_{ij}(x)\eta'_j + Q_{ij}(x)\eta_j \qquad (x \in \aleph;\ i = 1, \ldots, m) \tag{2.1}$$

Since the determinant $|R_{ij}(x)| \neq 0$ for $x \in \aleph$, (2.1) is equivalent to relations of the form

$$\eta'_i = a_{ij}(x)\xi_j + b_{ij}(x)\eta_j \qquad (x \in \aleph;\ i = 1, \ldots, m) \tag{2.2}$$

for suitable choices of C^0-mappings a_{ij} and b_{ij} of $\aleph$ into R. Subject to (2.1) or (2.2)

$$Q_{ji}(x)\eta'_j + P_{ij}(x)\eta_j = c_{ij}(x)\xi_j + d_{ij}(x)\eta_j \qquad (x \in \aleph) \tag{2.3}$$

for arbitrary m-tuples ξ and η, and suitable C^0-mappings c_{ij} and d_{ij} of $\aleph$ into R. The DE (1.5) are then equivalent to the DE

$$\begin{cases} \dfrac{d}{dx}\eta_i = a_{ij}(x)\xi_j + b_{ij}(x)\eta_j \\[2ex] \dfrac{d}{dx}\xi_i = c_{ij}(x)\xi_j + d_{ij}(x)\eta_j \end{cases} \qquad (i = 1, \ldots, m) \tag{2.4}$$

The classical Cauchy Lipschitz theorem on the existence of solutions of ordinary DE (Goursat, [1], p. 384) implies the following:

THEOREM 2.1. *Given a value* $c \in \aleph$ *and* m*-tuples* (η^0,ξ^0), *there exist unique* C^1*-mappings* (η,ξ) *of* $\aleph$ *into* R^m *which are solutions of* (2.4) *and such that*

$$(\eta(c),\xi(c)) = (\eta^0,\xi^0). \tag{2.5}$$

Theorem 2.1 has the following corollary:

COROLLARY 2.1. *The mapping* η *in Theorem* 2.1 *is a solution of the DE* (1.5). *Given a value* $c \in \aleph$ *and* m*-tuples* η^0,η'^0 *there is a unique* C^1*-solution* η *of the DE* (1.5) *such that*

$$\eta(c) = \eta^0, \qquad \eta'(c) = \eta'^0. \tag{2.6}$$

Notation for Lemma 2.1. There exists a $2m$-square determinant

$$Z(x) = \begin{vmatrix} p_{ij}(x), & q_{ij}(x) \\ \xi^p_{ij}(x), & \xi^q_{ij}(x) \end{vmatrix} \qquad (x \in \aleph) \qquad (i, j = 1, \ldots, m) \tag{2.7}$$

whose columns are solutions of the DE (2.4) and which is such that for a prescribed value $c \in \aleph$

$$\left\| \begin{matrix} p_{ij}(c), & q_{ij}(c) \\ \xi^p_{ij}(c), & \xi^q_{ij}(c) \end{matrix} \right\| = \left\| \begin{matrix} \delta^j_i, & 0 \\ 0, & \delta^j_i \end{matrix} \right\| \tag{2.8}$$

We introduce the $2m$-square determinant

$$P(x) = \begin{vmatrix} p_{ij}(x), & q_{ij}(x) \\ p'_{ij}(x), & q'_{ij}(x) \end{vmatrix} \qquad (x \in \aleph) \tag{2.9}$$

and prove the following lemma.

LEMMA 2.1. *For no value of* $x \in \aleph$ *does the determinant* $Z(x)$ *or* $P(x)$ *vanish.*

The Nonvanishing of $Z(x)$. If $Z(a) = 0$ for some $a \in \aleph$, a proper linear combination of the columns of the determinant (2.7) would vanish when $x = a$ and hence vanish identically by Theorem 2.1. This, however, is impossible, since $Z(c) \neq 0$.

The Nonvanishing of $P(x)$. If a proper linear combination (η, η') of the columns of the determinant $P(x)$ vanished when $x = a$, then $\eta(x) \equiv 0$ by Corollary 2.1. The same proper linear combination of the columns of the determinant (2.7) would be a pair (η, ξ) with the same η, and a mapping ξ which is the canonical associate of η. Since $\eta(x) \equiv 0$, $\xi(x) \equiv 0$, contrary to the fact that $Z(c) \neq 0$.

The following corollary is readily verified:

COROLLARY 2.2. *Each solution* η *of the JDE* (1.5) *is a mapping with component mappings* $x \to \eta_i(x)$ *of the form*

$$x \to A_j p_{ij}(x) + B_j q_{ij}(x) : \aleph \to R \qquad (i = 1, \ldots, m) \tag{2.1}$$

where A *and* B *are* m*-tuples in* R^m *uniquely determined by the solution* η.

3. Conjugate Points

We begin by verifying the following classical lemma.

LEMMA 3.1. *For* $(x,c) \in \aleph \times \aleph$ *and* i,j *on the range* $1, \ldots, m$, *there exists an* m*-square matrix* $\|v^j_i(x,c)\|$ *of elements in* R *such that the mappings*

$$(x,c) \to v^j_i(x,c) : \aleph \times \aleph \to R \tag{3.1}$$

are of class C^0 *and for fixed* c *and* j, *the partial mappings*,

$$x \to v_i^j(x,c) : \aleph \to R \qquad (i = 1, \ldots, m) \tag{3.2}$$

define a C^1*-solution of the JDE* (1.5) *such that*

$$v_i^j(c,c) = 0,\ v_{ix}^j(c,c) = \delta_i^j \qquad (i,j = 1, \ldots, m). \tag{3.3}$$

By virtue of Corollary 2.2 and Lemma 2.1 the mappings v_i^j can be obtained by solving the equations

$$\begin{aligned} 0 &= A_k^j p_{ik}(c) + B_k^j q_{ik}(c) \\ \delta_i^j &= A_k^j p'_{ik}(c) + B_k^j q'_{ik}(c) \end{aligned} \tag{3.4}$$

for m-tuples $A^j(c)$ and $B^j(c)$ for each j, and setting

$$v_i^j(x,c) = A_k^j(c) p_{ik}(x) + B_k^j(c) q_{ik}(x). \tag{3.5}$$

Conjugate points are defined for *free* JDE (1.5) as they were for derived JDE in Section 5 of Chapter 1. Distinct points $x = c$ and $x = d$ on $\aleph$ are thus termed *conjugate* if there exists a nontrivial solution η of the JDE such that $\eta(c) = 0$ and $\eta(d) = 0$. Equivalently $x = c$ is conjugate to $x = d$ if and only if the determinant

$$D(x,c) = |v_i^j(x,c)| \tag{3.6}$$

vanishes when $x = d$.

Notation for Lemma 3.2. Let τ be an open subinterval of $\aleph$ in which no point is conjugate to another point of τ. We introduce an open subset S_τ of $\tau \times \tau$ defined by setting

$$S_\tau = \{(c,d) \in \tau \times \tau \mid c \neq d\}. \tag{3.7}$$

The following lemma will be verified:

LEMMA 3.2. *For each triple* $(x : c,d) \in \tau \times S_\tau$ *there exists an* m*-square matrix* $\|z_i^j(x : c,d)\|$ *of elements in* R *such that the mappings*

$$(x : c,d) \to z_i^j(x : c,d) : \tau \times S_\tau \to R \qquad (i, j = 1, \ldots, m) \tag{3.8}$$

are of class C^0, *and for fixed* c, d *and* j, *the partial mappings*

$$x \to z_i^j(x : c,d) : \tau \to R \qquad (i = 1, \ldots, m) \tag{3.9}$$

define a C^1*-solution of the JDE* (1.5) *such that for distinct* c,d *in* τ

$$z_i^j(c : c,d) = 0;\ z_i^j(d : c,d) = \delta_i^j. \tag{3.10}$$

Let i, j, k have the range $1, \ldots, m$. The mappings z_i^j can be obtained by solving the equations

$$\delta_i^j = A_k^j v_i^k(d,c) \qquad (i, j = 1, \ldots, m) \tag{3.11}$$

for the m-square matrix $\|A_k^j\|$, given an arbitrary pair $(c,d) \in S_\tau$. This is possible since the determinant $|v_i^k(d,c)|$ never vanishes for $(c,d) \in S_\tau$. One obtains thereby an m-square matrix $\|A_k^j(c,d)\|$ and sets

$$z_i^j(x : c,d) = A_k^j(c,d)v_i^k(x,c) \qquad (x \in \tau) \tag{3.12}$$

thereby satisfying the lemma.

In setting up our free index forms in Section 15 the following corollary of Lemma 3.1 is needed.

COROLLARY 3.1. *Let* τ *be an open subinterval of* $\aleph$ *in which no point is conjugate to another point of* τ. *Let* a *and* b *be values in* τ with $a < b$ *and let*

$$(a, y_1, \ldots, y_m), \qquad (b, \bar{y}_1, \ldots, \bar{y}_m) \tag{3.13}$$

be points in E_{m+1}. *The* m *mappings*

$$x \to y_j z_i^j(x : b,a) + \bar{y}_j z_i^j(x : a,b) \qquad (x \in \aleph;\ i = 1, 2, \ldots, m) \tag{3.14}$$

of τ *into* R *define a unique* C^1*-solution* η *of the JDE* (1.5) *such that*

$$\eta(a) = y, \qquad \eta(b) = \bar{y}. \tag{3.15}$$

Note 1. The solution η of the JDE (1.5) affirmed to exist in Corollary 3.1 is a linear form in the ordinates

$$y_1, \ldots, y_m; \bar{y}_1, \ldots, \bar{y}_m \tag{3.16}$$

whose coefficients* define solutions of the JDE which vary continuously with a, b for $a < b$ and a and b in τ. This corollary should be compared with Theorem 5.1 of Chapter 1, a similar theorem concerning extremals.

Definition 3.1. *The* D^1*-mappings* Θ *of* $\aleph$ *into* R^m. Let $a < b < c$ be values in $\aleph$ such that each of the intervals $[a,b]$ and $[b,c]$ is free from mutual conjugate points. It follows from Lemma 3.2 that for each j on the range $1, \ldots, m$ there exists a D^1-mapping

$$x \to \Theta^j(x : a,b,c) : \aleph \to R^m \tag{3.17}$$

with the following properties:

(i) The value of Θ^j when $x \in \aleph - [a,c]$ is a null m-tuple.

(ii) The graph of Θ^j when $a \leq x \leq c$ is a sequence of two secondary extremals separated by the point $(x,y) = (b,\delta_1^j, \ldots, \delta_m^j)$.

Lemma 3.3 is concerned with the representation of a broken secondary extremal with x-domain $[a^1,a^2]$, framed in the sense of Section 15.

* For fixed j, for $i = 1, \ldots, m$, and for $x \in \aleph$.

Preparation for Lemma 3.3. Let there be given a sequence of points

$$c_{-1} < c_0 < c_1 < \cdots < c_p < c_{p+1} < c_{p+2} \tag{3.18}$$

in $\aleph$ with $[c_0,c_{p+1}] = [a^1,a^2]$ and such that if a, b is any pair of successive values in the sequence (3.18) the interval $[a,b]$ contains no mutual conjugate points. Let $P_0, P_1, \ldots, P_p, P_{p+1}$ be a sequence of points $(x,y) \in R^{m+1}$ with x-coordinates respectively

$$c_0, c_1, \ldots, c_p, c_{p+1} \tag{3.19}$$

and with y-coordinates given by the respective m-tuples,

$$y^0, y^1, \ldots, y^p, y^{p+1}. \tag{3.20}$$

The successive points of the sequence $\mathbf{P} = (P_0, \ldots, P_{p+1})$ can be joined by a broken secondary extremal $E(\mathbf{P})$ without corners between the successive points. $E(\mathbf{P})$ can be represented as follows.

LEMMA 3.3. *The broken secondary extremal* $\mathrm{E}(\mathbf{P})$ *is the graph of the mapping*

$$x \rightarrow \sum_{q=0}^{p+1} \Theta^j(x : c_{q-1},c_q,c_{q+1})y_j^q : [a^1,a^2] \rightarrow R^m. \tag{3.21}$$

Note 2. The mapping (3.21) is linear in the coordinates y_j of the successive m-tuples (3.20), with coefficients which are D^1-mappings of $[a^1,a^2]$ into R^m. If the end m-tuples y^0 and y^{p+1} are m-tuples η^1 and η^2 subject to the conditions

$$\eta_j^s = c_{jh}^s u_h \qquad (s = 1, 2; h = 1, \ldots, r) \tag{3.22}$$

of (15.0)″, then one can replace y_j^0 and y_j^{p+1} in the right member of (3.21), respectively, by $c_{jh}^1 u_h$ and $c_{jh}^2 u_h$. The right member of (3.22) then becomes linear in the variables

$$u_1, \ldots, u_r : y_1^1, \ldots, y_m^1; \ldots ; y_1^p, \ldots, y_m^p \tag{3.23}$$

with coefficients which are D^1-mappings of $[a^1,a^2]$ into R^m whose graphs are broken secondary extremals.

In the preceding lemmas conjugate points and secondary extremals can be replaced by *σ-conjugate* points and *σ-secondary extremals.* σ-Conjugate points and σ-secondary extremals are ordinary conjugate points and secondary extremals defined by solutions of the DE

$$\frac{d}{dx}\,\omega^\sigma_{\zeta_i}(x,\eta,\eta') = \omega^\sigma_{\eta_i}(x,\eta,\eta') \qquad (i = 1, \ldots, m). \tag{3.24}$$

In the following section we study the dependence of solutions of DE such as (3.24) upon a parameter such as σ. In Section 4 this parameter is denoted by α.

4. Second-Order *DE* with a Parameter α

Let α be a parameter with values in an open interval π. Let there be given a 2-parameter family of quadratic forms $2\omega(x,\eta,\zeta : \alpha)$, explicitly of the form

$$R_{ij}(x,\alpha)\zeta_i\zeta_j + 2Q_{ij}(x,\alpha)\zeta_i\eta_j + P_{ij}(x,\alpha)\eta_i\eta_j, \tag{4.1}$$

with coefficients which are continuous mappings of $(\aleph \times \pi)$ into R and which, for each fixed $\alpha \in \pi$, are conditioned as are the coefficients R_{ij}, Q_{ij}, P_{ij} in (1.0). For each α, DE

$$\frac{d}{dx}\,\omega_{\zeta_i}(x,\eta,\eta' : \alpha) = \omega_{\eta_i}(x,\eta,\eta' : \alpha) \qquad (i = 1, \ldots, m) \tag{4.2}$$

are defined as in Section 1.

For each $\alpha \in \pi$ the DE (4.2) are equivalent, as in Section 2 to DE

$$\begin{cases} \dfrac{d}{dx}\,\eta_j = a_{ij}(x,\alpha)\xi_j + b_{ij}(x,\alpha)\eta_j \\ \dfrac{d}{dx}\,\xi_j = c_{ij}(x,\alpha)\xi_j + d_{ij}(x,\alpha)\eta_j \end{cases} \qquad (i = 1, \ldots, m) \tag{4.3}$$

with coefficients conditioned, as in Section 2 for each α, and in addition, continuous for $(x,\alpha) \in (\aleph \times \pi)$.

Classical Cauchy Lipschitz existence theorems show that there exists, for each $\alpha \in \pi$, a $2m$-square determinant

$$Z(x,\alpha) = \begin{vmatrix} p_{ij}(x,\alpha), & q_{ij}(x,\alpha) \\ \xi^p_{ij}(x,\alpha), & \xi^q_{ij}(x,\alpha) \end{vmatrix} \qquad (i, j = 1, 2, \ldots, m) \tag{4.4}$$

whose columns are solutions of the DE (4.3), conditioned for each α, as are the solutions (2.7) of DE (2.4), and, in addition, continuous for $(x,\alpha) \in (\aleph \times \pi)$. The fundamental Corollary 2.2 here takes the form

THEOREM 4.1. *For each $\alpha \in \pi$, the components η_i of a solution η of the DE (4.2) are partial mappings of the form*

$$x \to A_j p_{ij}(x,\alpha) + B_j q_{ij}(x,\alpha) : \aleph \to R \tag{4.5}$$

where A *and* B *are* m*-tuples in* R^m *uniquely determined for the given* α *by* η.

On reviewing Section 3, we see that the m-square determinant $D(x,c)$ introduced in (3.6), if defined for each $\alpha \in \pi$, yields an m-square determinant $D(x,c : \alpha)$ which varies continuously with $(x,\alpha) \in \aleph \times \pi$. The following conclusion can be drawn.

THEOREM 4.2. *If points* $x = c$ *and* $x \in d$ *in* $\aleph$ *are distinct and nonconjugate when* $\alpha = \alpha_0$, *these points are nonconjugate for* $|\alpha - \alpha_0|$ *sufficiently small.*

The mode of proof of Lemma 3.2 suffices to verify the following simplified extension.

LEMMA 4.1. *Let* $x = c$ *and* $x = d$ *be distinct points in* $\aleph$ *which are nonconjugate when* $\alpha = \alpha_0$. *Let* $\aleph_0$ *be the closed subinterval of* $\aleph$ *whose end points, in some order, are* $x = c$ *and* $x = d$. *If* π_0 *is a sufficiently small open subinterval of* π *which contains the value* α_0, *the following is true.*

For $j = 1, \ldots, m$, *there exist continuous mappings*

$$(4.6)' \qquad (x,\alpha) \to z_i^j(x,\alpha) : \aleph_0 \times \pi_0 \to R \qquad (i = 1, \ldots, m)$$

$$(4.6)'' \qquad (x,\alpha) \to z_{ix}^j(x,\alpha) : \aleph_0 \times \pi_0 \to R \qquad (i = 1, \ldots, m)$$

which, for fixed $\alpha \in \pi_0$, *define the components* η_i *and slopes* η_i' *of* C^1*-solutions of the DE* (4.2) *and are such that*

$$(4.7) \qquad z_i^j(c,\alpha) = 0; \; z_i^j(d,\alpha) = \delta_i^j \qquad (i, j = 1, \ldots, m).$$

In this lemma c may be less or greater than d. Suppose that, when $x = c$ and $x = d$ are interchanged, the mappings (4.6)′ are replaced by the mappings

$$(4.8) \qquad (x,\alpha) \to w_i^j(x,\alpha) : \aleph_0 \times \pi_0 \to R \qquad (i, j = 1, \ldots, m)$$

with

$$(4.9) \qquad w_i^j(d,\alpha) = 0, \qquad w_i^j(c,\alpha) = \delta_i^j.$$

Lemma 4.1 then has the following corollary.

COROLLARY 4.1. *Under the hypotheses of Lemma* 4.1, *let*

$$(4.10) \qquad (c, y_1, \ldots, y_m); \qquad (d, \bar{y}_1, \ldots, \bar{y}_m)$$

be points prescribed in E_{m+1}. *For each* $\alpha \in \pi_0$ *the mappings*

$$(4.11) \qquad x \to y_j z_i^j(x,\alpha) + \bar{y}_j w_i^j(x,\alpha) : \aleph_0 \to R \qquad (i = 1, \ldots, m)$$

define the components η_i *of a* C^1*-solution of the DE* (4.2) *such that*

$$(4.12) \qquad \eta(c) = \bar{y}, \qquad \eta(d) = y.$$

Note. The coefficients of y_j and $\bar{y}_j$ in (4.11) are characterized as in Lemma 4.1. This characterization is basic in our application of Corollary 4.1.

5. The End Space $((W_r))$ of W_r

We refer to the system W_r, $r > 0$ of free canonical conditions (15.0). Such a system is called ND (nondegenerate) if it possesses no characteristic root $\lambda = 0$. When $\lambda = 0$ the conditions (15.0) include the conditions

$$\begin{cases} \dfrac{d}{dx}\,\omega_{\zeta_i}(x,\eta,\eta') = \omega_{\eta_i}(x,\eta,\eta') & (i = 1, \ldots, m) \qquad (5.1)' \\ \eta_i^s = c_{ih}^s u_h \quad (s = 1, 2;\ i = 1, \ldots, m). & \qquad (5.1)'' \end{cases}$$

In Definition 24.1 we have introduced the vector space $((W_r))$ over R of C^1-solutions of the DE (5.1)′ which satisfy the conditions (5.1)″ with some r-tuple u. $((W_r))$ is termed the *end space* of W_r.

The following lemma is essential in the development of Section 24.

LEMMA 5.1. *If the system* W_r, $r > 0$, *is* ND, *then* $\dim ((W_r)) = r$.

Since the rank of the $2m \times r$ matrix $\|c_{ih}^s\|$ is r by hypothesis, the conditions (5.1)″ on the $2m$-tuples

$$(\eta_1(a^1), \ldots, \eta_m(a^1);\ \eta_1(a^2), \ldots, \eta_m(a^2)) = (\eta(a^1),\eta(a^2)) \tag{5.2}$$

are equivalent to $2m - r$ linearly independent linear conditions

$$K_i(\eta(a^1),\eta(a^2)) = 0 \qquad (i = 1, \ldots, 2m - r) \tag{5.3}$$

The vector space H of solutions η of the JDE (5.1)′ of Appendix I which satisfy the conditions (5.3) is the space $((W_r))$. According to Lemma 11.3 the space $((W_r)) = H$ has a dimension

$$\rho \geq 2m - (2m - r) = r. \tag{5.4}$$

It remains to show that $\rho = \mathrm{r}$.

The conditions (15.0)‴, subject to the conditions (15.0)″, are equivalent to r additional linearly independent linear conditions

$$H_j(\eta(a^1),\eta(a^2);\xi(a^1),\xi(a^2)) = 0 \qquad (j = 1, \ldots, r) \tag{5.5}$$

on the $4m$-tuple $(\eta(a^1),\eta(a^2);\xi(a^1),\xi(a^2))$. (Cf. Exercises 11.1 and 11.2.) According to Lemma 11.4 the vector subspace $\hat{H}$ of H which satisfies the conditions (5.5) is such that

$$\dim \hat{H} \geq \rho - r. \tag{5.6}$$

By hypothesis, however, the system W_r is nondegenerate or, equivalently, $\dim \hat{H} = 0$. Hence $\rho = r$.

Thus Lemma 5.1 is true.

We prepare for Lemma 5.2.

With W_r, $r > 0$, there is associated a functional, denoted by $I_r^0 \# W_r$ in Section 15 with values

$$(5.6)' \qquad I_r^0(\eta) = b_{hk}u_h u_k + \int_{a^1}^{a^2} 2\omega(x,\eta,\eta')\,dx,$$

where the coefficients b_{hk} are taken from (15.0)$'''$ and where η is any D^1-mapping of $[a^1,a^2]$ into R^m which satisfies the conditions (5.1)$''$ with the r-tuple u. If

$$(5.6)'' \qquad \mathscr{B} = (\eta^{(1)}, \ldots, \eta^{(r)})$$

is a base of the end space $((W_r))$ of an ND W_r, the respective mappings, $\eta^{(1)}, \ldots, \eta^{(r)}$, satisfy the conditions (5.1) with r-tuples

$$(5.7) \qquad (u^{(1)}, \ldots, u^{(r)}).$$

An arbitrary mapping $\eta \in ((W_r))$ has the form

$$(5.8) \qquad \eta = w_h\eta^{(h)}$$

for some r-tuple w, uniquely determined by η, and satisfies the conditions (5.1)$''$ with an r-tuple

$$(5.9) \qquad u = w_h u^{(h)}.$$

The quadratic end form $D_{\mathscr{B}}(w)$ of an ND W_r. Corresponding to a prescribed base $\mathscr{B}$: (5.6)$''$ of $((W_r))$, a *quadratic end* form of W_r was defined in Section 24 by setting

$$(5.10) \qquad D_{\mathscr{B}}(w) = I_r^0(w_h\eta^{(h)}) = d_{hk}w_k w_h,$$

where the r-square matrix $\|d_{hk}\|$ can and will be taken as symmetric.

The following lemma is essential to the formulation of Oscillation Theorem 24.1. For the definition of R-equivalence see footnote to Lemma 1.1 of Appendix II.

LEMMA 5.2. *When* W_r, $r > 0$ *is ND, each quadratic end form* $D_{\mathscr{B}}(w)$ *of* W_r *is* R*-equivalent to each other quadratic end form of* W_r.

If $\mathscr{B}$ and $\hat{\mathscr{B}}$ are two bases of $((W_r))$ of form

$$(5.11) \qquad \mathscr{B} = (\eta^{(1)}, \ldots, \eta^{(r)}), \qquad \hat{\mathscr{B}} = (\hat{\eta}^{(1)}, \ldots, \hat{\eta}^{(r)})$$

then, for a suitable choice of an r-square nonsingular matrix $\|c_{hk}\|$ of real numbers,

$$(5.12) \qquad \hat{\eta}^{(h)} = c_{hk}\eta^{(k)} \qquad (h = 1, \ldots, r).$$

By definition, for arbitrary r-tuples w and $\hat{w}$

$$(5.13) \qquad D_{\mathscr{B}}(w) = I_r^0(w_k\eta^{(k)}), \qquad D_{\hat{\mathscr{B}}}(\hat{w}) = I_r^0(\hat{w}_h\hat{\eta}^{(h)}).$$

If (5.12) holds and if one subjects the r-tuple $\hat{w}$, to the nonsingular linear transformation

$$w_k = c_{hk}\hat{w}_h \qquad (k = 1, \ldots, r) \tag{5.14}^*$$

then, by virtue of (5.12), (5.13) and (5.14),

$$w_k\eta^{(k)} = \hat{w}_h\hat{\eta}^{(h)}, \qquad D_{\mathscr{B}}(w) = D_{\hat{\mathscr{B}}}(\hat{w}). \tag{5.15}$$

The quadratic forms $D_{\mathscr{B}}(w)$ and $D_{\hat{\mathscr{B}}}(\hat{w})$ are thus R-equivalent.

Thus Lemma 5.2 is true.

* Note that the summation is with respect to k in (5.12) and with respect to h in (5.14).

APPENDIX II

Subordinate Quadratic Forms and Their Complementary Forms

1. Auxiliary Theorem 25.1

Appendix II will give a proof of this theorem. This section is a continuation of Section 25.

Theorem 25.1 (i) is a consequence of the following lemma. It concerns the quadratic form $Q(z)$ of (25.1).

LEMMA 1.1. *If the matrix of the* n *linear conditions*

$$\frac{\partial Q}{\partial z_{r+1}}(z) = \cdots = \frac{\partial Q}{\partial z_\mu}(z) = 0 \qquad \text{(of (25.6))} \tag{1.1}$$

is of rank n, *then any two forms which are complementary** *to* P(**0**,s) *are* R-*equivalent*,† *and so have a common index and nullity.*

If B and $\hat{B}$ are two bases of form

$$B = (z^{(1)}, \ldots, z^{(r)}), \qquad \hat{B} = (\hat{z}^{(1)}, \ldots, \hat{z}^{(r)}) \tag{1.2}$$

of the vector space π_r : (25.6), then for a suitable choice of an r-square nonsingular matrix $\|c_{hk}\|$ of real numbers,

$$\hat{z}^{(h)} = c_{hk} z^{(k)} \qquad (h = 1, \ldots, r) \tag{1.3}$$

By definition (25.8)

$$H_B(w) = Q(w_k z^{(k)}); \qquad H_{\hat{B}}(\hat{w}) = Q(\hat{w}_h \hat{z}^{(h)}) \tag{1.4}$$

* Defined in Section 25

† Two quadratic forms are termed R-*equivalent* if one form is obtainable from the other by a suitable nonsingular linear transformation of the variables of the one form into the variables of the other.

where w and $\hat{w}$ are arbitrary r-tuples. If (1.3) holds and if one subjects the r-tuple $\hat{w}$ to the nonsingular linear transformation $w_k = c_{hk}\hat{w}_h$, $k = 1, \ldots, r$, then

$$w_k z^{(k)} = \hat{w}_h \hat{z}^{(h)}, \qquad H_B(w) = H_{\hat{B}}(\hat{w}).$$

The quadratic forms $H_B(w)$ and $H_{\hat{B}}(\hat{w})$ are thus R-equivalent, so that Lemma 1.1 is true.

Theorem 25 (i) *follows from* Lemma 1.1.

Proof of Theorem 25.1 (ii). The following lemma implies Theorem 25.1 (ii). It concerns the symmetric quadratic form $H_B(w)$ over the base B : (25.7) of π_r : (25.6). As introduced in (25.8), $H_B(w)$ is defined for arbitrary r-tuples w and for α, β on the range $1, \ldots, \mu$ has the values,

$$H_B(w) = Q(w_h z^{(h)}) = (a_{\alpha\beta} z_\alpha^{(h)} z_\beta^{(k)}) w_h w_k = d_{hk} w_h w_k \tag{1.5}$$

where Q is the basic symmetric quadratic form (25.1).

LEMMA 1.2. *If the quadratic form* Q : (25.1) *is nonsingular, the following is true:*

(a_1) *For* h, k, j *on the range* 1, . . . , r, *and* d_{hk} *defined by* (1.5), *then*

$$2d_{hk} = z_j^{(h)} Q_{z_j}(z^{(k)}). \tag{1.6}$$

(a_2) *The* r-*square determinant* $|Q_{z_j}(z^{(h)})| \neq 0$.
(a_3) *Nullity* P(**0**,s) = *nullity* $H_B(w)$.

Proof of (a_1). It follows directly from (1.5) and the formula, $\partial Q/\partial z_\alpha = 2a_{\alpha\beta} z_\beta$, that

$$2d_{hk} = z_\alpha^{(h)} \frac{\partial Q}{\partial z_\alpha}(z^{(k)}) \tag{1.7}$$

for α on the range $1, \ldots, \mu$. The equality (1.6) follows from (1.7) and the fact that each μ-tuple $z^{(k)}$ is in the base B and hence satisfies the conditions (1.1).

Proof of (a_2). Were (a_2) false there would exist a nonnull r-tuple $(c_1, \ldots, c_r)$ such that $c_k Q_{z_h}(z^{(k)}) = 0$ for $h = 1, \ldots, r$. The μ-tuple $\hat{z} = c_k z^{(k)}$ is in the r-plane π_r : (25.6), and not null, since c is a nonnull r-tuple. Note that

$$\frac{\partial Q}{\partial z_1}(\hat{z}) = \cdots = \frac{\partial Q}{\partial z_r}(\hat{z}) = 0. \tag{1.8}$$

The point $\hat{z}$ satisfies the conditions (1.1) as well as the conditions (1.8). Hence $\hat{z}$ is a nonnull critical point of Q, contrary to the hypothesis that Q is nonsingular.

Hence (a_2) is true.

Proof of (a_3). We show first that

$$\text{nullity } P(\mathbf{0},s) \le \text{nullity } H_B(w) \tag{1.9}$$

To that end set $P(\mathbf{0},s) = f(s)$ for each n-tuple s. Let ρ be the nullity of the quadratic form f. If $\rho = 0$, (1.9) is true. If $\rho > 0$ there exist ρ linearly independent critical points $s^{(1)}, \ldots, s^{(\rho)}$ of f. Note first that $\rho \le r$, since there are at most r linearly independent solutions of conditions (1.1). One can choose a base $B = (z^{(1)}, \ldots, z^{(r)})$ for π_r : (25.6) in which

$$z^{(1)}, \ldots, z^{(\rho)} = (\mathbf{0},s^{(1)}), \ldots, (\mathbf{0},s^{(\rho)}),$$

where $\mathbf{0}$ is a null r-tuple. For such a choice of B it is clear from (1.6) that $d_{hk} = 0$ for $h = 1, \ldots, \rho$, since $z_j^{(h)} = 0$ for $h = 1, \ldots, \rho$ and $j = 1, \ldots, r$. Hence the nullity of the matrix $\|d_{hk}\|$ is at least ρ.

Thus (1.9) is true. It remains to prove that

$$\text{nullity } P(\mathbf{0},s) \ge \text{nullity } H_B(w). \tag{1.10}$$

Set nullity $H_B(w) = \nu$. If $\nu = 0$, (1.10) is trivially true. If $\nu > 0$ there exist ν linearly independent r-tuples $\hat{c} = (\hat{c}_1, \ldots, \hat{c}_r)$ for each of which $\hat{c}_h d_{hk} = 0$, $k = 1, \ldots, r$. If d_{hk} is given by (1.6), then for j on the range $1, \ldots, r$,

$$\hat{c}_h z_j^{(h)} Q_{z_j}(z^{(k)}) = 0 \qquad (k = 1, \ldots, r). \tag{1.11}$$

Since the r-square determinant of (a_2) does not vanish, we infer that $\hat{c}_h z_j^{(h)} = 0$ for $j = 1, \ldots, r$.

Thus the ν r-tuples $\hat{c}$ give rise to ν μ-tuples $\hat{z} = \hat{c}_h z^{(h)}$ with null initial r-tuples and with linearly independent terminal n-tuples. Since the ν μ-tuples $\hat{z}$ satisfy (1.1) and are linearly independent, their ν terminal n-tuples $\hat{s}$ are linearly independent critical points of the quadratic form f. Hence nullity $P(\mathbf{0},s) \ge \nu$.

Thus Lemma 1.2(a_3) *is true and implies Theorem* 25.1 (ii).

In Section 2 we shall drop the hypothesis that Q is nonsingular. In Section 3 we shall assume again that Q is nonsingular.

2. The Case of a Nonsingular Form, $P(\mathbf{0},s)$, (Subordinate to $P(u,s) = Q(z)$)

In this section we make no assumption* as to the singularity of $Q(z)$.

The conditions (1.1) defining the r-plane π_r : (25.6) can be written in the form

$$P_{s_i}(u,s) = 0 \qquad (i = 1, \ldots, n) \tag{2.0}$$

* Except for Corollary 2.2 of this section.

since $Q(z) = P(u,s)$, subject to (25.2). Suppose that

$$P(\mathbf{0},s) \equiv e_{ij}s_i s_j, \tag{2.1}$$

where i and j have the range $1, \ldots, n$, and $e_{ij} = e_{ji}$. When $P(\mathbf{0},s)$ is nonsingular the n-square determinant $|e_{ij}| \neq 0$. There then exist n linear forms,

$$(L_1(u), \ldots, L_n(u)) = \mathbf{L}(u) \tag{2.2}$$

in the n-tuples u, affording solutions $s = \mathbf{L}(u)$ of the n equations (2.0). For arbitrary r-tuples w set

$$P(w,\mathbf{L}(w)) \equiv H^0(w). \tag{2.3}$$

One sees that when $P(\mathbf{0},s)$ is nonsingular, the r-plane

$$\pi_r = \{z \in R^\mu \mid z = (w,\mathbf{L}(w)),\ w \in R^r\} \tag{2.4}$$

is the r-plane π_r of (25.6).

The following lemma is formal and trivial in proof.

LEMMA 2.1. *If* $\mathrm{P}(\mathbf{0},\mathrm{s})$ *is nonsingular, the form* $\mathrm{H}^0(\mathrm{w})$: (2.3) *is a quadratic form* $\mathrm{H}_{\mathrm{B}_0}(\mathrm{w})$ *complementary to* $\mathrm{P}(\mathbf{0},\mathrm{s})$ *over a base* $\mathrm{B}_0 = (\mathrm{z}^{(1)}, \ldots, \mathrm{z}^{(r)})$ *for which the* μ*-tuple*

$$\mathrm{z}^{(\mathrm{h})} = (\mathrm{w}^{(\mathrm{h})},\mathbf{L}(\mathrm{w}^{(\mathrm{h})})) \qquad (\mathrm{h} = 1, \ldots, \mathrm{r}) \tag{2.5$'$}$$

where $\mathrm{w}^{(\mathrm{h})}$ *is the* h*-th row of the* r*-square matrix* $\|\delta_{\mathrm{k}}^{\mathrm{h}}\|$.

Theorem 25.1(iii) will be derived with the aid of the following theorem.

THEOREM 2.1. *If* $\mathrm{P}(\mathbf{0},\mathrm{s})$ *is nonsingular, then under a suitable nonsingular, homogeneous linear transformation* T *from the* μ *variables* $(\mathrm{u}_1, \ldots, \mathrm{u}_\mathrm{r} : \mathrm{s}_1, \ldots, \mathrm{s}_\mathrm{n}) = (\mathrm{u},\mathrm{s})$ *to the* μ *variables* $(\mathrm{w}_1, \ldots, \mathrm{w}_\mathrm{r} : \mathrm{t}_1, \ldots, \mathrm{t}_\mathrm{n}) = (\mathrm{w},\mathrm{t})$

$$\mathrm{P}(\mathrm{u},\mathrm{s}) = \mathrm{H}^0(\mathrm{w}) + \mathrm{P}(\mathbf{0},\mathrm{t}). \tag{2.5$''$}$$

We understand that under T

$$(w_1, \ldots, w_r : t_1, \ldots, t_n) = (T_1(u,s), \ldots, T_\mu(u,s)). \tag{2.6}$$

The Transformation T *Defined.* Let

$$(\theta_1(u,s), \ldots, \theta_n(u,s)) = \Theta(u,s) \tag{2.7}$$

be a set of n linear forms in the μ-tuples (u,s) obtained by solving the n linear equations

$$P_{s_i}(u,s) = 2e_{i1}t_1 + \cdots + 2e_{in}t_n \qquad (i = 1, \ldots, n) \tag{2.8}$$

for the variables $t_1, \ldots, t_n$. The solutions exist, since the n-square determinant $|e_{ij}| \neq 0$, or equivalently $P(\mathbf{0},s)$ is nonsingular. We define T by setting

$$(T_1(u,s), \ldots, T_\mu(u,s)) \equiv (u_1, \ldots, u_r : \theta_1(u,s), \ldots, \theta_n(u,s)). \tag{2.9}$$

Note that when $u = 0$, $(s_1, \ldots, s_n) = (t_1, \ldots, t_n)$ under T, in accord with (2.1) and (2.8), and that $T(u,s) = (u_1, \ldots, u_r : 0, \ldots, 0)$ whenever (u,s) satisfies conditions (2.0).

The transformation $(w,t) = T(u,s)$ is homogeneous, linear, and nonsingular. Under T, $(u_1, \ldots, u_r) = (w_1, \ldots, w_r)$. Moreover, under the transformation T, with h, k on the range $1, \ldots, r$, and i, j on the range $1, \ldots, n$, one has an identity

$$P(u,s) = \alpha_{hk}w_h w_k + \beta_{ik}w_k t_i + \beta_{ik}t_i w_k + \gamma_{ij}t_i t_j \tag{2.10}$$

in (u,s) or {alternately in (w,t)}. The coefficients on the right of (2.10) are constants of which $\gamma_{ij} = e_{ij}$. One can take $\alpha_{hk} = \alpha_{kh}$ so that the quadratic form on the right of (2.10) has a symmetric matrix. We shall prove the following.

(i). *The coefficients* $\beta_{\text{ik}} = 0$. To verify this let (u',s') be a second set of variables (u,s) and set $(w',t') = T(u',s')$. Under the transformations,

$$(w',t') = T(u',s'); \qquad (w,t) = T(u,s) \tag{2.11}$$

the bilinear form, in the μ-tuple (u,s) and the μ-tuple (u',s') whose matrix is the symmetric matrix of $P(u,s) = Q(z)$, equals the bilinear form, in the μ-tuple (w,t) and the μ-tuple (w',t') whose matrix is the symmetric matrix of the right member of (2.10). Subject to (2.11) one then has the identity

$$\tfrac{1}{2}(u_k' P_{u_k}(u,s) + s_i' P_{s_i}(u,s)) = \alpha_{hk}w_h' w_k + \beta_{ik}w_k' t_i + \beta_{ik}t_i' w_k + \gamma_{ij}t_i' t_j \tag{2.12}$$

in $(w,t : w',t')$ or {alternately in $(u,t : u',t')$}. If (2.12) is regarded as an identity in $(w,t : w',t')$, subject to (2.11), one can set each t_i and w_k' zero therein. Because (2.11) conditions the variables in (2.12), one must then set $u_k' = 0$, since $u_k' = w_k'$ under T. One must also set

$$P_{s_1}(u,s) = \cdots = P_{s_n}(u,s) = 0, \tag{2.13}$$

when $t_1 = \cdots = t_n = 0$, since (2.8) holds under T. By virtue of these substitutions, (2.12) reduces to the identity $\beta_{ik}t_i' w_k \equiv 0$.

Hence each coefficient $\beta_{ik} = 0$.

Proof of (2.5)″. If one sets $\bar{H}(w) \equiv \alpha_{hk}w_h w_k$, (2.10) now takes the form

$$P(u,s) = \bar{H}(w) + P(\mathbf{0},t) \qquad (\text{when } (w,t) = T(u,s)). \tag{2.14}$$

We shall verify (2.5)″ by showing that $\bar{H}(w) \equiv H^0(w)$, when $H^0(w)$ is defined by (2.3).

To that end we can regard (2.14) as an identity in the μ-tuples (u,s), provided $(w,t) = T(u,s)$. Recall that $u \to \mathbf{L}(u)$ is a linear mapping of R^r into R^n such that

$$P_{s_i}(u,\mathbf{L}(u)) = 0 \qquad (i = 1, \ldots, n). \tag{2.15}$$

Under T, as defined by (2.6) and (2.9), a μ-tuple $(u,\mathbf{L}(u))$ goes into a μ-tuple (w,t), in which $w = u$ and t is a null n-tuple. We infer then from (2.14) that

$$P(u,\mathbf{L}(u)) \equiv \bar{H}(u) + 0. \tag{2.16}$$

Thus $H^0(u) \equiv \bar{H}(u)$ by virtue of (2.3).

Hence (2.5)″ is true, and the proof of Theorem 2.1 is complete.

Theorem 2.1, aided by Lemmas 1.1 and 2.1, has the following corollary.

COROLLARY 2.1. *When the quadratic form* $P(\mathbf{0},s)$, *subordinate to the form* $P(u,s)$, *is nonsingular, then*

$$\textit{index } P(u,s) = \textit{index } P(\mathbf{0},s) + \kappa \tag{2.17}$$

$$\textit{nullity } P(u,s) = \nu \tag{2.18}$$

where κ *and* ν *are respectively the index and nullity of each quadratic form complementary to* $P(\mathbf{0},s)$.

Proof. The relations (2.17) and (2.18) follow from (2.5)″ of Theorem 2.1, since $H^0(w)$ is a quadratic form complementary to $P(\mathbf{0},s)$ by Lemma 2.1, and such quadratic forms have a common index and nullity by Lemma 1.1, when $P(\mathbf{0},s)$ is nonsingular.

Corollary 2.1 implies a second corollary of Theorem 2.1.

COROLLARY 2.2. *When the quadratic form* $P(u,s)$, *as well as its subordinate form* $P(\mathbf{0},s)$, *is nonsingular, then any form complementary to* $P(\mathbf{0},s)$ *is nonsingular and* (2.17) *holds, as in Corollary* 2.1.

By hypothesis nullity $P(u,s) = 0$. Hence by (2.18), $\nu = 0$.

3. Proof of Theorem 25.1 Completed

Theorem 25.1(iii) remains unproved when the quadratic form $P(\mathbf{0},s)$ is singular. Our proof of this theorem, without hypothesis as to the singularity of $P(\mathbf{0},s)$, will involve the family of quadratic forms

$$Q^{\lambda}(z) = Q(z) - \lambda \, \|z\|^2 \qquad (z \in R^{\mu}) \tag{3.1}$$

for values of λ near $\lambda = 0$. If one sets

$$Q^{\lambda}(z) = P^{\lambda}(u,s) \tag{3.2}$$

subject to (25.2), one can affirm that $P^{\lambda}(\mathbf{0},s)$ is "subordinate" to $P^{\lambda}(u,s)$. Note that

$$P^{\lambda}(\mathbf{0},s) \equiv P(\mathbf{0},s) - \lambda \, \|s\|^2. \tag{3.3}$$

A Constant $e > 0$. The quadratic form $Q(z)$ is nonsingular by hypothesis of Theorem 25.1. Hence for $-e < \lambda < e$ and for e sufficiently small, $Q^{\lambda}(z)$

will remain nonsingular. We can suppose that e is so small that for $-e < \lambda < e$ the quadratic form $P^\lambda(\mathbf{0},s)$ is singular at most when $\lambda = 0$. We see that index $P^\lambda(u,s)$ is then constant for $-e < \lambda < e$, while index $P^\lambda(\mathbf{0},s)$ is constant for $-e < \lambda < 0$, or for $0 < \lambda < e$. Moreover Corollary 2.2 implies the following.

LEMMA 3.1. *If the quadratic form* Q(z) *is nonsingular, if* e *is conditioned as in the preceding paragraph, and if* $-e < \lambda < 0$ *or* $0 < \lambda < e$, *then*

$$\text{(3.4)} \qquad \textit{index } P^\lambda(u,s) = \textit{index } P^\lambda(\mathbf{0},s) + \kappa^\lambda,$$

where κ^λ *is the index of any quadratic form (necessarily nonsingular) complementary to the form* $P^\lambda(\mathbf{0},s)$ *subordinate to* $P^\lambda(u,s)$.

It follows from (3.4) that κ^λ is constant for $-e < \lambda < 0$ and for $0 < \lambda < e$. It may have a discontinuity when $\lambda = 0$. Information about a possible discontinuity of κ^λ when $\lambda = 0$ will be derived from the following two propositions. Proposition 3.1 is of general character.

PROPOSITION 3.1. *For* $-\delta < t < \delta$ *let* $q^t(w)$ *be a real-valued symmetric quadratic form in* $w_1, \ldots, w_r$ *with coefficients which vary continuously with* t. *If* $q^t(w)$ *is singular at most when* $t = 0$ *and if index* $q^0(w) = \hat{\kappa}$ *and nullity* $q^0(w) = \hat{\nu}$ *then for* $-\delta < t < \delta$

$$\text{(3.5)} \qquad \textit{index } q^t(w) - \hat{\nu} \leq \hat{\kappa} \leq \textit{index } q^t(w).$$

The r characteristic roots $\lambda(t)$ of the matrix of $q^t(w)$ vary continuously with $t \in (-\delta,\delta)$. By classical theorems index $q^t(w)$ is the count of the roots which are negative for the given t, and nullity $q^t(w)$ is the count of the roots which vanish for the given t. In the notation of Proposition 3.1, when $t = 0$, $\hat{\kappa}$ of these roots are negative and $\hat{\nu}$ vanish. Moreover index $q^t(w)$ is constant, when $-\delta < t < 0$ or when $0 < t < \delta$. Finally, these roots vary continuously with t.

The relations (3.5) follow.

Introduction to Proposition 3.2. The relation (3.4) is the starting point of the proof of Theorem 25.1 (iii). However, (3.4) is not affirmed to hold when $\lambda = 0$, although each of its entries is defined when $\lambda = 0$. As we shall see, our information concerning index $P^\lambda(u,s)$ and index $P^\lambda(\mathbf{0},s)$ is adequate when $\lambda = 0$, but not that concerning κ^λ. Once Proposition 3.2 is proved, Proposition 3.1 will yield the required information concerning κ^λ.

PROPOSITION 3.2. *If the quadratic form* $Q(z) = P(u,s)$ *is nonsingular and if a positive constant* ε *is sufficiently small, then for* $-\varepsilon < \lambda < \varepsilon$, $P^\lambda(u,s)$ *is nonsingular and there exists a symmetric quadratic form*

$$\mathscr{H}^{\lambda}(\mathrm{w}) = \mathrm{d}_{\mathrm{hk}}(\lambda)\mathrm{w}_{\mathrm{h}}\mathrm{w}_{\mathrm{k}} \qquad (\mathrm{w} \in \mathrm{R}^{\mathrm{r}}) \tag{3.6}$$

which is complementary to the form $\mathrm{P}^{\lambda}(\mathbf{0},\mathrm{s})$ *and is such that the coefficients* $\mathrm{d}_{\mathrm{hk}}(\lambda)$ *vary continuously with* $\lambda \in (-\varepsilon,\varepsilon)$.

The Choice of ε. In Section 25 quadratic forms complementary to the quadratic form $P(\mathbf{0},s)$, subordinate to $P(u,s)$, have been defined. We follow this definition as a model, replacing $P(u,s)$ and $P(\mathbf{0},s)$ by $P^{\lambda}(u,s)$ and $P^{\lambda}(\mathbf{0},s)$, respectively, subject to the condition that $-\varepsilon < \lambda < \varepsilon$. A choice of ε is to be made explicit on considering the n-conditions

$$\frac{\partial Q^{\lambda}}{\partial z_{r+i}}(z) = 0 \qquad (r + i = r + 1, \ldots, \mu) \tag{3.7}$$

as a replacement for the conditions (1.1). For each α on the range $1, \ldots, \mu$ and i on the range $1, \ldots, n$ let $A^{\lambda}_{i\alpha}$ be the coefficient of z_{α} in $\partial Q^{\lambda}/\partial z_{r+i}$. The matrix $\|A^{\lambda}_{i\alpha}\|$ is n by μ. The elements $A^{\lambda}_{i\alpha}$ vary continuously with λ. The matrix $\|A^{0}_{i\alpha}\|$ has rank n, because $Q(z)$ is nonsingular. There accordingly exists an n-square submatrix $\|B^{0}_{ij}\|$ of $\|A^{0}_{i\alpha}\|$ of rank n. If $\varepsilon > 0$ is sufficiently small, the n-square submatrix $\|B^{\lambda}_{ij}\|$ of $\|A^{\lambda}_{i\alpha}\|$ which reduces to $\|B^{0}_{ij}\|$ when $\lambda = 0$, has the rank n for $-\varepsilon < \lambda < \varepsilon$.

We suppose ε *so chosen*, subject to the additional condition that $\varepsilon < e$, where e is the constant of Lemma 3.1.

When $-\varepsilon < \lambda < \varepsilon$, the n equations (3.7) can be solved by Cramer's rule for the n components z_j of the μ-tuple z which bear the indices j of the columns of the n-square matrix $\|B^{0}_{ij}\|$. The solutions will be linear in the r components of z which bear the r residual indices in the range $1, \ldots, \mu$. For each $\lambda \in (-\varepsilon,\varepsilon)$ one obtains thereby a base

$$B(\lambda) = (z^{(1)}(\lambda), \ldots, z^{(r)}(\lambda)) \tag{3.8}$$

for solutions z of (3.7). For $-\varepsilon < \lambda < \varepsilon$ and for prescribed r-tuples $(w_1, \ldots, w_r)$ the quadratic form

$$\mathscr{H}^{\lambda}(\mathrm{w}) = Q^{\lambda}(w_k z^{(k)}(\lambda)) \tag{3.9}$$

is "complementary" to the form $P^{\lambda}(\mathbf{0},s)$ and is representable as in (3.6).

This completes the proof of Proposition 3.2.

Proposition 3.2 and Proposition 3.1 have the corollary.

COROLLARY 3.1. *If, as in Theorem* 25.1, κ *and* ν *are, respectively, the index and nullity of* $\mathscr{H}^{0}(\mathrm{w})$, *then, when* $-\varepsilon < \lambda < 0$ *or when* $0 < \lambda < \varepsilon$

$$\kappa^{\lambda} - \nu \leq \kappa \leq \kappa^{\lambda} \tag{3.10}$$

where $\kappa^{\lambda} =$ *index* $\mathscr{H}^{\lambda}(\mathrm{w})$.

The proof of Theorem 25.1 (iii) **concluded.** Let ε be conditioned as in Proposition 3.2, with $\varepsilon < e$ of Lemma 3.1. We shall verify a sequence of five relations of which the last two, taken with Corollary 3.1, will imply Theorem 25.1(iii),

$$\text{index } Q^\lambda(z) = \text{index } Q(z) \qquad (-\varepsilon < \lambda < \varepsilon) \tag{3.11}$$

$$\text{index } P^\lambda(\mathbf{0},s) = \text{index } P(\mathbf{0},s) \qquad (-\varepsilon < \lambda < 0) \tag{3.12}$$

$$\text{index } P^\lambda(\mathbf{0},s) = \text{index } P(\mathbf{0},s) + \nu \qquad (0 < \lambda < \varepsilon) \tag{3.13}$$

$$\kappa^\lambda = \text{index } Q(z) - \text{index } P(\mathbf{0},s) = \kappa_- \qquad (-\varepsilon < \lambda < 0) \tag{3.14}$$

$$\kappa^\lambda = \text{index } Q(z) - \text{index } P(\mathbf{0},s) - \nu = \kappa_+ \qquad (0 < \lambda < \varepsilon) \tag{3.15}$$

introducing κ_- and κ_+.

Proof of (3.11). This relation is a consequence of the hypothesis that $Q^\lambda(z)$ is nonsingular for $-\varepsilon < \lambda < \varepsilon$.

Proof of (3.12) *and* (3.13). The constant ε has been chosen so that $P^\lambda(\mathbf{0},s)$ is nonsingular for $-\varepsilon < \lambda < \varepsilon$, except at most when $\lambda = 0$. By (3.3), $P^0(\mathbf{0},s) \equiv P(\mathbf{0},s)$. Nullity $P(\mathbf{0},s) = \nu$ by Theorem 25.1(ii). Relations (3.12) and (3.13) follow from classical theorems and (3.3).

Proof of (3.14). For $-\varepsilon < \lambda < 0$, $Q^\lambda(z)$ and $P^\lambda(\mathbf{0},s)$ are nonsingular by virtue of our choice of ε. Since $\varepsilon < e$ of Lemma 3.1, the relation (3.14) follows from (3.4), (3.11) and (3.12).

Proof of (3.15). For $0 < \lambda < \varepsilon$, (3.15) follows from (3.4), (3.11) and (3.13).

From (3.14) and (3.15) we find that $\kappa_- - \nu = \kappa_+$. From (3.10)*

$$\kappa_- - \nu \leq \kappa \leq \kappa_+ \qquad (\kappa = \text{index } \mathscr{H}^0(w)). \tag{3.16}$$

These two sets of relations imply that the equality prevails in (3.16). Thus $\kappa = \kappa_+$. Hence (3.15) gives the relation

$$\text{index } Q(z) - \text{index } P(\mathbf{0},s) - \nu = \kappa, \tag{3.17}$$

thereby establishing Theorem 25.1(iii).

This completes the proof of Theorem 25.1 (iii). *The proof of Theorem* 25.1 (i) *and* (ii) *is given in* Section 1.

* The left and right inequalities of (3.10) are applied separately, the left to κ^λ of (3.14), the right to κ^λ of (3.15).

APPENDIX III

Model Families of Quadratic Forms

1. Model Families F

Let there be given a subinterval Γ of R of values $\sigma \leq \gamma$ where γ is prescribed and fixed. For each $\sigma \in \Gamma$ let there be given a real valued symmetric quadratic form,

$$Q^{\sigma}(z) = a_{ij}(\sigma)z_i z_j, \tag{1.1}$$

in real variables $z_1, \ldots, z_\mu$. Such a family of quadratic forms will be called a *model family* F if it satisfies the following three conditions.

Condition I. For $\sigma \in \Gamma$ the coefficients $a_{ij}(\sigma)$ shall vary continuously with σ.

Condition II. For σ negative with $|\sigma|$ sufficiently large, Q^{σ} shall be positive definite.

Condition III. For each non-null μ-tuple z the value $Q^{\sigma}(z)$ shall decrease strictly as σ increases in Γ.

Example 1. If $c_{ij}z_i z_j$ is a symmetric quadratic form with real coefficients, then for $\sigma \in \Gamma$ and $z \in R^{\mu}$ the quadratic forms

$$c_{ij}z_i z_j - \sigma z_i z_i$$

define a special model family F of quadratic forms.

Conditions I, II, III are relatively easy to verify. However, there are important applications for which these conditions are too restrictive. This leads us to define model families G of quadratic forms (1.1) in which Condition III is replaced by less restrictive Conditions IV and V.

2. Model Families G

The families G are defined for $\sigma \in \Gamma : \sigma \leq \gamma$, as before, and for $z \in R^{\mu}$ and are subject to the following conditions.

Condition I, as on F of Section 1.

Condition II, as on F of Section 1.

Condition IV. For each $z \in R^{\mu}$ the value $Q^{\sigma}(z)$ shall decrease monotonically as σ increases in Γ.

Condition V. If for a value $\tau < \gamma$, y is a non-null critical μ-tuple of Q^{τ}, then $Q^{\lambda}(y) < 0$ whenever $\tau < \lambda \leq \gamma$.

Each quadratic form in the family F is in the family G, but the converse is not true, as the following example shows.

Example 2. Let F be a model family of quadratic forms Q^{σ}, satisfying Conditions I, II, III, for $z \in R^{\nu}$ and $\sigma \leq \gamma$. Let $K(u)$ be a positive definite symmetric quadratic form in the variables $u_1, \ldots, u_n$. Let G be a family of quadratic forms $y \to q^{\sigma}(y)$ defined for each $\sigma \leq \gamma$ and for $y \in R^{\mu}$ with $\mu = \nu + n$. One sets

$$q^{\sigma}(y) = Q^{\sigma}(z) + K(u) \tag{2.1}$$

when $y = (z_1, \ldots, z_{\nu}; u_1, \ldots, u_n)$. The family G satisfies Conditions I, II, IV, V, but not Condition III.

We shall use model families F, when possible, because of their relative simplicity. Families G include families F as special cases. For that reason we shall be concerned from this point on in Appendix III with model families G. The main result of Appendix III, namely Theorem 3.1 is valid for families G and hence for families F. We prepare for Theorem 3.1 by proving two lemmas.

LEMMA 2.1. *Suppose that a quadratic form* Q^{τ} *of the family* G *is such that nullity* $Q^{\tau} = \nu$, *possibly* 0. *If* $|\sigma - \tau|$ *is sufficiently small, then for fixed* τ

$$\text{index } Q^{\tau} \leq \text{index } Q^{\sigma} \leq \nu + \text{index } Q^{\tau}. \tag{2.2}$$

The proof of this lemma depends only on the fact that the forms of G satisfy Condition I. It is a classical theorem that the index of a real-valued symmetric quadratic form $c_{ij}z_iz_j$ in μ-variables $z_1, \ldots, z_{\mu}$ is the "count" of negative roots λ of the polynomial defined by the μ-square determinant $|c_{ij} - \lambda\delta_i^j|$. Lemma 2.1 is readily proved with the aid of this lemma on taking due account of the continuous variation with σ of the coefficients of the form Q^{σ}.

Our second lemma requires two definitions.

Definition 2.1. *Characteristic roots of* G. If for a fixed $\sigma \in \Gamma$, a non-null μ-tuple z is a critical points of Q^σ, then σ will be called a *characteristic root* of G. Such a characteristic root σ of G will be assigned a *multiplicity* equal to the number of independent μ-tuples which are critical points of Q^σ. Each characteristic root σ of G will be *counted* a number of times equal to its multiplicity.

Definition 2.2. *The vector subspace* K^λ *of* R^μ. Corresponding to a prescribed value $\lambda \in \Gamma$, let K^λ be the vector subspace of R^μ, over R, generated by the critical μ-tuples z of forms Q^σ of G for which $\sigma \leq \lambda$.

Our second lemma follows.

LEMMA 2.2. (a) *Let* $\lambda \in \Gamma$ *be prescribed in the parameter domain* Γ *of a model family* G. *Then* $\mathrm{Q}^\lambda(\mathrm{w}) < 0$ *for each non-null* μ*-tuple* $\mathrm{w} \in \mathrm{K}^\lambda$ *other than a critical* μ*-tuple* y *of* Q^λ.

(b) *For a critical* μ*-tuple* y *of* Q^λ, $\mathrm{Q}^\lambda(\mathrm{y}) = 0$.

(c) *Dim* K^λ *is the count of characteristic roots of* G *at most* λ.

Notation for proof of Lemma 2.2. Corresponding to a representation (1.1) of a quadratic form Q^σ of the family G, and to μ-tuples y and z we introduce the bilinear form,

$$B^\sigma(y,z) = a_{ij}(\sigma)y_i z_j. \tag{2.3}$$

If y is a critical μ-tuple of Q^σ, then $B^\sigma(y,z) = 0$. Hence

$$Q^\sigma(y) = B^\sigma(y,y) = 0 \tag{2.4$'$}$$

and for an arbitrary μ-tuple z,

$$Q^\sigma(y + z) = Q^\sigma(y) + 2B^\sigma(y,z) + Q^\sigma(z) = Q^\sigma(z). \tag{2.4$''$}$$

Proof of (a). Corresponding to λ of Lemma 2.2 let

$$\sigma_1 < \sigma_2 < \cdots < \sigma_n \leq \lambda \tag{2.5}$$

be an arbitrary sequence of n characteristic roots of G at most λ and let

$$w = z^{(1)} + z^{(2)} + \cdots + z^{(n)} \tag{2.6}$$

be a sum of n non-null critical μ-tuples of the respective quadratic forms

$$Q^{\sigma_1}, Q^{\sigma_2}, \ldots, Q^{\sigma_n} \tag{2.7}$$

of the model family G. We shall show that $Q^\lambda(w) < 0$, unless $n = 1$ and $\sigma_n = \lambda$.

If $n = 1$ and $\sigma_1 = \lambda$, $Q^\lambda(w) = 0$. If $n = 1$, and $\sigma_1 < \lambda$, $Q^\lambda(w) < 0$ by Condition V on G.

If $n > 1$ and w is given by (2.6), the following relations can be successively verified. One uses (2.4) to verify the first equality in each line.

$$
\begin{aligned}
&Q^{\sigma_1}(z^{(1)}) = 0 \\
&Q^{\sigma_2}(z^{(1)} + z^{(2)}) = Q^{\sigma_2}(z^{(1)}) <^* 0 \\
&Q^{\sigma_3}(z^{(1)} + z^{(2)} + z^{(3)}) = Q^{\sigma_3}(z^{(1)} + z^{(2)}) \leq^\dagger Q^{\sigma_2}(z^{(1)} + z^{(2)}) < 0 \\
&\quad \cdot \qquad \cdot \qquad \cdot \qquad \cdot \qquad \cdot \\
&\quad \cdot \qquad \cdot \qquad \cdot \qquad \cdot \qquad \cdot \\
&Q^{\sigma_n}(z^{(1)} + \cdots + z^{(n)}) = Q^{\sigma_n}(z^{(1)} + \cdots + z^{(n-1)}) \\
&\qquad\qquad\qquad \leq^\dagger Q^{\sigma_{n-1}}(z^{(1)} + \cdots + z^{(n-1)}) < 0
\end{aligned}
\tag{2.8}
$$

Since $\sigma^{(n)} \leq \lambda$ and $w = z^{(1)} + \cdots + z^{(n)}$ we conclude that

$$
Q^\lambda(w) \leq^\dagger Q^{\sigma_n}(w) < 0 \qquad (\text{if } n > 1). \tag{2.9}
$$

Lemma 2.2(a) follows. Its proof implies that the μ-tuples $z^{(1)}, \ldots, z^{(n)}$ are linearly independent, and that $n - 1 \leq \text{index } Q^\lambda \leq \mu$.

Lemma 2.2(b) follows from (2.4) with σ replaced by λ.

Lemma 2.2(c) is verified as follows. Dim K^λ is the number of elements of K^λ in a linearly independent set of generators of K^λ. Let $\nu_1, \nu_2, \ldots, \nu_s$ be the nullities of the respective characteristic roots $\tau_1 < \tau_2 < \cdots < \tau_s$ of G at most λ. There exists a set of generators of K^λ composed of $\nu_1 + \nu_2 + \cdots + \nu_s$ linearly independent critical μ-tuples of the respective quadratic forms

$$
Q^{\tau_1}, Q^{\tau_2}, \ldots, Q^{\tau_s}.
$$

It follows from (a) that the generators of this set are linearly independent.

Thus (c) is true and the proof of Lemma 2.2 is complete.

3. The Principal Theorem of Appendix III

THEOREM 3.1. *If* G *is a model family of quadratic forms the index* κ_τ *of a quadratic form* Q^τ *of the family* G *is the count of characteristic roots of* G *less than* τ.

By Condition II on G, Q^{σ_0} will be positive definite if $\sigma_0 < 0$ and $|\sigma_0|$ is sufficiently large. If Q^{σ_0} is represented as in (1.1), the determinant $|a_{ij}(\sigma_0)| \neq 0$ and index $Q^{\sigma_0} = 0$. Let σ increase from σ_0 to the prescribed $\tau \in \Gamma$. We distinguish two cases.

Case I. *There is no characteristic root of* G *less than* τ. *In* this case Q^σ will remain nonsingular as σ increases on $[\sigma_0, \tau)$ and hence remain of index

* By Condition V on the family G.
† By Condition IV on the family G.

zero. It follows from the left inequality in Lemma 2.1 that index $Q^\tau = 0$. Thus Theorem 3.1 is true in Case I.

Case II. *There exists a characteristic root of* G *less than* τ. As we have seen there is at most a finite number of characteristic roots of G less than τ, say $\lambda_1 < \lambda_2 < \cdots < \lambda_n$. Let $\nu_1, \ldots, \nu_n$ be the respective multiplicities of these roots. The nullities of the respective quadratic forms $Q^{\lambda_1}, \ldots, Q^{\lambda_n}$ are $\nu_1, \ldots, \nu_n$. Set $\nu_0 = 0$. We continue with an inductive proof that

$$\text{index } Q^{\lambda_n} \leq \nu_0 + \nu_1 + \cdots + \nu_{n-1} \qquad (n \geq 1). \tag{3.1}$$

Proof of (3.1). The truth of Theorem 3.1 in Case I implies that index $Q^{\lambda_1} = 0$. Thus (3.1) is true when $n = 1$.

Suppose then that $n > 1$. Proceeding inductively, let s be a positive integer such that $s < n$. We shall assume that (3.1) is true when n is replaced by s and prove (3.1) true, when n is replaced by $s + 1$. For $\lambda_s < \sigma < \lambda_{s+1}$, index Q^σ is constant, since Q^σ is nonsingular for such values of σ. Moreover for such values of σ

$$\text{index } Q^\sigma \leq \text{index } Q^{\lambda_s} + \nu_s = (\nu_0 + \cdots + \nu_{s-1}) + \nu_s \tag{3.2}$$

by the right inequality in (2.2) and the assumed validity of (3.1) when n is replaced by s. Since (3.2) holds for $\lambda_s < \sigma < \lambda_{s+1}$, the left inequality in (2.2) implies that

$$\text{index } Q^{\lambda_{s+1}} \leq \nu_0 + \nu_1 + \cdots + \nu_s. \tag{3.3}$$

This completes the inductive proof of (3.1).

Completion of Proof of Theorem 3.1. By hypothesis Q^σ is nonsingular for $\lambda_n < \sigma < \tau$, and hence index Q^σ constant. Hence for $\lambda_n < \sigma < \tau$ it follows, respectively, from the left and right* inequalities in (2.2) and from (3.1) that

$$\text{index } Q^\tau \leq \text{index } Q^\sigma \leq \text{index } Q^{\lambda_n} + \nu_n \leq \nu_1 + \cdots + \nu_n = k \tag{3.4}$$

introducing k. Thus $\kappa_\tau \leq k$.

It remains to show that $\kappa_\tau \geq k$. To that end let λ be any value such that $\lambda_n < \lambda < \tau$. Let $\hat{K}^\lambda$ be the vector subspace of R^μ generated by the critical μ-tuples of quadratic forms Q^σ of G for which $\sigma \leq \lambda$ or equivalently $\sigma < \lambda$. It follows from Lemma 2.2 that Q^λ is negative definite on $\hat{K}^\lambda$ and dim $\hat{K}^\lambda = k$. Since $\lambda < \tau$ and Condition IV holds, Q^τ is negative definite on $\hat{K}^\lambda$. Hence $\kappa_\tau \geq k$.

Thus index $Q^\tau = \kappa_\tau$ as affirmed in Theorem 3.1.

* The right inequality in (2.2) is applied with τ replaced by λ_n.

References

Birkhoff, G. D. and Hestenes, M. R.

1. Natural isoperimetric conditions in the calculus of variations. *Duke Math. J.*, **1** (1935), 198–286.

Bliss, G. A. and Schoenberg, I. J.

1. On separation, comparison and oscillation theorems for selfadjoint systems of linear second order differential equations. *Amer. J. Math.* **53** (1931), 781–800.

Bliss, G. A.

1. Jacobi's condition for problems of the calculus of variations in parametric form. *Trans. Amer. Math. Soc.* **17** (1916), 195–206.
2. *Lectures on the Calculus of Variations.* University of Chicago Press, Chicago, Ill., 1945.

Bôcher, M.

1. *Leçons sur les Méthodes de Sturm.* Gauthier-Villars, Paris, 1917.
2. *Introduction to Higher Algebra.* Macmillan, New York, 1938.

Bolza, O.

1. *Vorlesungen über Variationsrechnung.* B. G. Teubner, Leipzig and Berlin, 1909.
2. Über den "Abnormalen Fall" beim Lagrangeschen und Mayerschen Problem mit gemischter Bedingungen und variable Endpunkten. *Math. Ann.* **74** (1913), 430–446.

Carathéodory, C.

1. *Variationsrechnung und Partielle Differential Gleichungen erster Ordnung.* B. G. Teubner, Leipzig and Berlin, 1935.

Clebsch, A.

1. Über die Reduction der zweiten Variation auf ihre einfachste Form. *J. reine Angew. Math.* **55** (1858), 254–272.

Courant, R. and Hilbert, D.

1. *Methods of Mathematical Physics.* Wiley Interscience, New York, 1953.

Deheuvels, R.

1. Topologie d'une fonctionnelle. *Ann. Math.* **61** (1955), 13–72.

Diaz, J. B. and McLaughlin, J. R.

1. Sturm separation and comparison theorems for ordinary and partial differential equations. *Atti. Accad. Naz. Lincei Mem.*, Serie VIII, **9** (1969), 135–194.

Dickson, L. E.

1. *Modern Algebraic Theories.* Benj. H. Sanborn & Co., New York, 1930.

Eisenhart, Luther

1. *Differential Geometry.* Ginn & Company, New York, 1909.

Ettlinger, H. J.

1. Oscillation theorems for the real selfadjoint linear system of the second order. *Trans. Amer. Math. Soc.*, **22** (1921), 136–143.

Ewing, George M.

1. *Calculus of Variations with Applications.* W. W. Norton & Co., Inc., New York, 1969.

Gelfand, I. M. and Fomin, S. V.

1. *Calculus of Variations.* Prentice-Hall, Englewood Cliffs, N.J., 1963.

Goursat, Edouard

1. *Cours d'Analyse Mathématique.* Vol. II. Gauthier-Villars et Cie. Paris, 1929.

Hestenes, M. R.

1. *Calculus of Variations and Optimal Control Theory.* John Wiley & Sons, Inc., New York, London, Sydney, 1966.
2. See Birkhoff, G. D. and Hestenes, M. R.

Hickson, A. O.

1. An application of the calculus of variations to boundary value problems. *Trans. Amer. Math. Soc.*, **31** (1929), 563–579.

Hadamard, J.

1. *Leçons sur le Calcul des Variations, I.* A. Hermann et Fils, Paris, 1910.

Hahn, H.

1. Über raumliche Variationsprobleme. *Math. Ann.* **70** (1911), 110–142.

Halmos, Paul R.

1. *Finite Dimensional Vector Spaces.* Princeton University Press, Princeton, N.J., 1942.

Hartman, P.

1. On nonoscillatory linear differential equations of the second order. *Amer. J. Math.* **74** (1952), 389–400.

Hedlund, G. A.

1. Poincaré's rotation number and Morse's type number. *Trans. Amer. Math. Soc.* **34** (1932), 79–97.

Hermann, R.

1. Focal points of closed submanifolds of Riemannian spaces. *Nederl. Akad. Wetensch. Proc.* Series A66, **25** (1963), 613–628.

Hille, E.

1. Nonoscillation theorems. *Trans. Amer. Math. Soc.* **64** (1948), 234–252.

Ince, E. L.

1. *Ordinary Differential Equations.* Longmans Green, London, 1927.

Jordan, C.

1. *Cours d'Analyse.* Vol. 1, 3rd ed. Gauthier, Paris, 1909.

Kneser, A.

1. *Lehrbuch der Variationsrecknung, II.* F. Vieweg & Sohn, Braunschweig, 1925.
2. Untersuchungen über die reelen NullStellen der Integrale linearer Differentialgleichungen. *Math. Ann.* **42** (1893), 409–435.
3. Untersuchungen über die reelen NullStellen der Integrale linearer Differentialgleichungen. *J. reine Angew. Math.* **116** (1896), 178–212.

Kondrat'ev, V. A.

1. Sufficient conditions for nonoscillatory or oscillatory nature of solutions of the second-order equation $y'' + p(x)y = 0$. *Dokl. Akad. Nauk, SSSR* [N,S] **113** (1957), 742–745.

Leighton, W.

1. On selfadjoint differential equations of the second order. *J. Lond. Math. Soc.* **27** (1952), 37–47.
2. Comparison theorems for linear differential equations of second order. *Proc. Amer. Math. Soc.* **13** (1962), 603–610.

Leighton, W. and Martin, A. D.

1. Quadratic functionals with a singular end point. *Trans. Amer. Math. Soc.* 78 (1955), 98–128.

Leighton, W. and Morse, M.

1. Singular quadratic functionals. *Trans. Amer. Math. Soc.* **40** (1936), 252–286.

Ljusternik, L. A.

1. *The Topology of Calculus of Variations in the Large.* Translations of Mathematical Monographs, Vol. 16. American Mathematical Society, Providence, R.I., 1966.

Mason, M. and Bliss, G. A.

1. The properties of curves in space which minimize a definite integral. *Trans. Amer. Math. Soc.* **9** (1908), 440–466.

Milnor, J. W.

1. *Morse Theory.* Princeton University Press, Princeton, N.J., 1969.

Morse, M.

1. "The Calculus of Variations in the Large." *Colloquium Publications*, Vol. 18, 4th printing. American Mathematical Society, Providence, R.I., 1965.
2. A generalization of the Sturm separation and comparison theorems in n-space. *Math. Ann.* **103** (1930), 52–69.
3. The order of vanishing of the determinant of a conjugate base. *Proc. Nat. Acad. Sci., USA*, **17** (1931), 319–320.
4. Sufficient conditions in the problem of Lagrange with variable end conditions. *Amer. J. Math.* **53** (1931), 517–546.
5. Recent advances in variational theory in the large. *Proc. Int. Cong. Math.* **2** (1950), 143–156.
6. Model families of quadratic forms, *Proc. Nat. Acad. Sci., USA* **68** (1971), 914–915.
7. Subordinate quadratic forms and their complementary forms. *Rev. Roumaine Math., pures appl.* **16** (1971), 559–569.
8. Singular quadratic functionals. Math. Ann. To appear.

Morse, M. and Cairns, S. S.

1. *Critical Point Theory in Global Analysis and Differential Topology.* Academic Press, New York, 1969.
2. Singular homology over **Z** on topological manifolds. *J. Diff. Geom.* **3** (1969), 257–288.

Nehari, Z.

1. Oscillation criteria for second-order linear differential equations. *Trans. Amer. Math. Soc.* **85** (1957), 428–445.

Picone, M.

1. Sui valori eccezionali di un parametro da cui dipende un'equazione differenziale lineare ordinaria del second'ordine. *Ann. Scuola Norm. Sup. Pisa* **11** (1909), 1–141.

Pitcher, E., with Morse, Marston.

1. On certain invariants of closed extremals. *Proc. Nat. Acad. Sci., USA* **20** (1934), 282–288.

Reid, W. T.

1. A comparison theorem for selfadjoint differential equations of second order. *Ann. Math.* **65** (1957), 197–202.
2. Oscillation criteria for selfadjoint differential systems. *Trans. Amer. Math. Soc.* **101** (1961), 91–106.
3. *Ordinary Differential Equations*. John Wiley & Sons, Inc., New York, 1971.

Signorini, A.

1. Esistenza di un'estremale chiusa dentro un contorno di Whittaker. *Rend. Circ. Mat. Palermo* **33** (1912), 187–193.

Stein, Junior

1. Singular quadratic functionals. Dissertation, University of California, Los Angeles, 1971.

Smtu, Cr.

1. Sur les équations différentielle linéaires du second order. *J. Math. Pures Appl.* **1** (1836), 106–186.

Swanson, A.

1. *Comparison and Oscillation Theory of Linear Differential Equations*. Academic Press, New York and London, 1968.

Tonelli, L.

1. *Fondamenti Calcolo delle Variazioni, I, II*. Nicola Zanichelli, Bologna, 1921.

Wintner, A.

1. A comparison theorem for Sturmian oscillation numbers of linear systems of second order. *Duke Math. J.* **25** (1958), 515–518.

INDEX